Mathematical Engineering

Today, the development of high-tech systems is unthinkable without mathematical modeling and analysis of system behavior. As such, many fields in the modern engineering sciences (e.g. control engineering, communications engineering, mechanical engineering, and robotics) call for sophisticated mathematical methods in order to solve the tasks at hand.

The series Mathematical Engineering presents new or heretofore little-known methods to support engineers in finding suitable answers to their questions, presenting those methods in such manner as to make them ideally comprehensible and applicable in practice.

Therefore, the primary focus is—without neglecting mathematical accuracy—on comprehensibility and real-world applicability.

To submit a proposal or request further information, please use the PDF Proposal Form or contact directly: Dr. Thomas Ditzinger (thomas.ditzinger@springer.com).

Indexed by SCOPUS, zbMATH, SCImago.

Alexander Kolpakov

Thin-Walled Structures with Initial Stresses

Multiscale Analysis of Classical and Non-Classical Problems

Springer

Alexander Kolpakov
Niederbipp, Switzerland

ISSN 2192-4732 ISSN 2192-4740 (electronic)
Mathematical Engineering
ISBN 978-3-032-07246-7 ISBN 978-3-032-07247-4 (eBook)
https://doi.org/10.1007/978-3-032-07247-4

This Springer imprint is published by the registered company Springer Nature Switzerland AG
The registered company address is: Gewerbestrasse 11, 6330 Cham, Switzerland

Preface

Published 20 years ago, book *Stressed Composite Structures: Homogenized Models for Thin-Walled Heterogeneous Structures with Initial Stresses* by Kolpakov has presented a method for deriving models of thin-walled structures (plates, membranes, beams, strings) with initial stresses directly from three-dimensional of elasticity theory problem. The method developed in the book was based on the observation that for a body with initial stresses, the equations relating additional stresses and additional strains are similar to Hooke's law, but the coefficients of the equations do not possess all symmetries of the elastic constants. In thin and small diameter regions, this loss of the symmetries leads to the appearance in the limit equations the terms known in the classical theories of thin-walled structures as "additional transverse forces" or "additional moments" (Timoshenko 1961, 1983).

Over the last 20 years, the method proposed in the above book has been tested and applied to new problems of bodies and structures with initial stresses. Generalization and organization of new results has led to the appearance of the present book.

The author includes in the book the chapters devoted to uniform plates (flat plates made of homogeneous material) and uniform beams (cylindrical beams made of homogeneous material) with initial stresses. This is done in order to demonstrate the origin of the "additional transverse forces" in the asymptotic theory in its "pure" form, without regard to the inhomogeneity of the structural elements. These chapters present an alternative method to deriving the classical stability equations for uniform plates and beams.

The author includes chapters about structural elements with initial stresses greater than usually considered in classical theories (such stresses arise, for example, as a result of thermal heating). Also, the local stresses are written for all the considered cases.

As a result, this book, in couple with the book mentioned above, covers all ranges of the resulting initial stresses in plates and beams. Recall that three ranges of resulting initial stresses are distinguished: stresses corresponding to axial or in-plane forces, stresses corresponding to moments, and stresses of the order of elastic constants.

Typically, engineering structures consist of structural elements of various types, assembled into a unit by using connections (joints). Thus, analyzing a structure, one needs (1) to take into account the properties and characteristics of structural elements, and (2) to take into account the properties and structure of joints. It is obvious that the stress-strain state in the joints differs from the stress-strain state in the joined elements. However, the local stress-strain state in a joint is determined by that of all the structure. The classical structural mechanics, see, e.g., Hibbeler (2008), Ventsel and Krauthammer (2001), Washizu (1992) ignores specific properties of joints, operating by abstract notions of supports and connections. More precisely, the problems of computation of the local stress-strain states in joints and in the joined structural elements are quite freely divided between such disciplines as "Structural mechanics" (Hulse and Cain 1991) and "Mechanics of joints (connections, fasteners)" (Blake 1985; Norton 2008; Hudgins and James 2014). Currently, intensive research is being conducted to develop integrated approach(s) to calculating structural elements and connections (joints), see, e.g., Ciarlet (1990), Zhikov (2002), Zhikov and Pastukhova (2003), Gaudiello and Kolpakov (2011), Kolpakov (2011). This book treats the structural elements.

As it was noted, the theory of thin-walled heterogeneous elastic structures with initial stresses can be constructed on the basis of the observation of the loss of symmetry in the linearized model of an elastic body with initial stresses (Washizu 1992) and the application of two-scale expansion method (Sanchez-Palencia 2000). The beginning chapter of the book contains the condensed exposition of these theories. A more detailed exposition of the general theory of elastic bodies with initial stresses can be found in the book by Washizu "Variational Methods in the Theory of Elasticity and Plasticity" (Washizu 1992) and expositions of the homogenization method can be found in the book by Sanchez-Palencia "Non-Homogeneous Media and Vibration Theory" (Sanchez-Palencia 2000). The author also recommends books (Bakhvalov and Panasenko 1989; Bensoussan et al. 1978; Cioranescu and Donato 2000; Panasenko 2005; Trabucho and Viano 1996) as referenced books on the mathematical foundations of the homogenization method and books (Andrianov et al. 2004; Andrianov and Awrejcewicz 20024; Kalamkarov and Kolpakov 1997; Kolpakov 2004; Lewinski 2000; Mityushev 1997) as referenced books on the application of the homogenization method to the engineering problems. This list is not exhaustive, other works are mentioned in the text of the book, see also the bibliography.

The structure of the book is the following.

In Chap. 1, the aims and objectives of the book are discussed, in particular, the historical background of the problems of thin-walled structural elements with initial stresses is presented in condensed form.

Chapter 2 contains examples of thin-walled structural elements in which initial stresses are significantly used and/or manifested.

In Chap. 3, we briefly present the procedures of homogenization theory that are directly used in this book.

In Chap. 4, we discuss the elastic composite body model with initial stresses, which will be used as a first-principles model throughout the book.

In Chap. 5, we make a transition from the three-dimensional elasticity theory problem to the corresponding two-dimensional problem for uniform plates (plates of constant thickness made of homogeneous material) with initial stresses. We consider the problems corresponding to in-plane resultant initial forces and moments of initial forces.

In Chap. 6, we consider plates of periodic structure with initial stresses. We consider the case corresponding to in-plane resultant initial forces.

In Chap. 7, we make a transition from the three-dimensional elasticity theory problem with initial stresses in a thin layer to the two-dimensional problem in the case when the initial stresses are of the same order as the elastic constants of the material of the plate. Although the initial stresses are of the same order as the elastic constants, they are small compared with the elastic constants in absolute value. Using this fact, we construct the first-order (linear) correctors in the problem under consideration.

In Chap. 8, we consider uniform beams with initial stresses, which are traditional and widely used structural elements.

In Chap. 9, we consider non-uniform beams of arbitrary periodic structure with initial stresses corresponding to axial resultant forces. We consider the case corresponding to the axial resultant initial force.

In Chap. 10, we make a transition from the three-dimensional problem of elasticity theory with initial stresses in a small-diameter body to a problem of beam theory when the initial stresses are of the same order as the elastic constants of the material from which the beam is made.

In Chap. 11, we compare the hypothesis of the classical theories of plates and beams with the terms of the asymptotic expansions obtained using the two-scale method and the homogenization method.

The "References" section contains a list of sources used in the book. Although this list includes over 300 references, it is far from exhaustive of the literature on thin-walled structures and homogenization.

The author thanks Prof. Dr. Sci. I. V. Andrianov (RWTH Aachen) for reading the manuscript and providing valuable comments, and Prof. Dr. Sci. V. V. Mityushev (Cracow University of Technology) for insightful discussions on random composites.

Niederbipp, Switzerland
2025

Alexander Kolpakov

References

Andrianov IV, Awrejcewicz J (2024) Asymptotic methods for engineers. CRC Press, Boca Raton, FL

Andrianov IV, Awrejcewicz J, Manevich LI (2004) Asymptotic mechanics of thin-walled structures. Springer, Berlin

Bakhvalov NS, Panasenko GP (1989) Homogenization: averaging processes in periodic media. Kluwer Academic Publication, Dordrecht

Bensoussan A, Lions J-L Papanicolaou G (1978) Asymptotic analysis for periodic structures. North-Holland Publication, Amsterdam

Blake A (1985) Design of mechanical joints. CRC Press, Boca Raton, FL

Ciarlet G (1990) Plates and junctions in elastic multi-structures. An asymptotic analysis. Macon, Paris
Cioranescu D, Donato P (2000) An introduction to homogenization. Oxford University Press, Oxford
Gaudiello A, Kolpakov AG (2011) Influence of non degenerated joint on the global and local behavior of joined rods. Int J Engg Sci 49(3):295–309
Hibbeler RC (2008) Structural analysis. 7th Ed. Prentice-Hall, Englewood Cliffs, NJ
Hudgins A, James B (2014) Fatigue of threaded fasteners. Adv Mater Proces Tech Art 172(8):18–22
Hulse R, Cain J (1991) Structural mechanics. MacMillan Publication, London
Kalamkarov AL, Kolpakov AG (1997) Analysis, design and optimization of composite structures. John Wiley & Sons, Chichester
Kolpakov AG (2004) Stressed composite structures: homogenized models for thin-walled nonhomogeneous structures with initial stresses. Springer, Berlin
Kolpakov AG (2011) Influence of non degenerated joint on the global and local behavior of joined plates. Int J Engg Sci 49(11):1216–1231
Lewinski T, Telega JJ (2000) Plates, laminates and shells: asymptotic analysis and homogenization. World Scientific, Singapoure
Mityushev V (1997) A functional equation in a class of analytic functions and composite materials. Demost Math 30:63–70
Norton RL (2008) Design of machinery. 4th Ed. McGraw-Hill Higher Education, Boston, MA
Panasenko GP (2005) Multi-scale modeling for structures and composites. Springer, Berlin
Sanchez-Palencia E (2000) Non-homogeneous media and vibration theory. Springer, Berlin
Timoshenko S (1961) Theory of elastic stability. 2nd Ed. McGraw-Hill, New York
Timoshenko S (1983) History of strength of materials (with a brief account of the history of elasticity and theory of structures). Dover, New York
Trabucho L, Viano JM (1996) Mathematical modeling of rods. In Ciarlet PG, Lions J-L editors, Handbook of numerical analysis, volume IV, pp 487–974, Amsterdam, Elsevier
Ventsel E, Krauthammer Th (2001) Thin plates and shells. theory, analysis, and applications. Marcel Dekker, New York
Washizu K (1992) Variational methods in elasticity and plasticity. Pergamon Press, Oxford
Zhikov VV (2002) Homogenization of elasticity problems on singular structures. Izvestiay: Math 66(2):299–365
Zhikov VV, Pastukhova SE (2003) Homogenization for elasticity problems on periodic networks of critical thickness. Russian Acad Sci: Sb Math 194:697–732

Competing Interests The author have no conflicts of interest to declare that are relevant to the content of this book.

Contents

Chapter 1
Introduction

The mechanics of structures with initial stresses is a traditional part of structural mechanics. It is closely related to the problem of stability (buckling) of structures and structural elements. Note that the theory of stability of beams arose earlier than the theory of elastic bodies with initial stresses. The basic concepts of elastic stability of structures go back to works by Euler (1759) (beams) and Bryan (1891, 1888) (plates). Later, Hill (1958), Marguerre (1938), Prager (1947), Timoshemko and Goodier (1951), Trefftz (1930), Washizu (1992), and other researchers demonstrated that the problem of deformation of solids with initial stresses is related to variational principles and nonlinear problems in elasticity.

The engineering analysis of stability (buckling) problems for thin plates and beams with initial stresses is based on the computation of additional normal force arising as a result of additional deformation of stressed structural element. Ab alternative approach is based on the computation of work of initial stresses on the additional displacements (Amabili 2008; Ventsel and Krauthammer 2001).

In the 1950s, composite materials began to be widely used in engineering, especially in aeronautics and spacecraft. It quickly became apparent that traditional computations and design methods developed for homogeneous materials and uniform structures were not applicable to composite structural elements. This initiated intensive research in the field of composite material and structure theory (Broutman and Krock 1974; Kelly and Rabotnov 1988), in particular, research on the buckling of composite structures (Boscolo et al. 2013; Zhang et al. 2024; Ghorbani et al. 2024; Turvey and Marshall 2012; Miller et al. 2011; Lopez Jimenez and Triantafyllidis 2013). This research continues to this day. The main directions of the research are the theory of stability of complex/nonhomogeneous thin-walled structures (Kesavan 1979; Ciarlet and Kesavan 1981; Mignot et al. 1980; Wang and Abdalla 2002; Marcal and Yamagata 2016), design of materials and structures possessing required properties (Alaire 2002; Bendsøe and Sigmund 2004; Haug et al. 1986; Kolpakov and Rakin 1986; Foraboschi 2013): (in particular, unusual properties: negative Poisson's ratio Almgren 1985; Kolpakov 1985; Lakes 1987; Novikov and Wojciechowski

A. Kolpakov, *Thin-Walled Structures with Initial Stresses*, Mathematical Engineering,
https://doi.org/10.1007/978-3-032-07247-4_1

1999, negative thermal expansions Kolpakov and Rakin 1986; Lakes 1996; Miller et al. 2009; Kelly et al. 2005, etc.), vibration of complex structures (Kaplunov et al. 2002; Kolpakov 2001, 2005; Pichugin and Rodgerson 2004; Wozniak et al. 2004). The lists of the problems and the references are not exhausting. Additional references may be found in the cited literature.

This book is devoted to a particular problem—the construction of the low-dimensional and homogenized models of elastic composite thin-walled structural elements with initial stresses. The analysis is carried out in the framework of a linear (linearized Washizu 1992) approach. Both uniform and nonhomogeneous plates and beams are considered in the book.

No doubt, there are many interesting and important problems outside the frame of our consideration, for example, a local loss of stability (Li et al. 2015; Sun et al. 2023) or substantially non-linear problems of loss of stability (Chakraborty et al. 2024a, b; Lal and Markad 2021; Hu et al. 2021; Goriely et al. 2008; Machado 2008).

Historically, the basic concepts of structural mechanics were developed for uniform structural elements: flat plates, beams and membranes, cylindrical beams and strings, etc. made of homogeneous materials. When dealing with composite plates and beams or structural elements of complex geometry (such as ribbed plates or lattice masts), one finds that many concepts of classical theory cannot be transferred to the mechanics of heterogeneous structure. To illustrate this point, we discuss the following two concepts of classical theory.

1. *Cross-section*. The notion of cross-section plays a important role in the classical theory of beams and plates. In particular, the classical notions of the resultant forces and moments in a plate or beam assume the integration over cross-section. In an inhomogeneous plate/beam, the geometry and/or material characteristics of the cross-section change from one to another adjacent section, see Fig. 1.1. In Fig. 1.1, three cross-sections: $S(y_1)$, $S(y_2)$, and $S(y_3)$ of a plate of a complex structure are displayed. These cross-sections are very different. This example demonstrated that we are usually unable to introduce either a standard cross-section or a representative section for plates or beams of complex geometry.
2. *Neutral plane (axis)*. Considering a non-flat plate, we have two possibilities to introduce neutral surfaces. The first is to introduce a neutral plane as the median surface of the actual plate. In this case, the neutral plane may be a curved surface. With respect to this curved neutral surface, one can write the formulas of the classical theory. Note that this procedure is not possible for an arbitrary plate. For example, it is possible for a simple corrugated plate, but it is impossible for the plate displayed in Fig. 1.1, which consists of a corrugated plate connected with a carrier plane. An alternative way is to introduce a flat neutral plane for the plate as a whole. It may happen that a large part of such neutral plane will be located outside the physical plate, as shown in Fig. 1.1. In addition, it will be necessary to predict the position of the neutral plane. It is not easy to do for the plate displayed in Fig. 1.1.

As a result, we arrive at the conclusion that it would be useful to have a theory of thin-walled inhomogeneous structures developed on the basis of three-dimensional

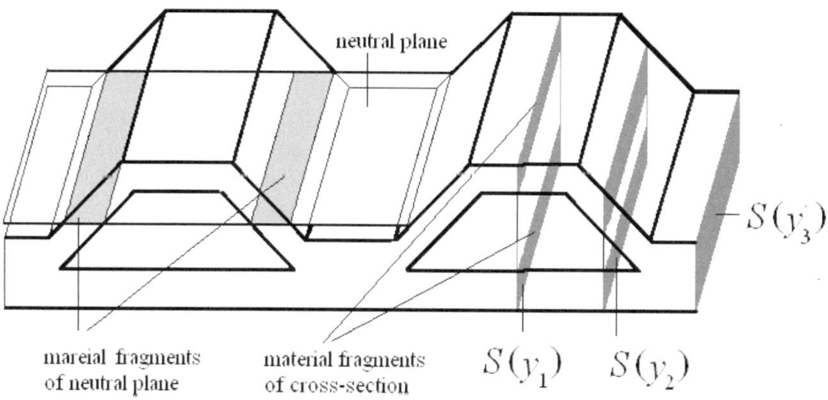

Fig. 1.1 A complex geometry plate

elasticity theory with no simplifying assumptions (in particular, with no a priori hypothesis). The need for such a theory was noted not only in connection with problems of stability, but also in connection with other problems with thin-walled structures, see e.g., Ball (2002), Ciarlet (1990), Kalamkarov (1992), Kaplunov et al. (1998), Smirnov and Tovstik (2001).

The first problem one encounters when analyzing a composite plate or a plate of complex structure is the problem of approximating such a plate with a suitable homogeneous plate. The reason for searching for an approximation is that a composite plate is formed from many constituent elements. A direct solution for many elements is impossible even on a modern computer. For example, the number of simple (basic) elements that form a plate with a honeycomb core varies from 10^4 to 10^6. Solving a problem of joint deformation of so many elements is difficult (often impossible) even with a modern computer. Direct numerical analysis of large systems may not give us an exact solution, even if we use a powerful computer. For example, numerical analysis of the strength of a composite in the key paper (Babŭshka et al. 1999) did not detect the energy localization effect between the closely placed particles (Kolpakov 2007; Kolpakov and Kolpakov 2007, 2010). It was because the finite elements mesh was regular and not condensed in the necks between closely placed particles. Obviously, one could have made the mesh finer in the necks if one had known about the (at that time unknown) localization effect. Note that the localization effect was discovered later numerically (Kolpakov 2007; Rakin 2014).

Returning to the two examples above, we can indicate the way of resolution to the problems by using the homogenization theory (the justifications and details of solutions will be presented below, see also Kolpakov 2004).

1. *Cross-section*. The resultant forces and moments in the plate or beam are determined by integration over the three-dimensional periodicity cell or representative cell of the plate or beam. In this case, the mechanical concept of a cross-section is absent altogether. It does not lead to a contradiction with classical theory

because for uniform (classical) plate the periodicity cell is cylinder with the base coinciding with the cross-section.
2. *Neutral plane/axis*. It was demonstrated (Kolpakov 2004) that there exist the "invariant" forms of the plate/beam model which do not depend (have the same form) on translation of the coordinate system in the direction normal to the plate plane or beam axis.

If we decide that it would be useful to have a theory of fine heterogeneous structures developed on the basis of three-dimensional elasticity theory without any a priori hypotheses, we are faced with the question:

(1) what can be adopted as the first-principles model(s) and
(2) what methods can be used to investigate the adopted model(s) to achieve the goal.

The author takes the linearized model of an elastic body with initial stresses (Washizu 1992) as the starting (the first principle) model. A weighty argument for this choice is that all classical models of structural elements subjected to initial stresses: composite solids, plates and membranes, beams and strings; can be derived from that model (it was done in Kolpakov 2004). All the derived models coincide in form with the corresponding engineering models, when the corresponding engineering model exists.

The author considers the asymptotic homogenization method as suitable for solving the second problem. The possibility of examining a homogeneous solid instead of examining the original inhomogeneous composite solid, was one of the principal results of the mathematical theory of homogenization, see Bensoussan et al. (1978), Oleinik et al. (1962), Jikov et al. (1994). Now, the asymptotic method is successfully applied to many problems in the theory of composite materials and complex structures. The advantage of asymptotic homogenization is that it uses three-dimensional elasticity models as the starting point, analyzes these models without any simplified hypothesis, and it can be applied to all types of structural elements: solids, plates, rods (see Caillerie 1984; Trabucho and Viano 1996; Kolpakov 1991). On the basis of the homogenization theory, engineering methods for analysis of composite materials and structures were developed, see, e.g., Annin et al. (1990), Kalamkarov (1992), Kalamkarov and Kolpakov (1997), Oskay and Pal (2010). The homogenization theory does not assume a direct full-size solution to the original problem. In the homogenization theory, it is sufficient to solve so-called periodicity cell problems (or local perturbation problems Gaudiello and Kolpakov 2011; Kolpakov 2011, if joints are considered).

It should be noted that the linearized elasticity problem, in the form as it was formulated in Washizu (1992) was used as a starting point for the analysis of bodies subjected to initial stresses, including the problem of the stability theory, for a long time (Timoshenko 1961; Biot 1939; Green et al. 1952; Machado and Cortinez 2007). The disadvantage of earlier approaches was that the linearized elasticity model was used in the couple with the Kirchhoff-Love hypothesis. Note that most works based at the linearized elasticity model (Washizu 1992) were devoted to volumetric bodies, see e.g., Guz (1999), Guz (1986), Pan'kov (2022).

In spite of the contradiction in the Kirchhoff-Love hypothesis (the main one of which is that the thickness of the plate does not change during deformation despite the Poisson effect Timoshenko and Woinowsky-Krieger 1959) they provide us with two-dimensional models, which are confirmed by centuries of engineering practice. A similar situation takes place for Euler-Bernoulli beam model. One meets here the situations where contradictory hypotheses lead to the correct final results. Note, that the fact that a wrong premise can lead to correct results is well-known from the course of logic (Walicki 2011). It is also known from a logic course that the incorrect premise can lead to both the correct and the incorrect result. Therefore, the argument "the theory contains contradictions, but it works", generally, cannot be accepted.

Once one wants to develop a consistent theory of thin-walled structures, the solution comes from the use of asymptotic methods. An early example of the application of asymptotic methods to problems with no initial stresses are the papers (Gol'denveizer 1961, 1962, 1963) devoted to the "asymptotic integration of the equations of the theory of elasticity". An early example of the application of asymptotic methods to the problem with initial stress is the paper (Alekseev 2002). In Alekseev (2002), an expansion in the Legendre series was used. The proposal to use a power series for the same purpose was expressed in Washizu (1992). The mentioned above approaches have a significant limitation, they are suitable for homogeneous plates of constant thickness, but not for non-homogeneous plates. Note that the use of asymptotic methods in engineering was, if not blocked, then restricted. There were weighty reasons for this restriction. On the one hand, conflicting hypotheses of Kirchhoff-Love, still lead to correct results for uniform plates. On the other hand, asymptotic methods are beyond the standard courses of mathematics for engineers. The problem became actual when the use of composite plates and plates of complex structure was started.

Historically, the first works in which a rigorous model of composite plate was derived from a three-dimensional elasticity theory problem by using a modification of the homogenization method were (Caillerie 1982, 1984). They were followed by works by other authors, see, e.g., Babŭshka and Li (1991), Destuynder (1981), Mielke (1995), Nazarov (2002). Note that the computations based on the asymptotic homogenization method were more compact and rigorous rather then the "asymptotic integration" computations. It is because the asymptotic homogenization method uses the methods of functional analysis and modern methods of partial differential equations (Bensoussan et al. 1978; Duvaut 1976; Jikov et al. 1994).

For plates with initial stresses, the homogenization method was used for the first time for the analysis of two-dimensional models (Kolpakov 1987; Mignot et al. 1980). The results of these studies are valid if the characteristic dimension of inhomogeneities is large as compared to the thickness of the plate. This condition is not fulfilled for many structures used in practice, for example, for lattice plates (Pshenichnov 1993) or perforated plates (Andrianov et al. 1995), but not for plates made of composite materials or plates with rapidly varying thickness. In the last cases, the three-dimensional elasticity problem should be taken as the original model.

Analysis of the inhomogeneous thin-walled structural elements subjected to initial stresses on the basis of the asymptotic homogenization method as applied to a three-dimensional elasticity theory problem started in 1990s. In Kolpakov (1992), Kolpakov (2001), asymptotic analysis was applied to three-dimensional composite solids of small diameter with initial stresses, and a one-dimensional model for a inhomogeneous beam with initial stresses was obtained. In Kolpakov (1995), Kolpakov and Sheremet (2004), the homogenization method was applied to a body of small thickness, and a two-dimensional model for a inhomogeneous plate with initial stresses was obtained. In Kolpakov and Sheremet (1997) two-dimensional membrane model and one-dimensional string model were derived from a three-dimensional elasticity theory problem. The results of the mentioned investigations corresponding to the classical thin-walled structures were summarized in Kolpakov (2004).

The new approach to inhomogeneous structures subjected to initial stresses leads to some significant changes in the concepts of thin-walled structures.

It should be noted that the analysis of problems in the mechanics of composites using the homogenization method requires knowledge of mathematics beyond what is typically covered in traditional textbooks on structural mechanics and the strength of materials. While the level of mathematics used in the book does not exceed the level of modern university mathematics courses, the reading of the book may require some mathematical efforts from an engineer. As a generous reward for some mathematical effort, the reader will gain access to a thorough theory of composite materials and structures with initial stresses. The examples given in the book will show readers that the behavior of composite materials and structures is not always similar to that of homogeneous materials and structures. However, the book also demonstrates that for homogeneous materials and structures, an approach based on homogenization theory leads to classical models. In addition to the fact that homogenization is a mathematically rigorous method, this will give the reader another reason to trust the methods and models presented in the book.

References

Alaire G (2002) Shape optimization by homogenization method. Springer, Berlin

Alekseev AE (2002) Linearized equations of nonlinear elastic deformation of thin plates. J Appl Mech Tech Phys 43(1):133–139

Almgren RF (1985) An isotropic three-dimensional structure with Poisson's ratio = –1. J Elast 15:427–430

Amabili M (2008) Nonlinear vibrations and stability of shells and plates. Cambridge University Press, Cambridge

Andrianov IV, Shevchenko VV, Kholod EG (1995) Asymptotic methods in the statics and dynamics of perforated plates and shells with periodic structures. Tech Mech 15(2):141–157

Annin BD, Kalamkarov AL, Kolpakov AG (1990) Analysis of local stresses in high modulus fiber composite. In: Localized damage computer-aided assessment and control, Vol 2. Comp. Mech. Publ., Southampton, pp 131–144

Babushka I, Anderson B, Smith P, Levin K (1999) Damage analysis of fiber composites. Part I. Statistical analysis on fiber scale. Comput Method Appl Mech Eng 172:27–77

Babushka I, Li L (1991) Hierarchic modelling of plates. Comput Struct 91:419–430

Ball JM (2002) Some open problems in elasticity. In: Newton P, Holmes P, Weinstein A (eds) Geometry mechanics and dynamics. Springer, New York, pp 3–59
Bendsøe MP, Sigmund O (2004) Topology optimization. Springer, Berlin
Bensoussan A, Lions J-L, Papanicolaou G (1978) Asymptotic analysis for periodic structures. North-Holland Publication, Amsterdam
Biot MA (1939) Non linear theory of elasticity and the linearized case for a body under initial stress. Philos Mag 27.89–115
Boscolo M, Fazzolari FA, Banerjee JR (2013) Buckling of composite plate assemblies using higher order shear deformation theory-an exact method of solution. Thin-Walled Struct 71:18–34
Broutman LJ, Krock RH (eds) (1974) Composite materials. In: Eight volumes. Academic, New York, London
Bryan GH (1891) On the stability of a plane plate under thrusts in its own plane, with applications to the "buckling" of the sides of a ship. Proc Lond Math Soc 22:54–67
Bryan GH (1888) On the stability of elastic systems. Proc Cambridge Philos Soc 6:199–210
Caillerie D (1982) Plaques elastiques minces a structute periodique de periode et depaisseur comparable. C R Acad Sci Ser 2 294(3):159–162
Caillerie D (1984) Thin elastic, periodic plates. Math Method Appl Sci 6:159–191
Chakraborty S, Dash S, Dey T (2024b) Non-linear stability characteristics of randomly distributed cnt-reinforced fiber composite beams under thermo-mechanical loadings. Mech Based Des Struct Mach Int J 52(9):7021–7044
Chakraborty S, Naskar S, Dey T, Kumar R, Mukhopadhyay T (2024a) Nonlinear stability of curved multiphase composite panels: Influence of agglomeration in randomly distributed carbon nanotubes with nonuniform in-plane loads. J Aerosp Eng 37(3)
Ciarlet G (1990) Plates and junctions in elastic multi-structures. An asymptotic analysis. Macon, Paris
Ciarlet PG, Destuynder P (1979) A justification of two-dimensional plate model. J Mecan 18:315–344
Ciarlet PG, Kesavan S (1981) Two-dimensional approximations of three-dimensional eigenvalue problems in plate theory. Comput Meth Appl Mech Eng 26(2):145–172
Destuynder P (1981) Unu Theorie Aymptotiques des Plaques Mincea en Elasticite Lenearie. Masson, Paris
Duvaut G (1976) Analyse fonctionnelle et mecanique des milieux continus—application a l'etude des materiaux composites elastiques a structure periodique—homogeneisation. In: Koiter W (ed) Proceedings 14th IUTAM congress. North Holland, Amsterdam, pp 119–132
Euler L (1759) Sur la force des colonnes. Histoire de l'Academie Royal des sciences et belle-lettres avec les Memoires. Tires des Registres de cette Academie 13:252–282
Foraboschi P (2013) Layered plate with discontinuous connection: exact mathematical model. Comput Part B: Eng 47:365–378
Gaudiello A, Kolpakov AG (2011) Influence of non degenerated joint on the global and local behavior of joined rods. Int J Eng Sci 49(3):295–309
Ghorbani A, Mirzaali MJ, Roebroek T, Coulais C, Bonn D, van der Linden E, Habibi M (2024) Suppressing torsional buckling in auxetic meta-shells. Nat Commun 15(6999)
Gol'denveizer AL (1961) Theory of elastic thin shells. Pergamon, Oxford
Gol'denveizer AL (1962) Derivation of an approximate theory of bending of a plate by the method of asymptotic integration of the equations of the theory of elasticity (in Russian). Prikl Math Mekh 26:668–686
Gol'denveizer AL (1963) Derivation of an approximate theory of shells by means of asymptotic integration of the equations of the theory of elasticity (in Russian). Prikl Math Mekh 27:593–608
Goriely A, Vandiver R, Destrade M (2008) Nonlinear Euler buckling. Proc R Soc A 464(2099):3003–3019
Green AE, Rivlin RS, Shield RT (1952) General theory of small elastic deformations superposed on finite elastic deformations. Proc R Soc Ser A 211(1104):128–154

Guz AN (1986) Elastic waves in bodies with initial stresses. In: Two volumes. Naukova Dumka, Kiev (in Russian)
Guz AN (1999) Fundamentals of the three-dimensional theory of stability of deformable bodies. Springer, Berlin
Haug EJ, Choi KK, Komkov V (1986) Design sensitivity analysis of structural systems. Academic, Orlando
Hill R (1958) A general theory of uniqueness and stability in elastic-plastic solids. J Mech Phys Solids 6(3):236–249
Hu Ch-F, Li Zh, Hu Q-S (2021) On non-linear behavior and buckling of arch-beam structures. Eng Struct 239:112214
Jikov VV, Kozlov SM, Oleinik OA (1994) Homogenization of differential operators and integral functionals. Springer, Berlin
Kalamkarov AL, Kolpakov AG (1997) Analysis, design and optimization of composite structures. Wiley, Chichester
Kalamkarov AL (1992) Composite and reinforcement of elements of construction. Wiley, Chichester
Kaplunov JD, Kossovich LY, Nolde EV (1998) Dynamics of thin walled elastic bodies. Academic, New York
Kaplunov JD, Nolde EV, Rogerson GA (2002) Short-wave motion of a pre-stressed plate. IMA J Appl Math 67:383–399
Kelly A, McCartney LN, Clegg WJ, Stearn RJ (2005) Controlling thermal expansion to obtain negative expansivity using laminated composites. Comput Sci Tech 65(1):47–59
Kelly A, Rabotnov YN (eds) (1988) Handbook of composites. North-Holland, Amstredam
Kesavan S (1979) Homogenization of elliptic eigenvalue problems. Parts I and II. Appl Math Optim 5:157–167 and 197–216
Kolpakov AG (1995) Asymptotic problem stability of beams. Buckling under torsion. Appl Mech Tech Phys 6:133–141
Kolpakov AG (2004) Stressed composite structures: homogenized models for thin-walled nonhomogeneous structures with initial stresses. Springer, Berlin
Kolpakov AA, Kolpakov AG (2010) Capacity and transport in contrast composite structures: asymptotic analysis and applications. CRC Press, Boca Raton, FL
Kolpakov AG, Sheremet IG (2004) 2-D model of stresses plate derived from the 3-D elasticity theory model. In: Kienzler R, Ott I, Altenbach H (eds) Theories of plates and shells. Critical review and new applications, volume 16 of Lecture notes in applied and computational mechanics. Springer, Berlin, pp 97–106
Kolpakov AG (1985) On determination of the average characteristics of elastic frameworks. J Appl Math Mech 49:739–745
Kolpakov AG (1987) Averaging in problems of the bending and oscillation of stressed inhomogeneous plate. J Appl Math Mech 51(1):44–49
Kolpakov AG (1991) Calculation of the characteristics of thin elastic rods of periodic structure. J Appl Math Mech 57:440–448
Kolpakov AG (1992) On problem of beam with initial stress. Appl Mech Tech Phys 6:139–144
Kolpakov AG (2001) On the calculation of rigidity characteristics of the stressed constructions. Int J Solids Struct 38(15):2469–2485
Kolpakov AG (2005) Effect of inflation of initial stresses on the homogenized characteristics of composite. Mech Mater 37(8):840–854
Kolpakov AA (2007) Numerical verification of existence of the energy concentration effect in high-contrast high-filled composite material. J Eng Phys Thermophys 4:812–819
Kolpakov AG (2011) Influence of non degenerated joint on the global and local behavior of joined plates. Int J Eng Sci 49(11):1216–1231
Kolpakov AA, Kolpakov AG (2007) Asymptotic of capacity of a system of closely placed bodies. Dokl Phys 415(2):188–192
Kolpakov AG, Rakin SI (1986) On problem of sintsesys of composite material of unidirectional structure with prescribed characteristics. J Appl Mech Tech Phys 6:143–150

Kolpakov AG, Sheremet IG (1997) Asymptotic model of membrane and string. J Appl Math Mech 61(5):845–85
Lakes R (1987) Foam structures with negative Poisson's ratio. Science 235:1038
Lakes R (1996) Cellular solids structures with unbounded thermal expansion. J Mater Sci Lett 15(5):475–477
Lal A, Markad K (2021) Static and dynamic nonlinear stability analyses of hybrid sandwich composite beams under variable in-plane loads. J Mech Sci Technol 35:3895–3908
Li GQ, Jiang J, Zhu Q (2015) Local buckling of compression flanges of h-beams with corrugated webs. J Constr Steel Res 112:69–79
Lopez Jimenez F, Triantafyllidis N (2013) Buckling of rectangular and hexagonal honeycomb under combined axial compression and transverse shear. Int J Solids Struct 50:3934–3946
Machado SP (2008) Non-linear buckling and postbuckling behavior of thin-walled beams considering shear deformation. Int J Non-Linear Mech 43:345–365
Machado S, Cortinez VH (2007) Free vibration of thin-walled composite beams with static initial stresses and deformations. Eng Struct 29(3):372 382
Marcal P, Yamagata N (2016) Buckling analysis of grid-stiffened composite shells. In: Design and analysis of reinforced fiber composites, pp 1–18. Springer, Cham
Marguerre K (1938) Die behandlung der stabilitätsproblemen mit hilfe der energetishen methoden. Z Angew Math Mech 18(1):57–73
Mielke A (1995) To the justification of plate theories in linear elasticity theory using exponential decay estimates. J Elast 38:165–208
Mignot F, Puel J-P, Suquet P-M (1980) Homogenization and bifurcation of perforated plates. Int J Eng Sci 18:409–414
Miller W, Smith CW, Mackenzie DS, Evans KE (2009) Negative thermal expansion: a review. J Mater Sci 44:5441–5451
Miller W, Smith CW, Evans KE (2011) Honeycomb cores with enhanced buckling strength. Comput Struct 93:1072–1077
Nazarov SA (2002) Asymptotic analysis of thin plates and rods. Nauchay Kniga, Novosibirsk (in Russian)
Novikov VV, Wojciechowski KW (1999) Negative Poisson's ratio of fractal structures. Phys Solid State 41(12):1970–1975
Oleinik OA, Shamaev AS, Yosifian GA (1962) Mathematical problems in elasticity and homogenization. North Holland, Amsterdam
Oskay C, Pal G (2010) Multiscale failure model for analysis of thin heterogeneous plates. Int J Damage Mech 19(5):575–610
Pan'kov AA (2022) Effect of the initial stress state on the effective properties of piezocomposite. Mech Comput Mater 58:733–746
Pichugin AV, Rodgerson GA (2004)An asymptotic model for long-wave symmetric motion of a prestressed incompressible plate with fixed faces. In: Kienzler R, Ott I, Altenbach H (eds) Theories of plates and shells. Critical review and new applications, volume 16 of Lecture notes in applied and computational mechanics. Springer, Berlin, pp 177–184
Prager W (1947) The general variational principle of the theory of structural stability. Q Appl Math 4(4):38–378
Pshenichnov GI (1993) Theory of laticced plates and shells. World Scientific, Singapore
Rakin SI (2014) Numerical verification of the existence of the elastic energy localization effect for closely spaced rigid disks. J Eng Phys Thermophys 87:246–252
Smirnov AL, Tovstik PE (2001) Asymptotic methods in the buckling theory of elastic shells. World Scientific, Singapore
Sun HJ, Guo YL, Wen CB, Zuo JQ (2023) Local and global buckling prevention design of corrugated steel plate shear walls. J Build Eng 68:106055
Timoshemko S, Goodier JN (1951) Theory of elasticity. McGrow-Hill, New York
Timoshenko S (1961) Theory of elastic stability, 2nd edn. McGraw-Hill, New York
Timoshenko S, Woinowsky-Krieger K (1959) Theory of plates and shells. McGraw-Hill, New York

Trabucho L, Viano JM (1996) Mathematical modeling of rods. In: Ciarlet PG, Lions J-L (eds) Handbook of numerical analysis, vol IV. Elsevier, Amsterdam, pp 487–974
Trefftz E (1930) Uber die ableitung der stabilitäs-kriterien des elastichen gleichgewichtes aus der elastitätstheorie endlicher deformationen. In: Proceedings of the 3rd international congress for applied mechanics, pp 44–50
Turvey GJ, Marshall IH (2012) Buckling and postbuckling of composite plates. Springer, Berlin
Ventsel E, Krauthammer T (2001) Thin plates and shells. Theory analysis, and applications. Marcel Dekker, New York
Walicki M (2011) Introduction to mathematical logic. World Scientific, Singapore
Wang D, Abdalla M (2002) Buckling analysis of composite plates. In: Durban D, Givoli D, Simmonds JG (eds) Advances in the mechanics of plates and shells. Solid mechanics and its applications, vol 88. Kluver, Dordrecht, pp 135–150
Washizu K (1992) Variational methods in elasticity and plasticity. Pergamon Press, Oxford
Wozniak M, Wierzbicki E, Wozniak C (2004) Macroscopic modelling of prestressed microperiodic elastic media. Acta Mech 173(1–4):107–117
Zhang G, Hu Y, Yan B, Tong M, Wang F (2024) Buckling and post-buckling analysis of composite stiffened panels: a ten-year review (2014–2023). Thin-Walled Struct 205(B):112525

Chapter 2
Thin-Walled Structural Elements Operating Under Initial Stresses

In this chapter, we provide examples of thin-walled structures and structural elements in which initial stresses are significantly used and/or manifested.

It is noted in literature, see, e.g., Shams et al. (2011), Kumar et al. (2021) that the presence of initial stresses in solid materials can have a substantial effect on their subsequent response to applied loads that is very different from the corresponding response in the absence of initial stresses. The effect of initial stresses is clearly manifested in the mechanics of textile materials.

Initial stresses may arise from various causes: from applied loads, as in the case of gravity; processes of growth and development in living tissue, the heating, and many other reasons of natural and artificial nature. In the case of engineering components, initial stresses may arise from the manufacturing process, either by design to improve the performance of the component or unintentionally (which can lead to defective behavior).

In this book, the term initial stress is used in its broadest sense, irrespective of how the stress arises. This includes situations where the stress is due to an applied load leading to an accompanying finite deformation from an unstressed configuration, in which case the term prestress is commonly used. If an initial stress is present in the absence of applied loads (body forces and surface stresses), it is referred to as residual stress.

There are two widely used approaches to the problems related to initial stresses. One approach is based on the finite deformations elasticity theory. The main motivation for this approach comes from soft tissue biomechanics and residual stresses in artery walls, see Holzapfel and Ogden (2010) and references therein. It is supported by the experimental observations of large deformations in the mentioned objects. The development of constitutive laws for residual stressed materials has been a focus of the work (Hoger 1985, 1986, 1993; Johnson and Hoger 1993), see also Saravanan (2008).

Another approach is based on the small deformation elasticity theory, a body with initial stresses. This approach is related to engineering structures and geophysics

A. Kolpakov, *Thin-Walled Structures with Initial Stresses*, Mathematical Engineering,
https://doi.org/10.1007/978-3-032-07247-4_2

problems. In the context of engineering structure problems, mention the brief but very insightful discussion on the initial stresses by Washizu (1992). In the context of acoustoelasticity, mention work by Biot developed within the geophysical context and is particularly concerned with the effect of initial stress on the propagation of small-amplitude elastic waves. Biot developed a static theory of small deformations influenced by initial stress (Biot 1939), followed by a corresponding theory for wave propagation (Biot 1940), see also Biot (1965).

In this book, the approach based on the theory of small-strain elasticity for a body with initial stresses predominates.

Present examples of initial stresses in some structural elements and discuss briefly the related problems.

2.1 Plates and Plate-Like Structural Elements with Initial Stresses

Thin-walled structures: plates, shells, membranes, etc. are found in many branches of engineering, such as civil, mechanical, aeronautical, marine and chemical engineering. Such widespread use of thin-walled structures is due to their intrinsic properties. With appropriate design, even very thin plates, and especially shells, can withstand heavy loads. Thus, they are used in structures such as aerospace vehicles, in which low weight is important. Composite technologies provide engineers with additional opportunities to both improve the properties of thin-walled structures and to impart new properties to them. Another application of shell engineering is in the field of biomechanics: shells are found in various biological forms, such as the eye and the skull, as well as in plant and animal forms.

Prestressing is an extra tool to influence the properties of plates and shells and improve their mechanical characteristics and consumer properties.

2.1.1 Prestressed Concrete Slabs

Prestressed concrete is defined as concrete that has been prestressed so that the induced internal actions resist the external load to the desired degree (King 1956; Masterman 1952). Prestressed concrete members are often produced by tensioning steel reinforcing bars, see Fig. 2.1, although other prestressed concrete technologies are available (Fotinos 1974).

There exist numerous textbooks and codes on prestressed concrete, see, e.g., Abeles and Bardhan-Roy (1981), Nawy (2005), Dolan and Hamilton (2018), Libby (1990) (this is a very short list of the textbooks and codes on the subject). Information on concrete prestressing technology and equipment can be found, for example, in

Fig. 2.1 Stress of reinforcement in concrete structure

James (2012), Kong and Evans (1987), Dolan and Hamilton (2019), Menn (1990), Raju (2006) and references in these books.

A characteristic feature of concrete is the presence of microcracks. Thus, when considering prestressed concrete, it is necessary to keep in mind not only the problem of initial stresses, but also the problem of a medium with a system of cracks.

Prestressed concrete structural members are usually manufactured in the form of beams, lattice beams, and slabs (Fig. 2.2). In thin-walled structures, the initial stresses and additional stresses can be (and often are) of the same order of magnitude. This determines the importance of taking into account the initial stresses in thin-walled prestressed concrete structures. The theme is widely discussed and there exists extensive literature on specific reinforced concrete plate-like structural elements as well as constructions made of reinforced concrete plate-like structural elements (precast concrete structures). We indicate the references Allen (1992), Bhatt (2019), Hurst (2019), Nawy (2000), Nicholson (1997), which include additional references to the theme.

2.1.2 Textiles

Textiles and textile structures are another type of material where initial stresses play a decisive role. This is due to the fact that the textile structural elements are often operated in a membrane mode at the macro level. Another characteristic feature of textiles is that the structural elements of textiles operate in contact interaction with each other at the micro level (Fig. 2.3).

A textile is a flexible material made by creating an interlocking network of yarns. Yarns are produced by various spinning technologies, from length-limited or endless fibers (from either natural or synthetic sources) to long and twisted lengths. Through a defined interlocking of one or many yarn systems, a two- or three-dimensional textile is formed by means of textile technologies (weaving, weft knitting, warp knitting, etc.). A technical textile is a textile product manufactured for non-aesthetic purposes, where function is the primary criterion. Introduction to technical textiles from a mechanical point of view can be found in Chou and Ko (1989), Byrne (2000), Milwich et al. (2006). Technical textiles include textiles for automotive applications, medical textiles (e.g., implants and bandages), geotextiles (reinforcement of embankments), agrotextiles (textiles for crop protection), protective clothing with thermophysiological effect (e.g., heat and radiation protection for firefighter clothing), functional and sports clothing, as well as light-weight constructions. The underlying textile structure, i.e., the woven or knitted fabric made of threads, determines the mechanical behavior and can be analyzed rigorously with the help of asymptotic analysis and homogenization theory (Griso et al. 2020a, b).

An application of homogenization and asymptotic analysis to the problems arising in the mechanics of textiles was presented in Griso et al. (2020b), Orlik et al. (2016), Cioranescu et al. (2013), Griso and Orlik (2018), Shiryaev and Orlik (2016), Shiryaev

Fig. 2.2 Reinforced concrete slab

Fig. 2.3 Textiles: glass fiber textile (left), aluminum fiber textile (right)

et al. (2018), Orlik et al. (2021), Kouznetsova et al. (2004), see also references in the mentioned papers. In the mentioned papers, it was noted that the main modeling issue of textiles is the contact between yarns.

2.2 Beams and Beam-Like Structural Elements with Initial Stresses

Beam and beam-like structures are widely used in mechanical, civil, and aerospace engineering. Many structural components can be approximated as beam elements, a few examples being aircraft wings, automobile drive shafts, and diving boards. For many types of beam problems, exact solutions based on solid mechanics theory exist. However, for many beams with complex geometry or made of heterogeneous materials, an exact solution is either absent or difficult to find. Numerical solutions for such cases can serve as a good alternative approach. Unfortunately, if the beam geometry is sufficiently complex or the material is highly heterogeneous, numerical methods may not be able to solve the problem.

2.2.1 *Prestressed Beams and Columns*

Above, we have given examples of prestressed concrete slabs. Many prestressed concrete structural elements are manufactured in the form of beams and columns. Everything that has been said about prestressed slabs is also valid as applied to beams and columns. Suppliers of prestressed beams offer structural engineers a huge variety of beams and columns for use in different situations, see, for example, Richardson (2003).

A specific example of beam-like structural elements is prestressed concrete pipes. Prestressed concrete pipes are large-diameter concrete pressure pipes used for transporting water and wastewater. Prestressed concrete pipe is a composite structure that is composed of a concrete core, high-tensile prestressed steel wires, a thin steel cylinder, and mortar coating (Kejie et al. 2021).

2.2.2 *Stressed Rails in Continuous Welded Rail Technology*

Continuous welded rails are track segments welded together to form a continuous miles-long track. With respect to mechanically-jointed rails, the continuous welded rails are stronger, smoother, require less maintenance, and can be traveled at higher speeds.

Despite these advantages, the continuous welded rails are prone to fracture during cold seasons and to instability during warm seasons, when the increase in the steel temperature induces extreme compression. Continuous welded rails are subjected to

Fig. 2.4 Rails stressing with hydraulic equipment

longitudinal stresses in rails due to temperature influence, as well as inhomogeneous stress accumulation due to train loadings. The stresses cause the accelerated track lateral irregularity accumulation that, without timely maintenance, can cause track buckling, see Fig. 2.5.

This type of railway track is designed to withstand significant (compared to traditional section track) temperature stresses that occur on the lines of rails when the ambient temperature fluctuates. Mechanical stresses that occur on the lines of rails during operation reach values of more than 100 tons on each of the lines.

Rails can be artificially stressed depending on the temperature during installation and the operating temperature. The stressing process involves either heating the rails or stretching the rails with hydraulic equipment, see Fig. 2.4. In the figure, the jacks (T) pull the rails together, and one section (W) is welded using thermite welding. Additional information in the field may be found in Enshaeian and Rizzo (2021), Atapin et al. (2022), Pucillo (2016), Yang and Bradford (2016), Lim et al. (2003).

We note that the track buckling is a deformation of a system "rail lines and sleepers", as it is seen in Fig. 2.5, not a buckling of a single rail line or an independent buckling of rail lines.

2.3 Initial Stresses in Biology and Medicine

Whatever the source of the initial stress and whether it is accompanied by a finite deformation, of particular interest in many applications is the effect that the initial stress has on small deformations (static or time-dependent). Specifically, deformations are linearized relative to the initial stress-strain state and are often referred to as incremental deformations or motions. Some examples of the application of incremental deformation theories in biology and medicine can be found in Erbay et al. (1987), Back et al. (2013), Parikh et al. (2023), Kassab (2006) and references therein.

2.3.1 Initial Stresses in Blood and Lymphatic Vessels

The role of initial (or residual) stresses is, for example, to ensure that in soft biological tissues in arterial walls the distribution of circumferential stress across the arterial wall thickness is close to uniform at typical physiological blood pressure values (Shams et al. 2011; Berntsson et al. 2016; Kozlov and Nazarov 2016a, b, 2013).

Figure 2.6 gives an image of the dimensions and structure of human aorta (Holzapfel et al. 2007; Du et al. 2019). In the picture, it is shown as unloaded from blood pressure and from axial and radial residual stresses. In the figure, the circumferential residual stress has caused a consequent contraction in the unloaded

Fig. 2.5 Rails buckling

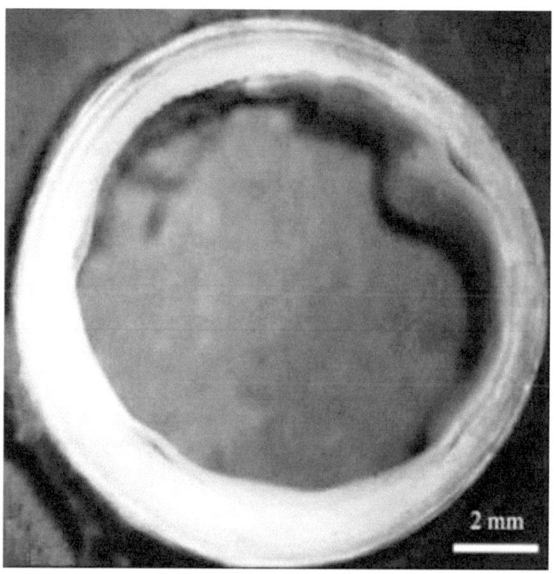

Fig. 2.6 Layer-specific structure of a cut human aorta ring

state, leading to buckling and delimitation on the inner surface. In the human body, the aorta is loaded from blood pressure, and from axial and residual stresses.

2.3.2 *Korotkoff Sounds*

Among the manifestations of initial stresses in medicine that are well known to the ordinary public, we name Korotkoff sounds, which occur with a certain combination of blood pressure load and external stress. The mechanism of Korotkoff sounds was considered and explained by numerous researchers, see, e.g., Baranger et al. (2023), Benmira et al. (2016) and references in these papers.

The mechanism of Korotkoff sounds is investigated until now. For example, the paper (Baranger et al. 2023) proposes the following scheme: (i) by decreasing the brachial artery stiffness, the pressure cuff slows down the pulse wave and amplifies the local arterial wall displacements; (ii) the resulting high velocity of the arterial wall compared to the propagation speed of the pulse wave results in a highly non-linear propagation regime; (iii) the nonlinear vibrations radiate into the surrounding tissues. These tissue vibrations propagate jointly with the arterial pulse wave; (iv) when these shear tissue displacements reach the stethoscope, they are converted into compressional displacements within the stethoscope diaphragm and create the Korotkoff sounds (see for detail the cited paper).

2.3.3 *Initial Stresses in Stomatology*

The role of initial stresses in stomatologic and related problems was discussed in the papers (Geramy 2002; Cobo et al. 1993; Fill et al. 2011). For example, models of the maxilla, teeth, periodontal ligament, lingual-orthodontic appliance and a micro-implant complex were constructed, and corresponding computations were carried out using finite element software. There were developed models of various implants and corresponding simulated models were constructed see for details Aghera et al. (2019) and references therein.

2.4 Initial Stresses in Plants and Vegetables

Biological, in particular, plant objects provide us with numerous examples of bodies with initial/residual stresses. Present below two frequently cited and pictorial examples.

2.4.1 *Initial Stresses in Trees and Wood*

Initial stresses are typical for both trees and wood. It is noted in Gril et al. (2017) that tree growth stress refers to the mechanical stress permanently supported by wood in a living tree during tree growth. In accordance with Gril et al. (2017), it results from the combined action of two mechanisms, i.e., cell wall maturation and the increase of dead weight. The growth stresses in trees are discussed in Archer (1987). In Matsuo-Ueda et al. (2022), one can find experimental information about the values of the residual stresses in the trees.

Residual stresses in wood are the result of biological and technological processes in the green logs (Gril et al. 2017; Muench and Wulf 2023; Kameyama et al. 2023; Gril et al. 2017). Residual stresses are almost always an annoyance for the timber industry. The researchers try to develop the technologies to diminish negative impacts on lumber quality. Many ideas have been proposed and applied for removing or diminishing the residual stress in logs, e.g., smoking, boiling, and steaming. These heating methods contribute to accelerating the growth stress release through increased molecular mobility of cell wall components. Each has merits and demerits from a cost-effectiveness perspective.

2.4.2 *Initial Stresses in Vegetables*

Another frequently cited example of initial stresses in a plant is the initial stresses in carrots, see Du et al. (2019), Vo-Bui et al. (2024). The manifestations of initial stress in carrots can be easily observed by cutting the carrot lengthwise, see Fig. 2.7.

The above brief overview convinces us that we are surrounded by numerous objects, structures and structural elements with initial stresses.

2.5 Stability of Thin-Walled Composite Structural Elements

It would seem that the stability problem should have been the first candidate for study in the theory of bodies with initial stresses. However, this did not happen. In the literature on initial stresses in elastic bodies before (Kolpakov 2004), geophysics and biomechanics were indicated as the main applications of models with initial stresses. An exception, very important but also very brief, is the discussion of the problem of elastic stability in Washizu (1992). The author explains this situation by the existence of recognized classical theories of plate and beam stability developed by outstanding scientists. Classical theories of plate and beam stability work well for homogeneous plates and beams. However, it is unclear how to solve the problem of stability of essentially non-homogeneous thin-walled structural elements using

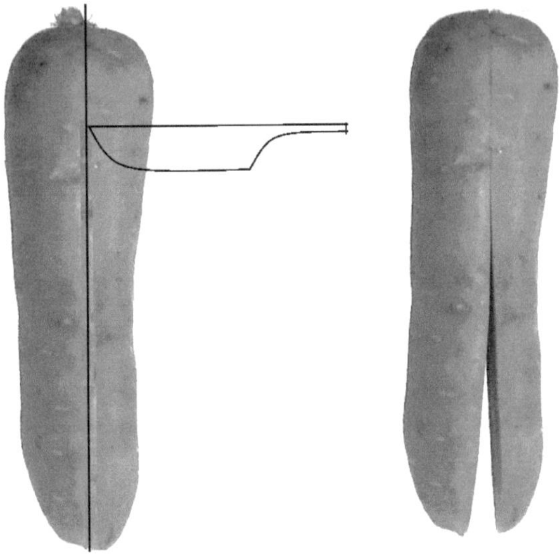

Fig. 2.7 Fresh carrot, cut in a straight line

classical theories. In the review (Sanchez-Palencia 1986), the buckling problem was called one of two "wide open problems" of homogenization in mechanics.

The studies presented in Kolpakov (2004) were initiated by problems that arose in composite materials, when it became clear that classical models were not suitable for composite materials and substantially inhomogeneous thin-walled structural elements.

Buckling modes in composite materials and heterogeneous structures are very diverse. They can demonstrate both similarities and differences with classical buckling modes. Figure 2.8 shows experimentally observed buckling modes of a three-layer plate with a honeycomb middle layer. The upper figure shows buckling in the global (classical) mode. The figures below show buckling in local modes.

Mention the work (Panasenko 1980) where it was discussed the loss of stability in the "ideal periodic" mode, when the period of the loss of stability coincides with the period of structure. It is evident that the "ideal periodic" buckling mode is theoretically available as well as that there exist many alternative local buckling modes. An example of buckling with a period not coinciding with the minimal period of structure can be found in Sanchez-Palencia (1986), see Fig. 2.9. In Fig. 2.9, the periodicity cell of the structure is Y, and the periodicity cell of the buckling is $4Y$.

The discussion in this book corresponds to buckling in global modes. The idea presented in Washizu (1992) opens up prospects for the analysis of other buckling modes in composite structural elements, but this idea is not discussed in this book.

Mention the paper on the homogenization eigenvalue problems (Kesavan 1979; Oleinik et al. 1962; Donato et al. 2021; Vanninati 1981; Nazarov et al. 2011; Pastukhova 2001; Chechkin et al. 2007; Nazarov and Piatnitski 2010), which are close but not equivalent to the problems discussed in this book.

Fig. 2.8 Various modes of the loss of stability of composite plate

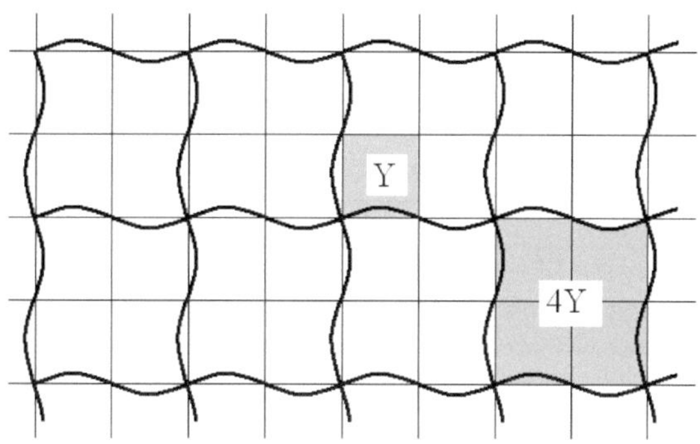

Fig. 2.9 Periodic buckling mode with period different from the period of structure

References

Abeles W, Bardhan-Roy BK (1981) Prestressed concrete designer's handbook. CRC Press, Boca Raton, FL

Aghera D, Shyagali TR, Kambalya P (2019) Evaluation of initial displacement and stresses in the maxillary dentition following en masse retraction using fixed lingual appliance and micro-implants: A finite element analysis. Int Orthod 17(3):451–460

Allen AH (1992) An introduction to prestressed concrete. British Cement Ass, Slough

Archer RR (1987) Internal residual stress patterns in tree stems. In: Growth stresses and strains in trees, volume 3 of Springer series in wood science. Springer, Berlin, pp 66–113

Atapin V, Bondarenko A, Sysyn M, Grun D (2022) Monitoring and evaluation of the lateral stability of CWR track. J Fail Anal Prev 22:319–332

Back M, Gasser TCh, Michel J-B, Caligiuri G (2013) Biomechanical factors in the biology of aortic wall and aortic valve diseases. Cardiovasc Res 99(2):232–41

Baranger J, Villemain O, Goudot G, Dizeux A, Le Blay H, Mirault T, Messas E, Pernot M, Tanter M (2023) The fundamental mechanisms of the Korotkoff sounds generation. Sci Adv 9(40)

Baranger J, Villemain O, Goudot G, Dizeux A, Le Blay H, Mirault T, Messas E, Pernot M, Tanter M (2023) The fundamental mechanisms of the Korotkoff sounds generation. Sci Adv 9(40)

Benmira A, Perez-Martin A, Schuster I, Aichoun I, Coudray S, Bereksi-Reguig F, Dauzat M (2016) From Korotkoff and Marey to automatic non-invasive oscillometric blood pressure measurement: does easiness come with reliability? Expert Rev Med Dev 13(2):179–189

Berntsson F, Karlsson M, Kozlov V, Nazarov SA (2016) A one-dimensional model of viscous blood flow in an elastic vessel. Appl Math Comput 274:125–132

Bhatt ThJ (2019) Prestressed concrete design to eurocodes. CRC Press, Boca Raton, FL

Biot MA (1965) Mechanics of incremental deformations: theory of elasticity and viscoelasticity of initially stressed solids and fluids, including thermodynamic foundations and applications to finite strain. Wiley, New York

Biot MA (1939) Non linear theory of elasticity and the linearized case for a body under initial stress. Philos Mag 27:89–115

Biot MA (1940) The influence of initial stress on elastic waves. J Appl Phys 11:522–530

Byrne C (2000) Technical textiles market—an overview. In: Horrocks AR, Anand SC (eds) Handbook of technical textiles, 1st edn. Woodhead Publication, Sawston, pp 1–23

Chechkin GA, Piatnitski AL, Shamaev SA (2007) Homogenization. Methods and applications. American Math. Soc. Providence

Chou T-W, Ko FK (eds) (1989) Textile structural composites. Elsevier Science Publiction, Amsterdam

Cioranescu D, Damlamian A, Orlik J (2013) Homogenization via unfolding in periodic elasticity with contact on closed and open cracks. Asympt Anal 82(3):201–232

Cobo J, Sicilia A, Arguelles J, Suarez D, Vijande M (1993) Initial stress induced in periodontal tissue with diverse degrees of bone loss by an orthodontic force: Thridimensional analysis by means of the finite element method. Am J Orthod Dentofac Orthop 104(5):448–454

Dolan CW, Hamilton HR (2019) Prestressed concrete. Springer Nature, Cham

Dolan ChW, Hamilton HR (2018) Prestressed concrete: building, design, and construction. Springer Nature, Cham

Donato P, Gemida E, Jose EC (2021) Homogenization of an eigenvalue problem in a two-component domain with interfacial jump. In: Donato P, Luna-Laynez M (eds) Emerging problems in the homogenization of partial differential equations. Springer Nature, Cham

Du Y, Lu C, Destrade M, Chen W (2019) Influence of initial residual stress on growth and pattern creation for a layered aorta. Sci Rep 9:Article number: 8232

Enshaeian A, Rizzo P (2021) Stability of continuous welded rails: a state-of-the-art review of structural modeling and nondestructive evaluation. Proc Inst Mech Eng Part F J Rail Rapid Transit 235(10):1291–1311

Erbay HA, Erbay S, Demiray H (1987) Pulse waves in prestressed arteries. Bull Math Biol 49:289–305

Fill TS, Carey JP, Toogood RW, Major PW (2011) Experimentally determined mechanical properties of, and models for, the periodontal ligament: critical review of current literature. J Dent Biomech 1:312980

Fotinos GC (1974) Design of prestressed concrete pile foundations. PCI J 19(5):40–53

Geramy A (2002) Initial stress produced in the periodontal membrane by orthodontic loads in the presence of varying loss of alveolar bone: a three-dimensional finite element analysis. Eur J Orthod 24(1):21–33

Gril J, Jullien D, Bardet S, Yamamoto H (2017) Tree growth stress and related problems. J Wood Sci 63:411–432

Gril J, Jullien D, Bardet S, Yamamoto H (2017) Unique characteristics of residual stress distribution of large-diameter keyaki (Zelkova serrata) logs and examination of their measurement method. J Wood Sci 63(5):411–432
Griso G, Orlik J (2018) Homogenization of contact problem with coulomb's friction on periodic cracks. Math Method Appl Sci 42(18):6435–6458
Griso G, Orlik J, Wackerle S (2020) Asymptotic behavior for textiles. SIAM J Math Anal 52:1639–1689
Griso G, Orlik J, Wackerle S (2020) Asymptotic behavior for textiles in von-Karman regime. J Math Pures Appl 144(1):164–193
Hoger A (1993) The elasticity tensors of a residually stressed material. J Elast
Hoger A (1985) On the residual stress possible in an elastic body with material symmetry. Arch Rational Mech Anal 88:271–289
Hoger A (1986) On the determination of residual stress in an elastic body. J Elast 16:303–324
Holzapfel GA, Ogden RW (2010) Modelling the layer-specific 3d residual stresses in arteries, with an application to the human aorta. J R Soc Interf 7:787–799
Holzapfel GA, Sommer G, Auer P, Regitnig M, Ogden RW (2007) Layer-specific 3D residual deformations of human aortas with non-atherosclerotic intimal thickening. Ann Biomed Eng 35:530–545
Hurst MK (2019) Prestressed concrete design, 2nd edn. CRC Press, Boca Raton, FL
James RL (2012) Modern prestressed concrete. Springer, New York
Johnson BE, Hoger A (1993) The dependence of the elasticity tensor on residual stress. J Elast 33:145–165
Kameyama N, Matsuo-Ueda M, Chen S, Jiang Z, Ichiyanagi T, Yoshida M, Yamamoto H (2023) Unique characteristics of residual stress distribution of large-diameter keyaki (Zelkova serrata) logs and examination of their measurement method. J Wood Sci 69(16)
Kassab GS (2006) Interface biomechanics of the cardiovascular system: the aorta as an illustratory example. J R Soc 3(11):719–740
Kejie Zh, Hongyuan F, Chengchao G, Pengpeng N, Bing F, Fuming W, Chao Zh (2021) Strengthening of PCCP with broken wires using prestressed CFR. Constr Build Mater 267(7):120903
Kesavan S (1979) Homogenization of elliptic eigenvalue problems. Parts I and II. Appl Math Optim 5:157–167 and 197–216
King J (1956) Prestressed concrete design. Nature 178:443–444
Kolpakov AG (2004) Stressed composite structures: homogenized models for thin-walled nonhomogeneous structures with initial stresses. Springer, Berlin
Kong FK, Evans RH (1987) Reinforced and prestressed concrete. Springer, New York
Kouznetsova VG, Geers MGD, Brekelmans WAM (2004) Multi-scale second-order computational homogenization of multi-phase materials: a nested finite element solution strategy. Comput Method Appl Mech Eng 193(48–51):5525–5550
Kozlov VA, Nazarov SA (2013) Asymptotic models of the blood flow in arteries and veins. J Math Sci 194(1):44–57
Kozlov VA, Nazarov SA (2016) Asymptotic models of anisotropic heterogeneous elastic walls of blood vessels. J Math Sci 213(4):561
Kozlov VA, Nazarov SA (2016) A simple one-dimensional model of a false aneurysm in the femoral artery. J Math Sci 214(3):287–301
Kumar P, Singh AK, Chattopadhyay A (2021) Influence of an impulsive source on shear wave propagation in a mounted porous layer over a foundation with dry sandy elastic stratum and functionally graded substrate under initial stress. Soil Dyn Earthq Eng 142:106536
Libby JR (1990) Modern prestressed concrete: design principles and construction methods, 4th edn. Springer, New York
Lim N-H, Park N-H, Kang YJ (2003) Stability of continuous welded rail track. Comput Struct 81(22–23):2219–2236
Masterman O (1952) Prestressed concrete. Nature 169:901–905

Matsuo-Ueda M, Tsunezumi T, Jiang Zh, Yoshida M, Yamashita K, Matsuda Y, Matsumura Y, Ikami Y, Yamamoto H (2022) Comprehensive study of distributions of residual stress and Young's modulus in large-diameter sugi (Cryptomeria japonica) log. Wood Sci Technol 56:573–588

Menn C (1990) Prestressed concrete bridges. Birkhauser, Basel

Milwich M, Speck T, Speck O, Stegmaier T, Planck H (2006) Biomimetics and technical textiles: solving engineering problems with the help of nature's wisdom. Am J Botany 93:1455–1465

Muench I, Wulf JB (2023) Residual stresses in green wood based on a phase field model. In: Merta I, Page J, Amziane S (eds) Bio-based building materials. ICBBM 2023, volume 45 of RILEM bookseries. Springer, Cham, pp 899–910

Nawy EG (2000) Prestressed concrete, 3rd edn. Prentice-Hall, Englewood Cliffs

Nawy EG (2005) Prestressed concrete: a fundamental approach, 5th edn. Pearson, Upper Saddle River, NJ

Nazarov SA, Piatnitski AL (2010) Homogenization of the spectral Dirichlet problem for a system of differential equations with rapidly oscillating coefficients and changing sign density. J Math Sci 169(2):212–248

Nazarov SA, Pankratova IL, Piatnitski AL (2011) Homogenization of the spectral problem for periodic elliptic operators with sign-changing density function. Arch Rational Mech Ana 200:747–788

Nicholson BA (1997) Simple bridge design using prestressed beams. Prestressed Concrete Ass, Leicester

Oleinik OA, Shamaev AS, Yosifian GA (1962) Mathematical problems in elasticity and homogenization. North Holland, Amsterdam

Orlik J, Krier M, Neusius D, Pietsch K (2021) Recent efforts in modeling and simulation of textiles. Textiles, MDPI 1(2):322–336

Orlik J, Panasenko G, Shiryaev V (2016) Optimization of textile-like materials via homogenization and beam approximations. Multisc Model Simul 14(2):637–667

Panasenko GP (1980) Averaging of periodic structures with highly conducting inhomogeneities (in Russian). Vest Mosk Univ Comput Math Cybern 3:4–11

Parikh Sh, Moerman KM, Ramaekers MJFG, Schalla S, Bidar E, Delhaas T, Reesink K, Huberts W (2023) Biomechanical characterisation of thoracic ascending aorta with preserved pre-stress. Bioengineering, MDPI 10(7):846

Pastukhova S (2001) Spectral asymptotics for a stationary heat conduction problem in a perforated domain. Math Notes 69(3–4):546–558

Pucillo GP (2016) Thermal buckling and post-buckling behavior of continuous welded rail track. Veh Syst Dyn 54(12):1785–1807

Raju NK (2006) Prestressed concrete. Tata McGraw-Hill Education, New Delhi

Richardson J (2003) Precast concrete structural elements. In: Newman J, Choo BS (eds) Advanced concrete technology, vol 4. Elsevier, Amsterdam, pp 3–46

Sanchez-Palencia E (1986) Homogenization in mechanics. A survey of solved and open problems. Rend Sem Mat Univ Polot Torino 44(1):251–270

Saravanan U (2008) Representation for stress from a stressed reference configuration. Int J Eng Sci 46:1063–1076

Shams M, Destrade M, Ogden RW (2011) Initial stresses in elastic solids: constitutive laws and acoustoelasticity. Wave Motion 48(7):552–567

Shiryaev V, Orlik J (2016) A one-dimensional computational model for hyperelastic string structures with Coulomb friction. Math Method Appl Sci 40(3):741–756

Shiryaev V, Neusius D, Orlik J (2018) Extension of one-dimensional models for hyperelastic string structures under coulomb friction with adhesion. Lubricants, MDPI 6(2):33

Vanninati M (1981) Homogenization of eigenvalue problems in perforated domains. Proc. Indian Acad. Sci. (Math. Sci.) 90(30):239–271

Vo-Bui NA, Rogers BA, Pegg EC (2024) Modelling of longitudinally cut carrot curling induced by the vascular cylinder-cortex interference pressure. R Soc Open Sci 11(1)

Washizu K (1992) Variational methods in elasticity and plasticity. Pergamon Press, Oxford

Yang G, Bradford MA (2016) Thermal-induced buckling and postbuckling analysis of continuous railway tracks. Int J Solids Struct 97–98:637–649

Chapter 3
Homogenization Method in the Linear Elasticity. Basic Approaches

There are now books available which contain excellent expositions of the theory of homogenization and its applications to the natural sciences and engineering. An exposition of the mathematical aspects of the homogenization theory can be found in Bensoussan et al. (1978), Jikov et al. (1994), Cioranescu and Donato (2000). Applications of the homogenization theory to problems in mechanics and physics can be found in Kalamkarov and Kolpakov (1997), Milton (2002). The homogenization method is applied to the design of materials with new and unusual properties (Bendsøe and Sigmund 2004). The books (Sanchez-Palencia 2025; Oleinik et al. 1962; Lewinski and Telega 2000; Gluzman et al. 2018; Drygas et al. 2019) present a combination of mathematical theory with the study of applied problems. There exist review papers on the subject, see, e.g., Sanchez-Palencia (1986), Kanout et al. (2009), Allaire et al. (2019), Charalambakis (2010), Gloria (2012), Nandakumaran (2007).

We will briefly present the procedures of the homogenization theory that are directly used in this book. The presentation will be an introduction to both the methods of the homogenization theory and the theory of elasticity of bodies with initial stresses.

3.1 The Homogenization Method

Homogenization is understood as the substitution of a real non-homogeneous body by a homogeneous body (referred to as the homogenized body) that is close in behavior to the original one. The mentioned homogeneous body can either physically exist or be an abstract (physically non-existent) body. A non-homogeneous body cannot be identical to a homogeneous body. This follows from both mathematical and mechanical reasons. The closeness of the behavior of homogeneous and inhomogeneous bodies can only take place if the characteristic size of the inhomo-

A. Kolpakov, *Thin-Walled Structures with Initial Stresses*, Mathematical Engineering,
https://doi.org/10.1007/978-3-032-07247-4_3

geneities (absolute or relative) of the inhomogeneous body is small. This means that homogenization should be understood as an asymptotic phenomenon.

Although the size of heterogeneities ε tends to zero, ε is always positive. That is, we encounter a situation in calculus courses (Rudin 1967): the argument of the function (it is ε) tends to zero, but does not take zero value. As far as the original function (the characteristic of a composite as a whole), it does not exist for $\varepsilon = 0$, but the limit of this function, as $\varepsilon \to 0$, does exist. Thus, if some characteristic of the composite has a limit, as $\varepsilon \to 0$, then we talk about asymptotic homogenization of this characteristic of the composite. This is the mathematical meaning of homogenization.

Physically, homogenization means that the size of the heterogeneities becomes so small that we begin to perceive the material as homogeneous. Of course, the material does not cease to be heterogeneous due to the limitations of our perception.

From the above explanations, it follows that approximation does not take place for all mechanical characteristics. This means that some mechanical characteristics in the original and homogenized bodies may not coincide, and the differences may be large.

The homogenization of composites of periodic structures is based on various mathematical techniques. The basic homogenization techniques are presented in Spagnolo (1968), Bensoussan et al. (1978), Sanchez-Palencia (2025), Spagnolo (1968), Boccardo and Marcellini (1976), Bakhvalov and Panasenko (1989), Jikov et al. (1994), Trabucho and Viano (1996), Tutek and Aganovich (1987), Alaire (1992), Panasenko (2022). We will give a brief presentation of two approaches used in this book: the asymptotic expansion method and G–limit approach. Note that for linear problems of elasticity theory, both these approaches are equivalent (Kalamkarov and Kolpakov 1997).

3.2 Basic Approaches of the Homogenization Theory

Let us consider a non-uniform elastic body occupying a domain G with a piece-wise smooth boundary ∂G and subjected to a mass force $\mathbf{F}^{\varepsilon}$. The body is assumed to be fixed on the surface $S_u \subset \partial G$, and subjected to a surface stress $\mathbf{H}^{\varepsilon}$ is applied on the surface $S_\sigma \subset \partial G$ (other types of boundary conditions can also be considered, see, e.g., Bensoussan et al. 1978).

We write the problem of elasticity theory in the form

$$L_{\varepsilon}\mathbf{u}^{\varepsilon} + \mathbf{F}^{\varepsilon}(\mathbf{x}) = 0 \text{ in } G, \tag{3.1}$$

$$l_{\varepsilon}\mathbf{u}^{\varepsilon} = \mathbf{H}^{\varepsilon}(\mathbf{x}) \text{ on } S_{\sigma}, \tag{3.2}$$

$$\mathbf{u}^{\varepsilon}(\mathbf{x}) = 0 \text{ on } \partial S_u,$$

where

$$L_\varepsilon \mathbf{u} = \frac{\partial}{\partial x_j}\left(c^\varepsilon_{ijkl}(\mathbf{x})\frac{\partial u_k}{\partial x_l}\right), \tag{3.3}$$

is the elasticity theory operator and (3.1), (3.2) are the boundary value conditions, where

$$l_\varepsilon \mathbf{u} = c^\varepsilon_{ijkl}(\mathbf{x})\frac{\partial u_k}{\partial x_l} n_j \tag{3.4}$$

is the operator describing the normal stresses. In (3.4) and throughout, $\mathbf{n} = (n_1, n_2, n_3)$ means the external normal to S_σ.

In (3.1), (3.2) and throughout, we write mass forces and surface normal stresses following Washizu (1992).

We consider the homogeneous boundary value condition $\mathbf{u}^\varepsilon(\mathbf{x}) = 0$ for the displacement for the sake of simplicity of presentation. Note that the elasticity problem with the general condition $\mathbf{u}^\varepsilon(\mathbf{x}) = \mathbf{u}^0(\mathbf{x})$ can always be transformed into a problem with a homogeneous boundary condition.

For a non-homogeneous body, elastic constants c^ε_{ijkl}, mass forces $\mathbf{F}^\varepsilon$, and surface stress $\mathbf{H}^\varepsilon$ depend on the spacial variable $\mathbf{x}$. They also depend on the characteristic size of the components of composite, which we denote by ε. The index ε marks the original elastic constants, mass forces, and surface stress and reminds that we deal with not a single problem, but with a sequence of problems.

The following standard conditions are applied to the elastic constant $c^\varepsilon_{ijkl}(\mathbf{x})$ (Fichera 1984):

$$\begin{aligned} &C1.\ c^\varepsilon_{ijkl}(\mathbf{x}) = c^\varepsilon_{klij}(\mathbf{x}) = c^\varepsilon_{jikl}(\mathbf{x}); \\ &C2.\ |c^\varepsilon_{ijkl}(\mathbf{x})| < M; \\ &C3.\ c^\varepsilon_{ijkl}(\mathbf{x})e_{ij}e_{kl} \geqslant \gamma|e_{ij}|^2 \text{ for any } e_{ij} \text{ such that } e_{ij} = e_{ji}; \end{aligned} \tag{3.5}$$

for any $\mathbf{x} \in G$, where the constants $0 < m, M < \infty$ do not depend on $\mathbf{x}$, ε, and e_{ij}.

Denote sets of finite infinitely differentiable in G functions

$$C_0^\infty(G) = \{\mathbf{u}(\mathbf{x}) \in C^\infty(G) : \mathbf{u}(\mathbf{x}) = 0 \text{ for } \mathbf{x} \in S_u\},$$

and denote V the set of functions obtained by closure of $C_0^\infty(G)$ in the norm

$$||\mathbf{u}||_1 = \sum_{i=1}^{3}\int_G \left(\sum_{j=1}^{3}\left|\frac{\partial u_i}{\partial x_j}\right|^2 + |u_i|^2\right) d\mathbf{x} \tag{3.6}$$

In the terms of mechanics, V is the set of admissible displacements (Washizu 1992; Sanchez-Palencia 2025).

Denote $\langle , \rangle$ and $\langle , \rangle_{S_\sigma}$ the operations of dual (bilinear Ekeland and Temam 1976) pairing of elements from V and V^* (V^* means dual space for V Ekeland and Temam 1976) and elements from $H^{1/2}(S_\sigma)$ and $H^{-1/2}(S_\sigma)$ (Lions and Magenes 1973) in the standard duality of these spaces, which coincide with the scalar products in $L_2(G)$ for smooth functions:

$$\langle \mathbf{F}, \mathbf{u} \rangle = \int_G \mathbf{F}\mathbf{u} d\mathbf{x}, \ \langle \mathbf{H}, \mathbf{u} \rangle_{S_\sigma} = \int_{S_\sigma} \mathbf{H}\mathbf{u} d\mathbf{x},$$

see Lions and Magenes (1973), Rudin (1992) for details.

Problem (3.1) and (3.2) admits the following variational formulation (see, e.g., Fichera 1984): find $\mathbf{u}^\varepsilon$ from the following minimization problem:

$$J_\varepsilon(\mathbf{u}) - \langle \mathbf{F}^\varepsilon, \mathbf{u} \rangle - \langle \mathbf{H}^\varepsilon, \mathbf{u} \rangle_{S_\sigma} \to \min, \mathbf{u} \in V; \tag{3.7}$$

where

$$J_\varepsilon(\mathbf{u}) = \frac{1}{2} \int_G c^\varepsilon_{ijkl}(\mathbf{x}) e_{ij}(\mathbf{u}) e_{kl}(\mathbf{u}) d\mathbf{x}, \tag{3.8}$$

is the elastic energy and the strains $e_{ij}(\mathbf{u})$ corresponding to the displacements $\mathbf{u}$ are defined by the formula (Timoshemko and Goodier 1951)

$$e_{ij}(\mathbf{u}) = \frac{1}{2}\left(\frac{\partial u_i}{\partial x_j} + \frac{\partial u_j}{\partial x_i} \right). \tag{3.9}$$

The equivalence of problem (3.1), (3.2) and (3.7) is a well-known result (Lions and Magenes 1973; Sanchez-Palencia 2025).

Problem (3.1) and (3.2) can be written as

$$\int_G \sigma^\varepsilon_{ij} \frac{\partial w_i}{\partial x_j} d\mathbf{x} - \int_G \mathbf{F}^\varepsilon \mathbf{w} d\mathbf{x} - \int_{S_\sigma} \mathbf{H}^\varepsilon \mathbf{w} d\mathbf{x} = 0 \text{ for any } \mathbf{w} \in V, \tag{3.10}$$

where stresses σ^ε_{ij} are related with the strains e_{ij} by Hook's low:

$$\sigma^\varepsilon_{ij} = c^\varepsilon_{ijkl}(\mathbf{x}) e_{ij}(\mathbf{u}^\varepsilon), \tag{3.11}$$

where $c^\varepsilon_{ijkl}(\mathbf{x})$ are elastic constants.

Since composite material is non-homogeneous, the elastic constants $c^\varepsilon_{ijkl}(\mathbf{x})$ depend on the spatial variable $\mathbf{x}$. The index ε indicates that the characteristics of composite change their values significantly at the distance of order of ε. Since $\varepsilon \ll 1$, we deal with the material with high-oscillating characteristics.

Taking into account definition (3.9) of strains and symmetries of the elastic constants, see (3.5), we can write the Hooke's low (3.11) in the form

$$\sigma_{ij}^{\varepsilon} = c_{ijkl}^{\varepsilon}(\mathbf{x})\frac{\partial u_i^{\varepsilon}}{\partial x_j}. \tag{3.12}$$

The equivalence of the boundary-value problems (3.1), (3.2), minimization problem (3.7) and variational principle (3.10) may be prove by varying the functional (3.7) and following integrating by parts (for more details, see Lions and Magenes 1973).

A homogenized material associated with a composite is described by the elasticity theory problem

$$L\mathbf{u} = \frac{\partial}{\partial x_j}\left(C_{ijkl}(\mathbf{x})\frac{\partial u_k}{\partial x_l}\right) + F_i = 0 \text{ in } G, \tag{3.13}$$

$$l\mathbf{u} = C_{ijkl}(\mathbf{x})\frac{\partial u_k}{\partial x_l}n_j = H_i \text{ on } S_\sigma, \tag{3.14}$$

$$\mathbf{u}(\mathbf{x}) = 0 \text{ on } \partial S_u, \tag{3.15}$$

with the elastic characteristics $C_{ijkl}(\mathbf{x})$ called the homogenized (also called effective, macroscopic, or average Aboudi 1991; Broutman and Krock 1974; Kalamkarov and Kolpakov 1997; Christensen 2005) elastic characteristics, which do not depend on the parameter ε. In many cases, $C_{ijkl}(\mathbf{x})$ are constants. If C_{ijkl} are constants, we talk about a homogenized material possessing these elastic constants. If $C_{ijkl}(\mathbf{x})$ are not constants, we talk about the homogenized body. The homogenized constants $C_{ijkl}(\mathbf{x})$ are determined by the local elastic constants $c_{ijkl}^{\varepsilon}(\mathbf{x})$ of the composite. Note that a real material or body with the constants $C_{ijkl}(\mathbf{x})$ may not exist at all.

The problem (3.15) can also be written in the following equivalent form: determine $\mathbf{u} \in V$ from the solution to the minimization problem

$$I(\mathbf{u}) - \langle \mathbf{F}, \mathbf{u}\rangle - \langle \mathbf{H}, \mathbf{u}\rangle_{S_\sigma} \to \min, \mathbf{u} \in V; \tag{3.16}$$

where

$$I(\mathbf{u}) = \frac{1}{2}\int_G C_{ijkl}(\mathbf{x})e_{ij}(\mathbf{u})e_{kl}(\mathbf{u})d\mathbf{x}, \tag{3.17}$$

$$\mathbf{F} = \text{w-}\lim_{\varepsilon\to 0}\mathbf{F}^{\varepsilon},\ \ \mathbf{H} = \text{w-}\lim_{\varepsilon\to 0}\mathbf{H}^{\varepsilon};$$

$\text{w-}\lim_{\varepsilon\to 0}$ means the weak limit in V or $H^{1/2}(S_\sigma)$, correspondingly. For smooth quickly oscillating functions, the indicated weak limits coincide with the average value (see for details Lions and Magenes 1973; Bensoussan et al. 1978).

The relationship between the operators L_ε, $L(\sigma)$ and the functionals I_ε and I is known Gajewskii et al. (1974): I_ε and I are potentials of the operators L_ε and $L(\sigma)$, the operators L_ε and L are derivatives in the sense of Gajewskii et al. (1974) of the functionals I_ε and I.

3.2.1 An Approach Based on Asymptotic Expansions for Media of Periodic Structure

Periodic materials and structures are widely used in engineering. In nature, periodic materials and structures are less common. In nature, materials with a random structure or a combination of periodic structure with random deviations from it are more common. The method of asymptotic expansion was originally developed and widely used for analysis of composite with periodic micro (local) structure. As will be demonstrated below, the idea of a representative element (representative volume) allows one to move from the analysis of a composite of random structure to the analysis of a proper composite of periodic structure.

Let a non-homogeneous linear elastic material have a periodic structure with periodicity cell (called also basic cell and periodic cell)

$$\varepsilon Y = [0, \varepsilon Y_1] \times [0, \varepsilon Y_2] \times [0, \varepsilon Y_3],$$

in coordinates x_1, x_2, x_3, see Fig. 3.1.

Hereafter, ε means the characteristic size of the periodicity cell (see Fig. 3.1). The material characteristics $c^{\varepsilon}_{ijkl}(\mathbf{x})$ of the medium under consideration are described by periodic functions of the form

$$c^{\varepsilon}_{ijkl}(\mathbf{x}) = c_{ijkl}(\mathbf{x}/\varepsilon), \tag{3.18}$$

where $c_{ijkl}(\mathbf{y})$ is a periodic function with periodicity cell Y. The mass forces and the surface stresses have the same form: $\mathbf{F}^{\varepsilon}(\mathbf{x}) = \mathbf{F}(\mathbf{x}/\varepsilon)$ and $\mathbf{H}^{\varepsilon}(\mathbf{x}) = \mathbf{H}(\mathbf{x}/\varepsilon)$, corre-

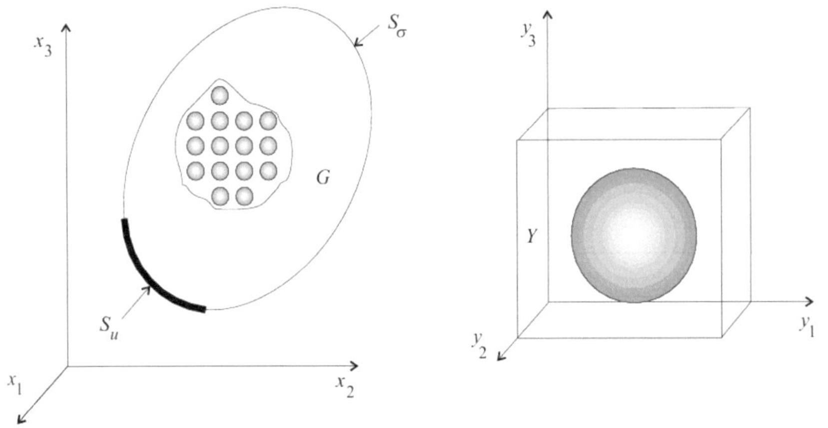

Fig. 3.1 Elastic body of periodic structure (left) and its periodicity cell (right)

spondingly. The mass forces and the surface stresses often have the form $\mathbf{F}(\mathbf{x}, \mathbf{x}/\varepsilon)$ and $\mathbf{H}(\mathbf{x}, \mathbf{x}/\varepsilon)$.

It is natural to distinguish the variables $\mathbf{x}$ associated with the body "as a whole" and the variables $\mathbf{x}/\varepsilon$ used to describe the microstructure of the body. This led to the introduction (see details and historical references in Sanchez-Palencia 2025, 1970) of two scales of variables: the "slow" variables $\mathbf{x}$ and the "fast" variables $\mathbf{y} = \mathbf{x}/\varepsilon$.

The method of asymptotic expansion is based on seeking a solution to the problem (3.1), (3.2) with rapidly oscillating coefficients $c_{ijkl}(\mathbf{x}/\varepsilon)$ (3.18) in the form of a series. Namely, the solution to the Eq. (3.1) is sought in the form of the asymptotic expansion

$$\mathbf{u}^{\varepsilon} = \mathbf{u}^{(0)}(\mathbf{x}) + \varepsilon\mathbf{u}^{(1)}(\mathbf{x}, \mathbf{y}) + \ldots = \mathbf{u}^{(0)}(\mathbf{x}) + \sum_{k=1}^{\infty} \varepsilon^k \mathbf{u}^{(k)}(\mathbf{x}, \mathbf{y}), \tag{3.19}$$

where $\mathbf{y} = \mathbf{x}/\varepsilon$ are "fast" variables, and $\mathbf{x}$ are "slow" variables, i.e. a two-scale expansion is considered. The functions $\mathbf{u}^{(k)}(\mathbf{x}, \mathbf{y})$ in (3.19) are assumed to be periodic in $\mathbf{y}$ with periodicity cell Y.

When differentiating functions of the form $Z(\mathbf{x}, \mathbf{y})$, where $\mathbf{y} = \mathbf{x}/\varepsilon$, we will replace the differentiation operators in accordance with the following differentiating rule (Sanchez-Palencia 2025):

$$\frac{\partial Z}{\partial x_n}(\mathbf{x}, \mathbf{y}) \to Z_{,nx} + \varepsilon^{-1} Z_{,ny}. \tag{3.20}$$

Hereafter, subscript $,nx$ means the derivative $\partial/\partial x_n$ with respect to the "slow" variables and, subscript $,ny$ means the derivative $\partial/\partial y_n$ with respect to the "fast" variables.

Using differentiating rule (3.20), we can write Eq. (3.1) as

$$L_{\varepsilon}\mathbf{u} = (\varepsilon^{-2} A_1 + \varepsilon^{-1} A_2 + A_3) \sum_{k=1}^{\infty} \varepsilon^k \mathbf{u}^{(k)}(\mathbf{x}, \mathbf{y}) = \mathbf{F}^{\varepsilon}(\mathbf{y}). \tag{3.21}$$

The operators A_1, A_2, A_3 act on a function of the form $\mathbf{v}(\mathbf{x}, \mathbf{y}) = \mathbf{v}(\mathbf{x}, \mathbf{x}/\varepsilon)$ according to the rule

$$\begin{aligned} A_1\mathbf{v} &= (c_{ijkl}(\mathbf{y}) v_{k,ly})_{,jy}; \\ A_2\mathbf{v} &= c_{ijkl}(\mathbf{y}) (v_{k,ly})_{,jx} + (c_{ijkl}(\mathbf{y}) v_{k,lx})_{,jy}; \\ A_3\mathbf{v} &= c_{ijkl}(\mathbf{y}) (v_{k,lx})_{,jx}. \end{aligned}$$

By equating in (3.21) the terms with the identical order of ε, we obtain an infinite sequence of problems, the first three of which (other ones doe not influence our investigation) are the following:

$$A_1\mathbf{u}^{(0)} = 0; \tag{3.22}$$

$$A_1\mathbf{u}^{(1)} + A_2\mathbf{u}^{(0)} = 0; \tag{3.23}$$

$$A_1\mathbf{u}^{(2)} + A_2\mathbf{u}^{(1)} + A_3\mathbf{u}^{(0)} + \mathbf{F}^{\varepsilon}(\mathbf{y}) = 0. \tag{3.24}$$

Equation (3.22) is satisfied because A_1 is an operator in variables $\mathbf{y}$ and the function $\mathbf{u}^{(0)} = \mathbf{u}^{(0)}(\mathbf{x})$ depends on the variables $\mathbf{x}$. Due to the independence of $\mathbf{u}^{(0)}(\mathbf{x})$ from $\mathbf{y}$, Eq. (3.23) takes the form

$$A_1 u^{(1)} + (c_{ijkl}(\mathbf{y}))_{,jy} u^{(0)}_{k,lx}(\mathbf{x}) = 0. \tag{3.25}$$

Separating the functions of the variables $\mathbf{x}$ and $\mathbf{y}$, we write the solution to the Eq. (3.25) as

$$\mathbf{u}^{(1)}(\mathbf{x}, \mathbf{y}) = \mathbf{N}^{kl}(\mathbf{y}) u^{(0)}_{k,lx}(\mathbf{x}) + \mathbf{V}(\mathbf{x}); \tag{3.26}$$

where function $\mathbf{N}^{kl}(\mathbf{y})$ is solution to the problem

$$\begin{cases} (c_{ijmn}(\mathbf{y}) N^{kl}_{m,ny} + c_{ijkl}(\mathbf{y}))_{,jy} = 0 \text{ in } Y, \\ \mathbf{N}^{kl}(\mathbf{y}) \text{ is periodic in } \mathbf{y} \text{ with periodicity cell } Y. \end{cases} \tag{3.27}$$

Problem (3.27) is called a periodicity cell problem (Sanchez-Palencia 2025; Bensoussan et al. 1978) (also names basic cell problem, cellular problem are used). Consider Eq. (3.24) in which the function $\mathbf{u}^{(2)}$ is unknown and $\mathbf{x}$ is a parameter. This problem has a periodic solution if the following condition is satisfied (Sanchez-Palencia 2025):

$$\langle A_2\mathbf{u}^{(1)} + A_3\mathbf{u}^{(0)} + \mathbf{F}^{\varepsilon} \rangle = 0. \tag{3.28}$$

where

$$\langle\rangle = \frac{1}{|Y|} \int_Y d\mathbf{y} \tag{3.29}$$

means the average value over the periodicity cell Y (do not confuse the symbol $\langle\rangle$ with the dual coupling symbol $\langle , \rangle$). Here and below $|Y|$ means the measure Y (volume in three-dimensional case, area in two-dimensional case, length in one-dimensional case).

From the Eq. (3.28), taking into account (3.26), we obtain the homogenized equations (3.13) for $\mathbf{u}^{(0)}(\mathbf{x})$:

$$\frac{\partial}{\partial x_j}\left(C_{ijmn} \frac{\partial u^{(0)}_m}{\partial x_n} \right) + \langle F^{\varepsilon}_i \rangle = 0. \tag{3.30}$$

From (3.26) and (3.28), we obtain the following formulas for the homogenized elastic constants C_{ijkl} in (3.30):

$$C_{ijkl} = \left\langle c_{ijmn}(\mathbf{y}) N^{kl}_{m,ny}(\mathbf{y}) + c_{ijkl}(\mathbf{y}) \right\rangle . \tag{3.31}$$

In the homogenization theory, the following fact about the average value of a function of two-scale variables is widely used.

Proposition 3.1 (Sanchez-Palencia 2025) *Let function* $F(\mathbf{x}, \mathbf{y})$ *of two variables* $\mathbf{x}$ *and* $\mathbf{y}$ *is periodic in* $\mathbf{y}$ *with the periodicity cell* Y *and* $|\partial F(\mathbf{x}, \mathbf{y})/\partial x_i| \le M < \infty$.
Then

$$\int_G F(\mathbf{x}, \mathbf{x}/\varepsilon) d\mathbf{x} \to \int_G \langle F(\mathbf{x}, \mathbf{y}) \rangle d\mathbf{x}, \ \text{as } \varepsilon \to 0. \tag{3.32}$$

The proof (Sanchez-Palencia 2025) of the proposition is based on replacing the integral on the left side (3.32) with the sum of integrals over the periodicity cells εY:

$$\sum_{\text{over } \varepsilon Y \text{ periods}} |\varepsilon Y| \left(\frac{1}{|\varepsilon Y|} \int_{\varepsilon Y} F(\mathbf{x}, \mathbf{x}/\varepsilon) d\mathbf{x} \right). \tag{3.33}$$

In every periodicity cell εY period, the first argument $\mathbf{x}$ in the integrand in (3.33) may be considered as a constant since the derivatives $F(\mathbf{x}, \mathbf{y})$ with respect to x_i are less M. Make the change of variables $\mathbf{y} = \mathbf{x}/\varepsilon)$ in every integral in (3.33). As a result, (3.33) becomes

$$\sum_{\text{over } \varepsilon Y \text{ periods}} |\varepsilon Y| \left(\frac{1}{|Y|} \int_{Y} F(\mathbf{x}, \mathbf{y}) d\mathbf{y} \right), \tag{3.34}$$

since Jacobian of the transformation $\mathbf{y} = \mathbf{x}/\varepsilon$ is equal to $|\varepsilon Y|/|Y|$. After that, we arrive at the required result.
Note. It is evident that Proposition 3.1 remains valid if the first argument of the function $F(\mathbf{x}, \mathbf{y})$ is changed for x_1, x_2 or x_1 and function is periodic in y_1, y_2 or y_1.

3.2.2 *Relationship Between the Solutions to the Original and the Homogenized Problems*

In the theory of homogenization, it is proved that (see for details Bensoussan et al. 1978):

$$\mathbf{u}^{\varepsilon}(\mathbf{x}) \to \mathbf{u}^{(0)}(\mathbf{x}) \text{ as } \varepsilon \to 0 \text{ weak in } H^1(G), \tag{3.35}$$

where $\mathbf{u}^{\varepsilon}(\mathbf{x})$ is solution to the original problem (3.1), (3.2), and $\mathbf{u}^{(0)}(\mathbf{x})$ is the solution to the homogenized problem (3.13)–(3.15). Moreover, it is proved that (Bensoussan et al. 1978),

$$\mathbf{u}^{\varepsilon}(\mathbf{x}) - (\mathbf{u}^{(0)}(\mathbf{x}) - \varepsilon \mathbf{N}^{kl}(\mathbf{x}/\varepsilon) u^{(0)}_{k,lx}(\mathbf{x})) \to 0 \text{ in } H^1(G), \text{ as } \varepsilon \to 0. \quad (3.36)$$

For the local stresses $\sigma^{\varepsilon}_{ij} = c_{ijmn}(\mathbf{x}/\varepsilon) u^{\varepsilon}_{k,l}$, we obtain from (3.12) and (3.36) that

$$\sigma^{\varepsilon}_{ij} - (c_{ijmn}(\mathbf{x}/\varepsilon) N^{kl}_{m,ny}(\mathbf{x}/\varepsilon) + c_{ijkl}(\mathbf{x}/\varepsilon)) u^{(0)}_{k,lx}(\mathbf{x}) \to 0 \text{ in } L_2(G), \quad (3.37)$$

as $\varepsilon \to 0$.

For the mechanics needs, it is required to know if the convergence in (3.37) takes place in space $C(G)$ (this means the pointvice convergence). At present, convergence in $C(G)$ in a fairly general case has been proved for ordinary differential equations, only (Kolpakov 1982).

The following consequence of (3.37) plays an important role in our considerations below: *local stresses in an inhomogeneous body are rapidly oscillating functions of the argument* $\mathbf{x}/\varepsilon$ *with the periodicity cell* εY.

Integrating (3.37) over the periodicity cell Y and taking into account (3.31), we obtain the following equality (see Oleinik et al. 1962 for details)

$$\left\langle \sigma^{\varepsilon}_{ij} \right\rangle = \sigma_{ij}, \quad (3.38)$$

where σ_{ij} are the stresses determined from the solution to the homogenized problem:

$$\sigma_{ij} = C_{ijkl} \frac{\partial u^{(0)}_k}{\partial x_l} = C_{ijkl} e_{ij}, \quad (3.39)$$

where

$$e_{ij} = e_{ij}(\mathbf{u}^{(0)}) = \frac{1}{2} \left(\frac{\partial u^{(0)}_i}{\partial x_j} + \frac{\partial u^{(0)}_j}{\partial x_i} \right). \quad (3.40)$$

The stresses σ_{ij} determined by (3.39) and strains e_{ij} determined by (3.40) are called the homogenized stresses and strains, correspondingly.

The limit (3.37) and equality (3.38) lead to the limit

$$\sigma^{\varepsilon}_{ij} \to \sigma_{ij} \text{ weak in } L_2(G), \text{ as } \varepsilon \to 0. \quad (3.41)$$

Elastic energies of the original and the homogenized bodies. The elastic energy of the original composite body is

$$E^{\varepsilon} = \frac{1}{2} \int_G \sigma^{\varepsilon}_{ij}(\mathbf{x}) e^{\varepsilon}_{ij}(\mathbf{x}) d\mathbf{x}, \quad (3.42)$$

where, in accordance with (3.37) and (3.36),

$$\sigma_{ij}^{\varepsilon} \approx (c_{ijmn}(\mathbf{x}/\varepsilon)N_{m,ny}^{kl}(\mathbf{x}/\varepsilon) + c_{ijkl}(\mathbf{x}/\varepsilon))u_{k,lx}^{(0)}(\mathbf{x}), \tag{3.43}$$

and

$$e_{ij}^{\varepsilon} \approx (N_{i,jy}^{mn}(\mathbf{x}/\varepsilon) + \delta_{im}\delta_{jn})u_{m,nx}^{(0)}(\mathbf{x}). \tag{3.44}$$

Substituting (3.43) and (3.44) into (3.42), we obtain

$$\begin{aligned} E^{\varepsilon} \approx \int_G &[c_{ijmn}(\mathbf{x}/\varepsilon)N_{m,ny}^{kl}(\mathbf{x}/\varepsilon) + c_{ijkl}(\mathbf{x}/\varepsilon))]u_{k,lx}^{(0)}(\mathbf{x}) \\ &\times[N_{i,jy}^{mn}(\mathbf{x}/\varepsilon) + \delta_{im}\delta_{jn}]u_{m,nx}^{(0)}(\mathbf{x})d\mathbf{x}. \end{aligned} \tag{3.45}$$

The integral

$$\begin{aligned} \int_G &[c_{ijmn}(\mathbf{x}/\varepsilon)N_{m,ny}^{kl}(\mathbf{x}/\varepsilon) + c_{ijkl}(\mathbf{x}/\varepsilon))]u_{k,lx}^{(0)}(\mathbf{x}) \\ &\times N_{p,qy}^{mn}(\mathbf{x}/\varepsilon)d\mathbf{x} = 0, \end{aligned} \tag{3.46}$$

because

$$\int_Y [c_{ijmn}(\mathbf{y})N_{m,ny}^{kl}(\mathbf{y}) + c_{ijkl}(\mathbf{y})]N_{i,jy}^{mn}(\mathbf{y})d\mathbf{y} = 0.$$

To test the last equality, one can multiply the equation in the periodicity cell problem (3.27) by $N_p^{mn}(\mathbf{y})$ and integrate the result by part taking into account the boundary conditions in (3.27) and periodicity of the function $N_p^{mn}(\mathbf{y})$.

From (3.45) and (3.46), it follows that the energy (3.45)

$$\begin{aligned} E^{\varepsilon} \approx &\int_G [c_{ijpq}(\mathbf{x}/\varepsilon)N_{m,ny}^{kl}(\mathbf{x}/\varepsilon) + c_{ijkl}(\mathbf{x}/\varepsilon))]u_{k,lx}^{(0)}(\mathbf{x})u_{i,jx}^{(0)}(\mathbf{x})d\mathbf{x} \\ &= \int_G C_{ijkl}u_{k,lx}^{(0)}(\mathbf{x})u_{i,jx}^{(0)}(\mathbf{x})d\mathbf{x}. \end{aligned} \tag{3.47}$$

The last equality in (3.47) follows from the definition of the homogenized elastic constants C_{ijmn} (3.31).

The integral on the right-side of (3.47) is the elastic energy E of the homogenized body.

As follows from the formulas above, in the limit (as $\varepsilon \to 0$), the displacements $\mathbf{u}^{\varepsilon}(\mathbf{x})$ and elastic energy E^{ε} in the composite material are close to the displacements $\mathbf{u}^{(0)}(\mathbf{x})$ and elastic energy E in a homogeneous material with elastic constants C_{ijkl} given by (3.31). At the same time, as is evident from (3.37) and (3.39), the local stresses $\sigma_{ij}^{\varepsilon}$ and the local strains e_{ij}^{ε} in the composite material can differ (and, as a rule, are different) from the homogenized stresses σ_{ij} and the homogenized strains e_{ij}. According to (3.38) the average values of the local stresses in composite material (but not the local stresses themselves) coincide with the homogenized stresses. Note that the actual stresses in a composite are local stresses, while the homogenized stresses are some fictitious quantities. The same is correct for the local and homogenized strains.

The issues of using local and homogenized stresses for calculations, in particular, for obtaining strength criteria for composite materials and structures, are considered in Panasenko (1983), Annin et al. (1990), Annin et al. (2022), Kalamkarov and Kolpakov (1997), Kolpakov and Rakin (2023), Kolpakov and Rakin (2020), Kolpakov and Rakin (2021). A comparative analysis of homogenized stresses and the corresponding local stresses can be found in Rakin (2024), Kolpakov and Rakin (2022), Gluzman et al. (2018), Drygas et al. (2019), Mishnaevsky (2007), Barbero (2013), Matthews et al. (2020), Panasenko (1988), see also references in these works. Numerical calculations verify the relationship (3.38) and show that local stresses $\sigma_{ij}^{\varepsilon}$ usually differ greatly from homogenized stresses σ_{ij} (3.39). The similar situation takes place for the local and the homogenized strains.

3.2.3 The Concept of G-Limit

We present briefly the basic ideas of G-limit approach to the homogenization problems. The concept of G-convergence of operators and functionals is an abstract mathematical theory, usually used in mathematical research. However, it most clearly expresses the idea of the proximity of the displacements and energies of the original composite body and the homogenized body. Recently, articles have appeared that use G-limit in applied research (Bauer et al. 2020; Neukamm and Velcic 2013; Mora and Scardia 2012).

We give some definitions from the convex analysis (Rockafellar 1970; Ekeland and Temam 1976) and the definition of G-convergence of operators and functionals following Spagnolo (1968), Marcellini (1973), Marcellini (1975), Boccardo and Marcellini (1976).

Let V denote a Banach reflexive space (below V is a subspaces of $H^1(G)$). Let V^* denote the space topologically dual to V, see Lions and Magenes (1973). Following Marcellini (1973), let $C_0(V)$ denote the set of convex lower semicontinuous functionals on V that take values in $(-\infty, +\infty]$ and are not identically equal to $+\infty$. Denote $C(\alpha, v_0, M) = \{F \in C_0(V) : F(v) \geqslant \alpha(v) \text{ for all } v \in V, F(v_0) \leqslant M < \infty\}$, where the functional $\alpha \in C_0(V)$ is such that $\alpha(v) - \langle v^*, v\rangle$ attains a minimum on V for any $v^* \in V^*$.

Definition 1 (Rockafellar 1970) The functional $F^* : V^* \to R$ defined by the equality $F^*(v^*) = \sup_{v \in V}(\langle v^*, v\rangle - F(v))$ is called the convex conjugate of the functional $F : V \to R$.

Definition 2 (Marcellini 1973) The sequence of functionals $\{F_\varepsilon\} \subset C(\alpha, v_0, M)$ G-converges to the functional F, as $\varepsilon \to 0$, if $\lim_{\varepsilon \to 0} F_\varepsilon^*(v^*) = F^*(v^*)$ for any $v^* \in V^*$.

Definition 3 (Spagnolo 1968) A sequence of operators $\{L_\varepsilon : V \to V^*\}$ G–converges to an operator $L : V \to V^*$, as $\varepsilon \to 0$, if the operators $\{L_\varepsilon\}$ and L are invertible and $L_\varepsilon^{-1} v^* \to L^{-1} v^*$ weakly in V, as $\varepsilon \to 0$, for any $v^* \in V^*$.

Under broad assumptions, satisfying in particular for the linear differential operators and quadratic functionals considered in this book, the equivalence of Definition 2 and Definition 3 was proved in Boccardo and Marcellini (1976).

The notion of G–convergence is directly related to the mechanics of composite materials. Namely, it is related to the convergence of displacements and energy. To demonstrate this, we consider the definition of the convex conjugate functional (Rockafellar 1970; Rudin 1992). If in Definition 1,

$$F_\varepsilon(\mathbf{v}) = \frac{1}{2}\int_G c_{ijkl}^\varepsilon(\mathbf{x}) e_{ij}(\mathbf{v}) e_{kl}(\mathbf{v}) d\mathbf{x}, \tag{3.48}$$

is the elastic energy of composite body, and $\mathbf{v}^*$ describes the forces acting on the body:

$$\langle \mathbf{v}^*, \mathbf{v} \rangle = \int_G \mathbf{F}^\varepsilon \mathbf{v} d\mathbf{x} + \int_{S_\sigma} \mathbf{H}^\varepsilon \mathbf{v} d\mathbf{x}, \tag{3.49}$$

then the adjoint functional $J^*(v^*)$ is equal to the additional energy (Washizu 1992; Ekeland and Temam 1976) of the elastic body subjected to the loads $\mathbf{v}^*$.

Denote $\mathbf{v}^\varepsilon$ solution to the elasticity theory problem for the composite body, and $\mathbf{v}$ solution to the elasticity theory problem for the homogenized body. If the sequence of functionals F_ε G-convergence to F, as $\varepsilon \to 0$, then $\mathbf{v}^\varepsilon \to \mathbf{v}$ weakly in V, thus, strong in L_2 Kolmogorov and Fomin (1970). Thus, $\langle \mathbf{v}^\varepsilon, \mathbf{F}^\varepsilon \rangle \to \langle \mathbf{v}, \mathbf{F} \rangle$ and $\langle \mathbf{v}^\varepsilon, \mathbf{H}^\varepsilon \rangle \to \langle \mathbf{v}, \mathbf{H} \rangle$, as $\varepsilon \to 0$. Then, $F^\varepsilon(\mathbf{v}^\varepsilon) \to F(\mathbf{v})$, as $\varepsilon \to 0$. It means that the elastic energy of the composite body tends to the elastic energy of the homogenized body.

Mention Γ–convergence method close to G–convergence (Del Maso 1993; De Giorgi 1977; Lewicka et al. 2010). Both the methods are applied to both volumetric and thin-walled structures, see, e.g., Cardone et al. (2002), Marcellini and Sbordone (1977), Buttazzo (1989), Neff et al. (2010). An attempt to apply Γ–convergence methods modelling thin-walled plares and films with initial tresses may be found in Padilla-Garza (2022), Bhattacharya et al. (2016), Lewicka (2023).

3.2.4 Homogenization of Boundary Value Problems Whose Coefficients Depend on Both "Slow" and "Fast" Variables

In many cases, the coefficients of the original problem depend on both the "fast" $\mathbf{x} = \mathbf{x}/\varepsilon$ and the "slow" variable $\mathbf{x}$ and have the form $c_{ijmn}(\mathbf{x}, \mathbf{x}/\varepsilon)$. Generally, the problem may have many scales and the general form of the function c_{ijkl} is: $c^{\varepsilon}_{ijkl} = c_{ijkl}(\mathbf{x}, \mathbf{x}/\varepsilon, \mathbf{x}/\varepsilon^2, ...)$ (Alaire 1992). If the function $c_{ijkl}(\mathbf{x}, ...)$ has bounded derivatives with respect to the variables x_i, then the construction of the homogenized operator or functional can be performed using the procedure known as reiterated homogenization (Allaire and Briane 1996). If the problem has only two scales: $c^{\varepsilon}_{ijkl} = c_{ijkl}(\mathbf{x}, \mathbf{x}/\varepsilon)$, the procedure of reiterated homogenization can be carry out by using the procedure known as the "freezing" of the "slow" variables. Namely, in $c_{ijkl}(\mathbf{x}, \mathbf{x}/\varepsilon)$ the first argument $\mathbf{x}$ is considered as a constant ("frozen" variable), after that the homogenization is performed with respect to the second argument—$\mathbf{x}/\varepsilon$. After the homogenization has been done, $\mathbf{x}$ again is considered as variable and the resulting homogenized coefficients depend on $\mathbf{x}$: $C_{ijkl} = C_{ijkl}(\mathbf{x})$.

The justification of the reiterated homogenization method, including the "freezing" procedure, can be found in Bakhvalov and Panasenko (1989), Alaire (1992), Allaire and Briane (1996).

3.3 Composites of Random Structure

The results obtained for periodic structures can be used for analysis of composites of random structure. There exists intensive literature on the composites of random structures. Many publications have been devoted to the problem of random homogenization. The mathematical aspects of the problem were discussed in Papanicolaou and Varadhan (1981), Papanicolaou (1995), Kozlov (1980, 1978), Zhikov (1993), Yurinski (1986, 1980). The problems related to the homogenization of media of random structure were discussed in Willis (2002), Molchanov (1991, 1994), Sab (1992), Berlyand and Kolpakov (2001), Ostoja-Starzewski (2006), Torquato (2002). The problem was discussed for both volumetric and thin-walled structural elements (Sab 2011; Gusella et al. 2019).

In works on random homogenization (Jikov et al. 1994; Kozlov 1980), the existence of the homogenization problems was proved, and it was demonstrated that in many cases the homogenized problems have a form similar to the original problems. The homogenization theorems for random composites were proved under condition of stationary spatial distribution of heterogeneity or similar conditions. In Mityushev and Rylko (2025), it was noted that the stationary of the spatial distribution of heterogeneity is the necessary condition for the existence of the homogenized characteristics for random composites. From an engineering point of view, a disadvantage of the existing theoretical research on random homogenization is the lack of a practically feasible method for calculating the homogenized coefficients.

One of the methods proposed for calculating the homogenization coefficients is the method of representative volume. In Hill (1963), a representative volume was defined as a sample of heterogeneous material with the following properties:

1. It is entirely typical of the whole mixture, on average,
2. It contains a sufficient number of inclusions for the apparent properties to be independent of the surface values of traction and displacement, so long as these values are macroscopically uniform.

Another definition (Drugan and Willis 1996; Kanit et al. 2003; Sun and Vaidya 1996): The representative element is the smallest material volume element of the composite for which the usual spatially constant (overall modulus) macroscopic constitutive representation is a sufficiently accurate model to represent the mean constitutive response.

The choice of the representative element size for practical calculation of the homogenization characteristics is still an open problem. In addition to this problem, if starting real calculations, one immediately encounters the problem of boundary conditions for the representative element.

One way to use the representative element is to combine it with the idea of self-consistency (about the self-consistency approach, see, e.g., Kanaun and Levin 2008; Hill 1965; Willis 1977; Cleary et al. 1987 for an introduction to self-consistency methods for composites). Briefly, the mentioned combination consists of placing the representative element in a homogeneous material with unknown elastic constants and seeking for the effective (averaged) constants from the conditions of coincidence of the averaged characteristics of the representative element and the composite.

An alternative way is modeling a random body as a periodic system of representative volumes. In this case, the representative volume of the random body is a periodicity cell Y and the periodicity cell problems of the standard homogenization theory are solved on the Y.

For a long time, there were no strong arguments (or arguments were not rigorous enough) in favor of one or another approach. The rigorous arguments were presented recently by Mityushev with co-authors (Gluzman et al. 2018; Drygas et al. 2019; Mityushev 2022).

In the works (Mityushev and Rylko 2022; Mityushev 2022) it is explained that the self-consistent concept is methodologically incorrect. It is also explained that the methodologically correct is to replace the random composite with a periodic composite with a quadruple cell constructed using representative volumes of the random composite. Although this conclusion is limited to the second order of approximation in the concentration of inclusions, it demonstrates that there is still a difference between the self-consistent and homogenization approaches. Interestingly, the difference is related to the shape of the representative volume. The difference is observed, for example, for spherical shapes of the representative volume. For rectangular shapes of the representative volume, the self-consistent approach can be reduced to problems justified by the homogenization theory.

Let us consider a representative volume of the random composite, Fig. 3.2 (the left picture). Repeating this representative volume, we obtain a new composite of periodic structure, Fig. 3.2 (the right figure). From the mechanical point of view, we agree that the original random composite body and "new" periodic composite body have the same homogenized constants if the representative element is large

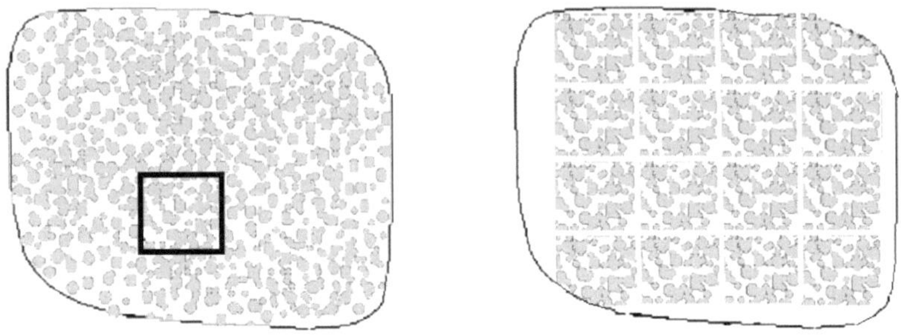

Fig. 3.2 Composite of random structure

enough. In common words, we agree that the "large bricks" are equivalent from the macroscopic point of view, and we can examine one of the "bricks" in order to determine the macroscopic (homogenized) constants. Additional arguments in the favor of this computational scheme (with detailed explanations and examples) may be found in Mityushev et al. (2025).

For the reason of numerical solution, the periodicity cell problem for the representative volume is usually a large, complex problem, whose solution may require large computational resources. The advantage of the discussed method is that it allows us to transfer all the qualitative results obtained for a composite of periodic structure to random composites.

3.4 Micro-To-Macro Coupling and the "Self-consistent" Method in the Mechanics of Composites

Close to the "self-consistent" method is the problem of micro-to-macro coupling. In the papers devoted to the problem of micro-to-macro coupling, the boundary layer between the homogenized and not homogenized fragments of the body is often reported (Ben Dhia 2008; Lubineau et al. 2012; Bauman et al. 2008; Pfaller et al. 2012; Wagner and Liu 2003; Xiao and Belytschko 2004; Parks et al. 2008; the Arlequin framework 2006; Hu et al. 2011). As an example, we present the numerically computed local stress-strain state in the neighbor of an interface of a layered elastic composite and the homogeneous elastic material, see Fig. 3.3. The elastic characteristics of the homogeneous material on the right in Fig. 3.3 are exactly equal to the homogenized elastic characteristics of the layered material on the left in Fig. 3.3.

There are two possible reasons for the appearance of a boundary layer:

- arising of physical boundary layer,
- arising of boundary layer in computation process as a result of methodological error.

A physical boundary layer occurs when both the laminated composite and the homogeneous material are physically existing materials, and they are connected. If the

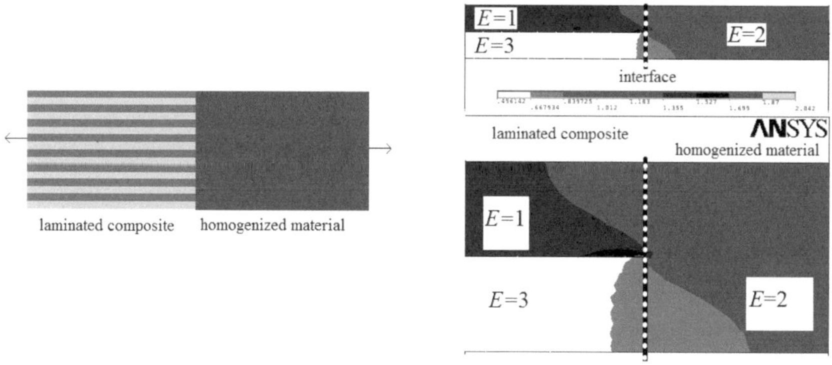

Fig. 3.3 Composite coupled to the homogenized material

laminated composite is a physically existing material, but the homogeneous material is an imaginary homogenized model, the boundary layer is a result of methodological error. Evidently, a physical boundary layer—a perturbation of the local stress-strain state—cannot arise as a result of imaginary manipulation. Additional details concerning the coupling of the original and homogenized materials may be found in Andrianov et al. (2018).

In the present solution, the classical interface conditions were accepted. As a result, the obtained solution corresponds to the interface of real materials, and we see the boundary layer in Fig. 3.3.

For the interface between the layered material and the homogenized material, it would be written interface condition

$$\sigma_{ij}^{\varepsilon} n_j = (c_{ijmn}(\mathbf{x}/\varepsilon) N_{m,ny}^{kl}(\mathbf{x}/\varepsilon) + c_{ijkl}(\mathbf{x}/\varepsilon)) u_{k,lx}^{(0)}(\mathbf{x}) n_j, \tag{3.50}$$

following from (3.37).

3.5 Homogenization for One and Two-Dimensional Problems

One- and two-dimensional problems can be homogenized using special methods associated with these dimensions. Namely, one-dimensional problems can be analyzed using special methods of ordinary differential equations. Two-dimensional problems can be analyzed using the technique of complex variables. The use of these methods is typical for one- and two-dimensional problems in mathematics and natural sciences.

3.5.1 Homogenization of Ordinary Differential Equations

A problem close to averaging for ordinary differential equations was analyzed in Senatorov (1971). Later, averaging problems for ordinary differential equations and systems of ordinary differential equations were usually reported as applications and illustrations of multidimensional averaging methods. Note that more detailed information can be obtained about the asymptotic of the solution of one-dimensional problems compared to multidimensional problems. We briefly present results concerning the homogenization of systems of ordinary differential equations (Kolpakov 1982).

Consider a system of ordinary differential equations, which depends on the parameter $\varepsilon \to 0$:

$$\sum_{k=0}^{n}(-1)^k(p^{\varepsilon}_{ik}(\mathbf{u}^{\varepsilon}(\mathbf{n}), ..., \mathbf{u}^{\varepsilon}, t))^{(k)} = 0, \;\; a \le t \le b, \;\; i = 1, ..., m; \tag{3.51}$$

with the boundary conditions

$$\mathbf{u}^{\varepsilon(l)}(a) = \mathbf{A}, \;\; \mathbf{u}^{\varepsilon(l)}(b) = \mathbf{B}, \;\; l = 0, ..., n-1. \tag{3.52}$$

We assume that the following conditions are satisfied: there exists $0 < M, m_1$, $m_{<}\infty$ such that for any $\varepsilon > 0$

1. the system of algebraic equations $p^{\varepsilon}_{in}(\mathbf{x}_n, ..., bfx_0, t) = y_i$ is solvable with respect to $\mathbf{x}_n$ for any $\mathbf{x}_{n-1}, ..., \mathbf{x}_0, t$. The solution function, $q^{\varepsilon}_i(\mathbf{y}, \mathbf{x}_{n-1}, ...,$ $\mathbf{x}_0, t)$ is continuous in t and Lipschitzian in $\mathbf{y}, \mathbf{x}_{n-1}, ..., \mathbf{x}_0$ uniformly in $\varepsilon > 0$ and t;
2. $p^{\varepsilon}_{ik}(\mathbf{x}_n, ..., \mathbf{x}_0, t) = y_i$ $(k < n)$ are continuous in t and Lipschitzian in $\mathbf{y}, \mathbf{x}_{n-1}, ...,$ $\mathbf{x}_0$ uniformly in $\varepsilon > 0$ and t;
3. for any $\mathbf{x}_1, \mathbf{x}_2 \in R^m$ and for any $\mathbf{y}_1, \mathbf{y}_2 \in (R^m)^{n+1}$

$$M||\mathbf{x}_1 - \mathbf{x}_2||^2 \ge \sum_{i=0}^{m}(p^{\varepsilon}_{in}(\mathbf{x}_1, ...) - p^{\varepsilon}_{in}(\mathbf{x}_2, ...))(\mathbf{x}_1 - \mathbf{x}_2)_i \ge m_1||\mathbf{x}_1 - \mathbf{x}_2||^2,$$

$$M||\mathbf{y}_1 - \mathbf{y}_2||^2 \ge \sum_{i=1}^{m}\sum_{k=0}^{n}(p^{\varepsilon}_{ik}(\mathbf{y}_1, t) - p^{\varepsilon}_{ik}(\mathbf{y}_2, t))(\mathbf{y}_1 - \mathbf{y}_2)_{ik} \ge m_1||\mathbf{y}_1 - \mathbf{y}_2||^2.$$

Proposition 3.2 *If for $\varepsilon \to 0$ there exist functions p_{ik} satisfying the above conditions 1–3, and such that for any continuous functions $\mathbf{x}(t)$ and $\mathbf{y}(t)$*

$$\begin{gathered} q^{\varepsilon}_i(\mathbf{x}(t), \mathbf{y}(t), t) \to q_i(\mathbf{x}(t), \mathbf{y}(t), t) \\ p^{\varepsilon}_{ik}(..., \mathbf{q}^{\varepsilon} \circ \mathbf{p}_n(\mathbf{x}(t), \mathbf{y}(t), t), ..., \mathbf{y}(t), t) \to p_{ik}(\mathbf{x}(t), \mathbf{y}(t), t) \end{gathered} \tag{3.53}$$

weakly in $L_2([a, b])$, then

$$\begin{aligned} \mathbf{u}^{\varepsilon}(t) &\to \mathbf{u}(t) \ in \ C([a,b]), \\ p^{\varepsilon}_{ik}(\mathbf{u}^{\varepsilon(n)}, ..., \mathbf{u}^{\varepsilon}, t) &\to p_{ik}(\mathbf{u}^{\varepsilon(n)}, ..., \mathbf{u}^{\varepsilon}, t) \ in \ C([a,b]), \ as \ \varepsilon \to 0, \end{aligned} \tag{3.54}$$

where is the solution to the problem obtained from (3.51) and (3.52) by replacing the functions p^{ε}_{ik} by p_{ik}.

Here $\mathbf{q}$ is the mapping constructed on the basis of the functions p_{in}, similarly for $\mathbf{q}^{\varepsilon}$, and $(\mathbf{q}^{\varepsilon} \circ \mathbf{p}_n(\mathbf{x}(t), \mathbf{y}(t), t))_i = q^{\varepsilon}_i(\mathbf{p}_n(\mathbf{x}(t), \mathbf{y}(t), t), \mathbf{y}(t), t)$.

3.5.2 Homogenization of Planar Problems Using Complex Variables

Complex variables for the purposes of (nowadays terms) homogenization were first used for perforated plates, see Grigolyuk and Filshtinskij (1970, 1992) and references therein. Later, complex variables were used for the homogenization of composites reinforced by a system of unidirectional fibers, see Mityushev and Rogozin (2000), Rodriguez-Ramos et al. (2020) and references in these works. These were analyzed both periodic (Makaruk et al. 2006; Mityushev 1997; Mityushev and Adler 2002; Mityushev 1997; Mityushev and Adler 2002; Mityushev 2001; Drygas et al. 2019) and random fiber systems (Drygas et al. 2019; Mityushev and Rylko 2022; Mityushev 2022; Mityushev et al. 2025).

The use of complex variables for the homogenization was based on the idea of Kolosov (1909), Kolossoff (1914) representation of solution to the elasticity theory problem using complex potentials (often referred to Kolosov-Muskhelishvili representation Muskhelishvili 1933, 1977) and doubly periodic Weierstrass elliptic functions (Lu 1995; Carrier et al. 1966; Grigolyuk and Filshtinskij 1992). As a result, it succeeds in obtaining the analytical solution which predicts both the local and overall properties of the composite material, see, e.g., Meguid and Kalamkarov (1994). The solution is obtained in the form of a series of complex variables.

Weierstrass function $\wp(z)$ of the complex variable $z = x + iy$ is defined by the series

$$\wp(z) = \frac{1}{z^2} + \sum_{m_1, m_2 \neq 0} \left(\frac{1}{(z-P)^2} - \frac{1}{P^2} \right) \tag{3.55}$$

where $P = m_1\omega_1 + m_2\omega_2$.

A solution to the double-periodic planar elasticity theory is constructed in the form of the sum of Weierstrass functions with undetermined coefficients. As a result, it is necessary to carry out additional massive computations to obtain final numerical values of the local and overall properties of the composite material or plate. Numerous examples of such computations may be found in Grigolyuk and Filshtinskij (1992). Now some calculations can be implemented on the computer, see, e.g., Gluzman et al. (2018). Detailed information, including both the state of art and the historical review, may be found in Mityushev et al. (2022).

References

Aboudi J (1991) Mechanics of composite material: a unified michromecanics approach. Elsevier Science Publication, Amsterdam

Alaire G (1992) Homogenization and two-scale convergence. SIAM J Math Anal 23(5):1482–1518

Allaire G, Briane M (1996) Multiscale convergence and reiterated homogenization. Proc R Soc Edinburg 126A:297–342

Allaire G, Cavallina L, Miyake N, Oka T, Yachimura T (2019) The homogenization method for topology optimization of structures: old and new. Interdisc Inf Sci 25(2):75–146

Andrianov IV, Kolpakov AG, Prikazchikov DA (2018) Asymptotic strategy for matching homogenized structures. Conductivity problem. Q J Mech Appl Math 71(4):519–535

Annin BD, Kalamkarov AL, Kolpakov AG (1990) Analysis of local stresses in high modulus fiber composites. In: Localized damage computer-aided assessment and control, vol 2. Comp. Mech. Publication, Southampton, pp 131–144

Annin BD, Kolpakov AG, Rakin SI (2022) Interface strength criterion for elastic bodies. Continuum Mech Thermodyn 35(4):1–13

Bakhvalov NS, Panasenko GP (1989) Homogenization: averaging processes in periodic media. Kluwer Academic Publication, Dordrecht

Barbero EJ (2013) Finite element analysis of composite materials using ANSYS. CRC Press, Boca Raton, FL

Bauer R, Neukamm S, Schaffner M (2020) Derivation of a homogenized bending-torsion theory for rods with micro-heterogeneous prestrain. J Elast 141:109–145

Bauman T, Ben Dhia H, Elkhodja N, Oden JT, Prudhomme S (2008) On the application of the Arlequin method to the coupling of particle and continuum models. Comput Mech 42:511–530

Ben Dhia H (2006) Global-local approaches: the Arlequin framework. Eur J Comput Mech 15:67–80

Ben Dhia H (2008) Further insights by theoretical investigations of the multiscale Arlequin method. Int J Multisc Comput Eng 6:215–232

Bendsøe MP, Sigmund O (2004) Topology optimization. Springer, Berlin

Bensoussan A, Lions J-L, Papanicolaou G (1978) Asymptotic analysis for periodic structures. North-Holland Publication, Amsterdam

Berlyand L, Kolpakov A (2001) Network approximation in the limit of small interparticle distance of the effective properties of a high-contrast random dispersed composite. Arch Rational Mech Anal 36(5):179–227

Bhattacharya K, Lewicka M, Schaffner M (2016) Plates with incompatible prestrain. Arch Rational Mech Anal 221(1):143–181

Boccardo L, Marcellini P (1976) Sulla convergenza della soluzioni di disequazioni variazionali. Ann Mat Pure Appl 110(1):137–159

Broutman LJ, Krock RH (eds) (1974) Composite materials. In eight volumes. Academic, New York, London

Buttazzo G (1989) Thin structures in elasticity: the variational method. Rendiconti del Seminario Matematico e Fisico di Milano 59:149–159

Cardone G, Corbo-Esposito A, Zhikov VV (2002) An integral representation result for the Γ-limit of functionals with non-standard growth conditions in the case of elasticity. Appl Anal 81(5):1179–1195

Carrier GF, Krook M, Pearson CE (1966) Functions of a complex variable: theory and technique. McGraw-Hill, New York

Charalambakis N (2010) Homogenization techniques and micromechanics. a survey and perspectives. Appl Mech Rev 63(3):030803–1–030803–10

Christensen RM (2005) Mechanics of composite materials. Dover, New York

Cioranescu D, Donato P (2000) An introduction to homogenization. Oxford University Press, Oxford

Cleary M, Chen I, Lee S (1987) Self-consistent techniques for heterogeneous solids. ASCE J Eng Mech 106:861–887

De Giorgi E (1977) Γ-convergenza e G-convergenza. Boll Unio Mat Ital 5:213–220

Del Maso G (1993) An introduction to Γ-convergence. Birkhäuser, Boston, Basel, Stuttgart

Drugan WJ, Willis JR (1996) A micromechanics-based nonlocal constitutive equation and estimates of representative volume element size for elastic composites. J Mech Phys Solids 44(4):497–524

Drygas P, Gluzman S, Mityushev V, Nawalaniec W (2019) Applied analysis of composite media. Analytical and computational results for materials scientists and engineers. Elsevier, Amsterdam

Ekeland I, Temam R (1976) Convex analysis and variational problems. North-Holland, Amsterdam

Fichera G (1984) Existence theorems in elasticity. Springer, Berlin

Gajewskii H, Groger K, Zacharias K (1974) Nichtlineare Operator- gleichungen und Operatordifferentialgleichungen. Akademie-Verlag, Berlin

Gloria A (2012) Numerical homogenization: survey, new results, and perspectives. ESAIM: Proc 37:50–116

Gluzman S, Mityushev V, Nawalaniec W (2018) Computational analysis of structured media. Academic, Amsterdam

Grigolyuk EI, Filshtinskij LA (1970) Perforated Platesand Shells. Nauka, Moscow (in Russian)

Grigolyuk EI, Filshtinskij LA (1992) Periodical piecewise homogeneous elastic structures. Nauka, Moscow (in Russian)

Gusella F, Cluni F, Gusella V (2019) Homogenization of dynamic behavior of heterogeneous beams with random young's modulus. Eur J Mech—A/Solids 73:260–267

Hill R (1963) Elastic properties of reinforced solids: some theoretical principles. J Mech Phys Solids 11:357–372

Hill R (1965) Self-consistent mechanics of composite materials. J Mech Phys Solids 13:213–222

Hu H, Damil N, Potier-Ferry M (2011) A bridging technique to analyze the influence of boundary conditions on instability patterns. J Comput Phys 230:3753–3764

Jikov VV, Kozlov SM, Oleinik OA (1994) Homogenization of differential operators and integral functionals. Springer, Berlin

Kalamkarov AL, Kolpakov AG (1997) Analysis, design and optimization of composite structures. Wiley, Chichester

Kanaun SK, Levin VM (2008) Self-consistent methods for composites. In two volumes. Dordrecht, Springer, New Delhi

Kanit T, Forest S, Galliet I, Mounoury V, Jeulin D (2003) Determination of the size of the representative volume element for random composites: statistical and numerical approach. Int J Solids Struct 40(13–14):3647–3679

Kanout D, Boso P, Chaboche JL, Schrefler BA (2009) Multiscale methods for composites: a review. Arch Comput Meth Eng 16:31–75

Kolmogorov AN, Fomin SV (1970) Introductory real analysis. Prentice-Hall, Englewood Cliffs

Kolosov GN (1909) On application of the theory of complex functions to plane problem of the mathematical theory of elasticity. Yur'ev, Mattisen

Kolossoff G (1914) Uber einige eigenschaften des ebenen problems der elastizitatstheorie. Zeitschr Math Phys 62:384–409

Kolpakov AG (1982) Averaging of some systems of ordinary differential equations. Math USSR—Sbornik 119(161):4(12):534–547

Kolpakov AG, Rakin SI (2022) Comparative analysis of local stresses in unidirectional and cross-reinforced composites. In: Andrianov I, Mityushev V, Gluzman S (eds) Mechanics and physics of structured media, vol 265. Academic, London, pp 395–416

Kolpakov AG, Rakin SI (2023) Local stresses in the reinforced plate with orthogonal systems of fibers. The homogenized strength criterion. Comput Struct 275:106922

Kolpakov AG, Rakin SI (2020) Homogenized strength criterion for composite reinforced with orthogonal systems of fibers. Mech Mater 148:103489

Kolpakov AG, Rakin SI (2021) Local stresses in the reinforced plate with orthogonal systems of fibers. Comput Struct 265:113772

Kozlov SM (1978) Averaging random structures. Dokl Akad Nauk SSSR 241(5):1016–1019

Kozlov SM (1980) Averaging of random operators. Math USSR—Sbornik 37(2):167–180

Kozlov SM (1980) Averaging of random operators. Math USSR Sbornik 37:167–180

Lewicka M (2023) Calculus of variations on thin prestressed films, asymptotic methods in elasticity. Birkhauser, Cham

Lewicka M, Mora MG, Pakzad M (2010) Shell theories arising as low energy Γ-limit of 3d nonlinear elasticity. Ann Scuola Norm Sup Pisa, Cl Sci 9(5):1–43

Lewinski T, Telega JJ (2000) Plates, laminates and shells: asymptotic analysis and homogenization. World Scientific, Singapore

Lions J-L, Magenes E (1973) Non-homogeneous boundary value problems and applications, Vol 1. Springer, Berlin

Lu J-K (1995) Complex variable methods in plane elasticity. World Scientific, Singapore

Lubineau G, Azdoud Y, Han F, Rey C, Askari A (2012) A morphing strategy to couple non-local to local continuum mechanics. J Mech Phys Solids 60:1088–1102

Makaruk SF, Mityushev VV, Rogosin SV (2006) An optimal design problem for two-dimensional composite materials. A constructive approach. In: Kilbas AA, Rogosin SV (eds) Analytic methods of analysis and differential equations. AMADE 2003. Cambridge Scientific Publishers, Cottenham, Cambridge, pp 153–167

Marcellini P (1973) Su una convergenza di funzioni convesse. Boll Uni Mat Ital 8(1):137–158

Marcellini P (1975) Un teorema di passagio de limite per la somma di convesse. Boll Uni Mat Ital 4:107–124

Marcellini P, Sbordone C (1977) Sur quelque de G-convergence et d'homogenisation non-lineare. C R Acad Sci Paris 284:535–537

Matthews FL, Davies GAO, Hitchings D, Soutis C (2020) Finite element modelling of composite materials and structures. CRC Press, Boca Raton, FL

Meguid SA, Kalamkarov AL (1994) Asymptotic homogenization of elastic composite materials with a regular structure. Int J Solids Struct 31(3):303–316

Milton GW (2002) The theory of composites. Cambridge University Press, Cambridge

Mishnaevsky Jr LL (2007) Computational mesomechanics of composites: numerical analysis of the effect of microstructures of composites of strength and damage resistance. Wiley, Chichester

Mityushev V (2022) Effective properties of two-dimensional dispersed composites. Part II. revision of self-consistent methods. Comput Math Appl 121:74–84

Mityushev V (1997) A functional equation in a class of analytic functions and composite materials. Demost Math 30:63–70

Mityushev VV (1997) Transport properties of doubly-periodic arrays of circular cylinders. Z Angew Math Mech 77:115–120

Mityushev V (2001) Transport properties of doubly periodic arrays of circular cylinders and optimal design problems. Appl Math Optim 44:17–31

Mityushev V, Adler PM (2002) Longitudinal permeability of a doubly periodic rectangular array of cylinders. I Z Angew Math Mech 82:335–345

Mityushev V, Nawalaniec W, Rylko N, Kycia R (2025) Introduction to mathematical modeling and computer simulations, 2nd edn. CRC - Taylor & Francis, Boca Raton, FL, Amsterdam

Mityushev V, Adler PM (2002) Longitudinal permeability of a doubly periodic rectangular array of cylinders. II. An arbitrary distribution of cylinders inside the unit cell. Z Angew Math Phys 53:486–517
Mityushev V, Andrianov I, Gluzman S (2022) L.a. filshtinsky's contribution to applied mathematics and mechanics of solids. In: Mityushev V, Andrianov I, Gluzman S (eds) Mechanics and physics of structured media. Academic, London, pp 1–40
Mityushev V, Drygas P, Gluzman S, Nauryzbayev N, Nurtazina K (2025) Computational strategy for analyzing effective properties of random composites—Part I: Conductivity. Appl Sci MDPI 15:3861
Mityushev V, Rogozin SV (2000) Constructive methods for linear and nonlinear boundary value problems of the analytic functions theory. Chapman&Hall/CRC, Boca Raton, FL
Mityushev V, Rylko N (2022) Effective properties of two-dimensional dispersed composites. Part I. Schwarz's alternating method. Comput Math Appl 111:50–60
Mityushev V, Rylko N (2025) Constructuve homogenization of disperced random composites. Part II. Computer simulation. Conference ECMS, in press
Molchanov S (1994) Lectures on random media. In: Lectures on probability theory, volume 1581 of Lecture notes in mathematics. Springer, New York, pp 242–411
Molchanov S (1991) Ideas in the theory of random media. Acta Appl Math 22(2/3):139–282
Mora MG, Scardia L (2012) Convergence of equilibria of thin elastic plates under physical growth conditions for the energy density. J Differ Eqn 252(1):35–55
Muskhelishvili NI (1933) Some mathematical problems of the plane theory of elasticity (in Russian) Akad. Nauk SSSR, Moscow-Leningrad
Muskhelishvili NI (1977) Some mathematical problems of the plane theory of elasticity. Springer, Dordrecht
Nandakumaran AK (2007) An overview of homogenization. J Indian Inst Sci 87(4)
Neff P, Hong K-I, Jeong J (2010) The reissner-mindlin plate is the Γ-limit of cosserat elasticity. Math Model Method Appl Sci 20:1553–1590
Neukamm S, Velcic I (2013) Derivation of a homogenized von-Karman plate theory from 3D nonlinear elasticity. Math Models Methods Appl Sci 23(14):2701–2748
Oleinik OA, Shamaev AS, Yosifian GA (1962) Mathematical problems in elasticity and homogenization. North Holland, Amsterdam
Ostoja-Starzewski M (2006) Material spatial randomness—from statistical to representative volume element. Probab Eng Mech 21(2):112–132
Padilla-Garza D (2022) Dimension reduction through gamma convergence for general prestrained thin elastic sheets. Calc Var 61(187):186–226
Panasenko GP (1988) Numerical solution of cell problems in averaging theory. USSR Comput Math Math Phys 28(1):183–186
Panasenko GP (1983) Strength of spatially reinforced composite materials. Vest. Moscow Univ. 15(2):20–27
Panasenko G (2022) Introduction to multiscale mathematical modeling. World Sci, Europe, London
Papanicolaou GC (1995) Diffusion in random media. In: Keller JB, McLaughlin D, Papanicolaou G (eds) Surveys in applied mathematics. Plenum Press, New York, pp 205–255
Papanicolaou GC, Varadhan SRS (1981) Boundary value problems with rapidly oscillating random coefficients. In: Colloquia Mathematica Societatis János Bolyai, vol 27. North-Holland, Amsterdam, pp 835–873
Parks ML, Bochev PB, Lehoucq RB (2008) Connecting atomistic-to-continuum coupling and domain decomposition. Multiscale Model Simul 7:362–380
Pfaller S, Possart G, Steinmann P, Rahimi M, Muller-Plathe F, Buhm MC (2012) A comparison of staggered solution schemes for coupled particle-continuum systems modeled with the Arlequin method. Comput Mech 49:565–579
Rakin SI (2024) Calculation of stress-strain state of fiber composite adhesive. Appl Mech Tech Phys 65(2):290–299
Rockafellar RT (1970) Convex analysis. Princeton University Press

Rodriguez-Ramos R, Ramirez-Torres A, Bravo-Castillero J, Guinovart-Diaz R, Guinovart-Sanjuan D, Cruz-Gonzalez OL, Sabina FJ, Merodio J, Penta R (2020) Multiscale homogenization for linear mechanics. In: Merodio J, Ogden R (eds) Constitutive modelling of solid continua volume 262 of Solid mechanics and its applications. Springer, Cham, pp 357–389
Rudin W (1967) Principles of mathematical analysis, 3rd edn. McGraw-Hill, New York
Rudin W (1992) Functional analysis. McGrow-Hill, New York
Sab K (1992) On the homogenization and the simulation of random materials. Eur J Mech—A/Solids 11(5):585–607
Sab K (2011) Homogenization of random plates. Int J Multisc Comput Eng 9(5):503–513
Sanchez-Palencia E (1970) Equations aux deivees partielles. Solutions periodiques par repport aux variables d'espace et application. C R Acad Sci Paris Ser A 271:1129–1132
Sanchez-Palencia E (1986) Homogenization in mechanics. A survey of solved and open problems. Rend Sem Mat Univ Polot Torino 44(1):251–270
Sanchez-Palencia E (2025) Non-homogeneous media and vibration theory. Springer, Berlin
Senatorov PK (1971) Coefficient stability of the solutions of second order ordinary differential equations and of parabolic equations on the plane. Differ Eqn 7(4):754–758
Spagnolo S (1968) Sulla convergenza di equazioni paraboliche et elittiche. Ann Scuola Norm Sup Pisa 22:577–597
Sun CT, Vaidya RS (1996) Prediction of composite properties from a representative volume element. Comput Sci Tech 56(2):171–179
Timoshemko S, Goodier JN (1951) Theory of elasticity. McGrow-Hill, New York
Torquato S (2002) Random heterogeneous materials. Springer, Berlin
Trabucho L, Viano JM (1996) Mathematical modeling of rods. In: Ciarlet PG, Lions J-L (eds) Handbook of numerical analysis, vol IV. Elsevier, Amsterdam, pp 487–974
Tutek Z, Aganovich I (1987) A justification of the one-dimensional linear model of elastic beam. Math Method Appl Sci 8:502–515
Wagner GJ, Liu WK (2003) Coupling of atomistic and continuum simulations using a bridging scale decomposition. J Comput Phys 249–274190
Washizu K (1992) Variational methods in elasticity and plasticity. Pergamon Press, Oxford
Willis JR (2002) Lectures on mechanics of random media. cism lecture notes. In: Jeulin D, Ostoja-Starzewski M (eds) Mechanics of random and multiscale structures. Springer, Wein, pp 221–267
Willis J (1977) Bounds and self-consistent estimates for the overall properties of anisotropic composites. J Mech Phys Solids 25:185–202
Xiao SP, Belytschko T (2004) A bridging domain method for coupling continua with molecular dynamics. Comput Methods Appl Mech Eng 193:1645–1669
Yurinski VV (1980) Average of an elliptic boundary problem with random coefficients. Siberean Math J 21:470–482
Yurinski VV (1986) Averaging of symmetric diffusion in random medium. Siberean Math J 27(4):167–180
Zhikov VV (1993) On homogenization in random perforated domains of general type. Matem Zametki 53(1):41–58

Chapter 4
Homogenization in Linear Elasticity Theory Problems with Initial Stresses

In this chapter, we discuss the model of elastic composite body with initial stresses, which will be used as a first-principles model throughout the book.

Following a common point of view (Washizu 1992), we understand as initial stresses σ_{ij}^* the stresses existing in a body in its initial state, i.e. before the deformation under consideration. These stresses can be caused by deformation incompatibilities (dislocations, etc.), by heating, or (the most common case, which is considered in this book) by applying forces. For a problem with initial stresses, the stressed state is taken as the reference state.

4.1 Linear (Linearized) Model of Elastic Body with Initial Stresses

Before moving on to the consideration of composite bodies with initial stresses, we should consider the classical linear elasticity theory problem, which describes elastic bodies with stresses from the general point of view.

In our consideration, we will follow Washizu (1992). Let the body occupying a domain G is clamped on a surface S_u and loaded on a surface S_σ. We assume that the initial stresses σ_{ij}^* in a body are the result of applying to the body mass forces $\mathbf{F}$, normal stress $\mathbf{H}$ on the surface S_σ, and displacement $\mathbf{u}^0(\mathbf{x})$ on the surface S_u, see Fig. 4.1. In this case, the initial stresses σ_{ij}^* are determined from the solution to the problem of elasticity theory (the calculations in this section do not directly address the microstructure of the body, so the index ε is not used in this section)

A. Kolpakov, *Thin-Walled Structures with Initial Stresses*, Mathematical Engineering,
https://doi.org/10.1007/978-3-032-07247-4_4

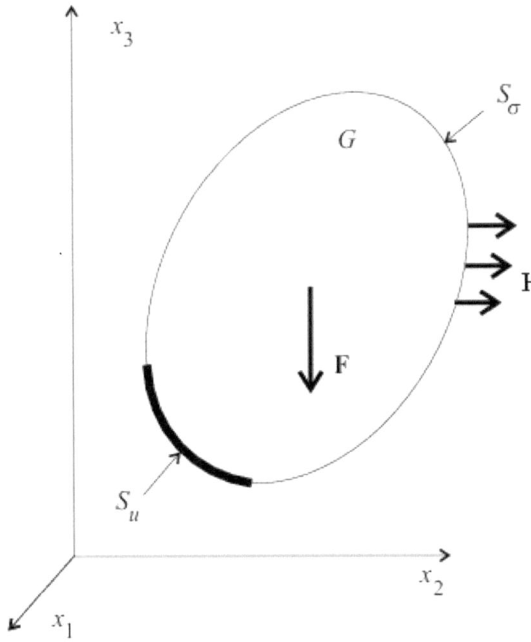

Fig. 4.1 Elastic body

$$\frac{\partial \sigma_{ij}^*}{\partial x_j} + F_i(\mathbf{x}) = 0 \text{ in } G, \tag{4.1}$$

$$\sigma_{ij}^* = c_{ijkl}(\mathbf{x}) \frac{\partial u_k}{\partial x_l}, \tag{4.2}$$

$$\sigma_{ij}^* n_j = H_i(\mathbf{x}) \text{ on } S_\sigma. \tag{4.3}$$

$$\mathbf{u} = \mathbf{u}^0(\mathbf{x}) \text{ on } S_u, \tag{4.4}$$

where $c_{ijkl}(\mathbf{x})$ are elastic constants of the body

Following Washizu (1992), we call the initial stresses satisfying (4.1)–(4.4) self-balanced.

When additional mass **f** and surface **h** forces are applied to the body, a problem of deformation of the body with initial stresses arises. We denote the additional stresses (i.e. the stresses corresponding to additional displacements **u**) by σ_{ij}.

We apply the variational principle of virtual work corresponding to the stresses $\sigma_{ij}^* + \sigma_{ij}$ (Washizu 1992):

$$\int_G (\sigma_{ij} + \sigma_{ij}^*)\delta e_{ij} d\mathbf{x} - \int_{S_\sigma} (\mathbf{H} + \mathbf{h})\delta \mathbf{u} d\mathbf{x} - \int_G (\mathbf{F} + \mathbf{f})\delta \mathbf{u} d\mathbf{x} = 0, \tag{4.5}$$

where the strains e_{ij} are calculated from the displacements **u** in the form (Washizu 1992)

$$e_{ij}(\mathbf{u}) = \frac{1}{2}\left(\frac{\partial u_i}{\partial x_j} + \frac{\partial u_j}{\partial x_i} + \frac{\partial u_k}{\partial x_i}\frac{\partial u_k}{\partial x_j}\right), \tag{4.6}$$

and δ denotes a variation of a function (Courant and Hilbert 1953; Washizu 1992).

Substituting (4.6) into (4.5) with account (4.15) and (4.3) yields

$$\int_G \left(\sigma_{ij}\delta e_{ij}(\mathbf{u}) + \sigma^*_{ij}\frac{\partial u_k}{\partial x_i}\frac{\partial \delta u_k}{\partial x_j}\right) d\mathbf{x} - \int_{S_\sigma} \mathbf{h}\delta\mathbf{u} d\mathbf{x} - \int_G \mathbf{f}\delta\mathbf{u} d\mathbf{x} = 0. \tag{4.7}$$

Provided that the additional displacements $\mathbf{u}$ are small, and the initial stresses σ^*_{ij} are not small, we obtain from (4.7) the variational principle for a body with initial stresses, see Washizu (1992),

$$\int_G \left(\sigma_{ij}\delta e_{ij}(\mathbf{u}) + \sigma^*_{ij}\frac{\partial \delta u_k}{\partial x_i}\right) d\mathbf{x} - \int_{S_\sigma} \mathbf{h}\delta\mathbf{u} d\mathbf{x} - \int_G \mathbf{f}\delta\mathbf{u} d\mathbf{x}, \tag{4.8}$$

where

$$e_{ij}(\mathbf{u}) = \frac{1}{2}\left(\frac{\partial u_i}{\partial x_j} + \frac{\partial u_j}{\partial x_i}\right) \tag{4.9}$$

are additional strains related to the additional stresses σ_{ij} by Hooke's law

$$\sigma_{ij} = c_{ijkl}(\mathbf{x})e_{ij}(\mathbf{u}) = c_{ijkl}(\mathbf{x})\frac{\partial u_k}{\partial x_l}. \tag{4.10}$$

The variational principle (4.8) leads to a model of a linearly elastic body with initial stresses, see Washizu (1992), consisting of the equilibrium equations

$$\frac{\partial}{\partial x_j}\left(\sigma_{ij} + \sigma^*_{kj}\frac{\partial u_i}{\partial x_k}\right) + f_i = 0 \text{ in } G, \tag{4.11}$$

constitutive equations (4.10) and boundary conditions

$$\sigma_{ij}n_j + \sigma^*_{kj}\frac{\partial u_i}{\partial x_k}n_j = h_i(\mathbf{x}) \text{ on } S_\sigma, \tag{4.12}$$
$$\mathbf{u} = 0 \text{ on } S_u.$$

Condition $\mathbf{u} = 0$ on S_u means that we assume no additional displacements on S_u. This assumption is done for the simplicity.

4.2 Transformation of the Governing Equations of Elastic Body with Initial Stresses

Let us consider the expression $\sigma_{ij} + \sigma^*_{kj}\dfrac{\partial u_i}{\partial x_k}$ in the brackets in the Eq. (4.11) and in (4.12). In accordance with (4.10) it can be rewritten as

$$\sigma_{ij} + \sigma^*_{lj}\delta_{ik}\frac{\partial u_k}{\partial x_l} = c_{ijkl}\frac{\partial u_k}{\partial x_l} + \sigma^*_{lj}\delta_{ik}\frac{\partial u_k}{\partial x_l} = (c_{ijkl} + \sigma^*_{lj}\delta_{ik})\frac{\partial u_k}{\partial x_l}, \tag{4.13}$$

where $\delta_{ii} = 1$ and $\delta_{ik} = 0$ if $i \neq k$ is Kronecker's delta.

Introduce the following quantities

$$h_{ijkl}(\mathbf{x}, \mathbf{x}/\varepsilon) = c_{ijkl}(\mathbf{x}) + \sigma^*_{jl}(\mathbf{x}, \mathbf{x}/\varepsilon)\delta_{ik}, \tag{4.14}$$

using which the problem (4.11) and (4.12) can be written as:
equilibrium equations

$$\frac{\partial}{\partial x_j}\left(h_{ijkl}(\mathbf{x}, \mathbf{x}/\varepsilon)\frac{\partial u_k}{\partial x_l}\right) + f_i = 0 \text{ in } G; \tag{4.15}$$

boundary conditions (do not confuse the coefficients h_{ijkl} with the surface force h_i)

$$h_{ijkl}(\mathbf{x}, \mathbf{x}/\varepsilon)\frac{\partial u_k}{\partial x_l}n_j = h_i(\mathbf{x}) \text{ on } S_\sigma;$$
$$\mathbf{u} = \mathbf{u}^0(\mathbf{x}) \text{ on } S_u. \tag{4.16}$$

Model (4.14)–(4.16) will be the basis of subsequent analysis.

4.3 Loss of Symmetry in the Equations Relating Additional Stresses and Additional Strains

By using Eq. (4.13), we introduce σ_{ij} as follows:

$$\sigma_{ij} = h_{ijkl}(\mathbf{x})\frac{\partial u_k}{\partial x_l}, \tag{4.17}$$

where $h_{ijkl}(\mathbf{x})$ is introduced by (4.14).

The boundary value problem (4.15) and (4.16) looks similar to the problem of elasticity theory. In particular, the equality (4.17) looks similar to the constitutive equations of an elastic body (the Hooke's low), compare with (4.10). The only difference between (4.10) and (4.17) is that the elastic constants c_{ijkl} in (4.10) are replaced by the coefficients h_{ijkl} (4.14) in (4.17).

The fundamental point from which the asymptotic theory of thin-walled structures with initial stresses arises is the loss by the coefficients h_{ijkl} (4.14) *some* symmetries inherent in the elastic constants c_{ijkl}. Namely, the tensor h_{ijkl} loses the minor symmetries (Knops and Payne 1971) inherent in the elastic constants, and generally,

$$h_{ijkl} \neq h_{ijlk}, \tag{4.18}$$

while the major symmetries

$$h_{ijkl} = h_{klij} \tag{4.19}$$

are preserved.

The equality (4.19) follows directly from the definitions of h_{ijkl} (4.14), and the symmetry of the Kronecker's delta δ_{ij} and the initial stresses σ^*_{ij} in i and j. Due to these symmetries, we have

$$h_{ijkl} = c_{ijkl} + \delta_{ik}\sigma^*_{jl} = c_{klij} + \delta_{ki}\sigma^*_{lj} = h_{klij}. \tag{4.20}$$

To verify the inequality (4.18), we consider the terms $h_{ijkl} = \delta_{ik}\sigma^*_{jl}$ and $h_{ijlk} = \delta_{il}\sigma^*_{jk}$ from (4.18). Take the indices i and l to be equal one to another and denote $i = l = A$ (A is an arbitrary number taking value 1, 2 or 3). For this case, terms $\delta_{ik}\sigma^*_{jl}$ and $\delta_{il}\sigma^*_{jk}$ take the form $\delta_{Ak}\sigma^*_{jA}$ and σ^*_{jk}, respectively. For $A \neq k$, the first term is equal to zero, while the second term may take non-zero values.

Note that this loss of symmetry takes place only in the presence of initial stresses. If the initial stresses absent: $\sigma^\varepsilon_{ij} = 0$, this asymmetry disappears.

Asymmetry (4.18) plays a decisive role in the theory of the thin-walled structures with initial stresses presented in this book.

4.4 Homogenization Method in the Mechanics of Composites with Initial Stresses

Now, we consider a composite body with initial stresses. As noted in Sect. 3.1, when considering a homogenization problem, we consider a problem depending on a parameter. In composite materials, there is a natural parameter, which is the characteristic size ε (absolute or relative) of inhomogeneities. Accordingly, the material characteristics and, as a consequence, the stress-strain state in composite depend on the parameter ε. To emphasize this fact, we mark the material characteristics of composite and stress-strain state in composite by the superscript ε. We will omit the superscript ε if parameter ε is specified in any other way (for example, if the function is specified to depend on the argument $\mathbf{x}/\varepsilon$).

Linear elasticity theory problem for a body of periodic structure with the initial stresses σ^*_{ij} has the form

$$\begin{gathered}\frac{\partial}{\partial x_j}\left((c_{ijkl}(\mathbf{x}/\varepsilon)+\delta_{ik}\sigma^{*\varepsilon}_{jl}(\mathbf{x},\mathbf{x}/\varepsilon))\frac{\partial u^\varepsilon_k}{\partial x_l}\right)+f^\varepsilon_i(\mathbf{x})=0 \text{ in } G,\\ (c_{ijkl}(\mathbf{x}/\varepsilon)+\delta_{ik}\sigma^{*\varepsilon}_{jl}(\mathbf{x},\mathbf{x}/\varepsilon))\frac{\partial u^\varepsilon_k}{\partial x_l}n_j=h^\varepsilon_i(\mathbf{x}) \text{ in } S_\sigma,\\ \mathbf{u}^\varepsilon(\mathbf{x})=0 \text{ on } S_u.\end{gathered} \tag{4.21}$$

By using the coefficient h_{ijkl} (4.14) and (4.17), which is analog but not equivalent if Hooke's low, we write (4.21) in the form

$$\begin{gathered}\frac{\partial}{\partial x_j}\left(h_{ijkl}(\mathbf{x},\mathbf{x}/\varepsilon)\frac{\partial u^\varepsilon_k}{\partial x_l}\right)+f^\varepsilon_i(\mathbf{x})=0 \text{ in } G,\\ h_{ijkl}(\mathbf{x},\mathbf{x}/\varepsilon)\frac{\partial u^\varepsilon_k}{\partial x_l}n_j=h^\varepsilon_i(\mathbf{x}) \text{ on } S_\sigma,\\ \mathbf{u}^\varepsilon(\mathbf{x})=0 \text{ on } S_u.\end{gathered} \tag{4.22}$$

Problem (4.22) is similar to the problem of elasticity theory (3.1) and (3.2), and the methods used for composite materials without initial stresses, described in Chap. 3, see Sect. 3.2.1, can be applied to it. All calculations from Sect. 3.2.1 can be repeated for problem (4.22). Having done this, we come to the conclusion that the homogenized material that we associate with the composite is described by the problem

$$L(\sigma)\mathbf{u}+\langle\mathbf{f}^\varepsilon\rangle=0 \text{ in } G, \tag{4.23}$$

$$l(\sigma)\mathbf{u}=\langle\mathbf{h}^\varepsilon\rangle \text{ on } S_\sigma, \tag{4.24}$$

$$\mathbf{u}(\mathbf{x})=0 \text{ on } S_u, \tag{4.25}$$

where

$$L(\sigma)\mathbf{u}=\frac{\partial}{\partial x_j}\left(C_{ijkl}(\sigma)\frac{\partial u_k}{\partial x_l}\right), \tag{4.26}$$

is the homogenized operator, and

$$l(\sigma)\mathbf{u}=C_{ijkl}(\sigma)\frac{\partial u_k}{\partial x_l}n_j \tag{4.27}$$

is the homogenized boundary normal stress operator. The homogenized elastic constants $C_{ijkl}(\sigma)$ in the operators defined by (4.29).

The periodicity cell problem corresponding to the problem (4.22) has the form

$$\begin{cases}(h_{ijmn}(\mathbf{x},\mathbf{y})K^{kl}_{m,ny}+h_{ijkl}(\mathbf{y}))_{,jy}=0 \text{ in } Y,\\ \mathbf{K}^{kl}(\mathbf{y}) \text{ is periodic in } \mathbf{y} \text{ with periodicity cell } Y,\end{cases} \tag{4.28}$$

and the homogenized coefficients $C_{ijkl}(\sigma)$ of composite with initial stresses are given by the formula (compare with (4.33))

$$C_{ijkl}(\sigma) = \left\langle h_{ijmn}(\mathbf{x}, \mathbf{y}) K^{kl}_{m,ny}(\mathbf{y}) + h_{ijkl}(\mathbf{x}, \mathbf{y}) \right\rangle . \tag{4.29}$$

The homogenized constants $C_{ijkl}(\sigma)$ describes the homogenized material. As follows from (4.28) and (4.29), the constants $C_{ijkl}(\sigma)$ are determined by both the local elastic constants as well as the local initial stresses. To emphasize this fact, we write the argument σ in $C_{ijkl}(\sigma)$.

The periodicity cell problem (4.28) looks similar to (3.27), but these problems differ in their coefficients. For this reason, we denote the solution of problem (4.28) by $\mathbf{K}^{kl}(\mathbf{y})$ to distinguish it from the solution $\mathbf{N}^{kl}(\mathbf{y})$ of the periodicity cell problem (3.27). Obviously, $\mathbf{K}^{kl}(\mathbf{y}) = \mathbf{N}^{kl}(\mathbf{y})$ if the initial stresses are zero.

The problem (4.23)–(4.27) may be written in the form of the minimization problem:

$$J_\sigma(\mathbf{u}) - \langle \mathbf{f}, \mathbf{u} \rangle - \langle \mathbf{h}, \mathbf{u} \rangle_{S_\sigma} \to \min , \mathbf{u} \in H^{01}(G), \tag{4.30}$$

where

$$J_\sigma(\mathbf{u}) = \frac{1}{2} \int_G C_{ijkl}(\sigma) e_{ij}(\mathbf{u}) e_{kl}(\mathbf{u}) d\mathbf{x},$$
$$\mathbf{f} = \text{w-}\lim_{\varepsilon \to 0} \mathbf{f}^\varepsilon, \quad \mathbf{h} = \text{w-}\lim_{\varepsilon \to 0} \mathbf{h}^\varepsilon .$$

$\text{w-}\lim\limits_{\varepsilon \to 0}$ means the weak limit in $H^{01}(G)$ and $H^{1/2}(S_\sigma)$, correspondingly. For smooth quickly oscillating functions, the indicated weak limits coincide with the average values on domain G and surface S_σ, correspondingly (see for details Lions and Magenes 1973; Bensoussan et al. 1978).

4.5 Small Initial Stresses

Stresses in elastic bodies, including the initial stresses, are limited by the ultimate strengths σ_S of the components of components composite. The ultimate strength of real materials does not exceed several percents of the Young's modulus E of the corresponding material, see Table 4.1 (more data on the mechanical constants of elastic materials can be found, for example, in Ashby and Jones 2005; Brandes and Brook 1992).

We suppose that the composite materials under consideration stay within the elastic regime. Then the coefficients h_{ijkl} (4.14) can be represented as

$$h_{ijkl}(\mathbf{x}, \mathbf{y}) = c_{ijkl}(\mathbf{y}) + \mu b_{ijkl}(\mathbf{x}, \mathbf{y}), \tag{4.31}$$

Table 4.1 Young's moduli E and strength limits σ_S

Material	Aluminum	Cooper	Steel	Nylon	Epoxy resin	Carbon tubes
E GPa ($10^9\,Pa$)	70	117	200	4	3	400
σ_s MPa ($10^6\,Pa$)	110	220	800	90	60	4000
σ_s/E	0.0016	0.0019	0.004	0.0225	0.02	0.01

where

$$b_{ijkl}(\mathbf{x},\mathbf{y}) = \delta_{ik}\sigma^*_{jl}(\mathbf{x},\mathbf{y}), \tag{4.32}$$

and μ is a small parameter, which has the order of ratio of characteristic value of the initial stresses to the elastic constants. This ratio does not excite the ratio $\dfrac{\sigma_S}{E}$. In practice, $\dfrac{\sigma_S}{E} < 0.02$, see Table 4.1.

The periodicity cell problem (4.28) for the coefficients (4.31) takes the form

$$\begin{cases} ((c_{ijmn}(\mathbf{y}) + \mu b_{ijmn}(\mathbf{x},\mathbf{y}))K^{kl}_{m,ny} + c_{ijkl}(\mathbf{y}) + \mu b_{ijkl}(\mathbf{x},\mathbf{y}))_{,jy} = 0 \text{ in } Y, \\ \mathbf{K}^{kl}(\mathbf{y}) \text{ is periodic in } \mathbf{y} \text{ with periodicity cell } Y. \end{cases} \tag{4.33}$$

and formula (4.29) for computation of the homogenized coefficients $C_{ijkl}(\sigma)$ takes the form

$$C_{ijkl}(\sigma) = \left\langle (c_{ijmn}(\mathbf{y}) + \mu b_{ijmn}(\mathbf{x},\mathbf{y}))K^{kl}_{m,ny}(\mathbf{y}) + h_{ijkl}(\mathbf{y})\right\rangle. \tag{4.34}$$

One can use the classical method of small parameter (concerning the method of small parameter and those application in mechanics of solids, see, e.g., Murray 1984; Lagerstrom 1988; Andrianov et al. 2004) to investigate the problem (4.33) and compute the homogenized coefficients $C_{ijkl}(\sigma)$ (4.34).

By using the classical method of small parameter, we seek a solution to the problem (4.33) in the form of series

$$\mathbf{K}^{kl}(\mathbf{y}) = \sum_{s=0}^{\infty} \mu^s \mathbf{K}^{kls}(\mathbf{y}). \tag{4.35}$$

Substituting (4.35) into the periodicity cell problem (4.33) and equating the term with identical power of μ, we obtain a series of the problems, the first two of them are the following:

$$\begin{cases} (c_{ijmn}(\mathbf{y})K^{kl0}_{m,ny} + c_{ijkl}(\mathbf{y}))_{,jy} = 0 \text{ in } Y, \\ \mathbf{K}^{kl0}(\mathbf{y}) \text{ is periodic in } \mathbf{y} \text{ with periodicity cell } Y, \end{cases} \tag{4.36}$$

$$\begin{cases} (c_{ijmn}(\mathbf{y})K^{kl1}_{m,ny} + c_{ijmn}(\mathbf{y})K^{kl0}_{m,ny} + \mu b_{ijkl}(\mathbf{x}, \mathbf{y}))_{,jy} = 0 \text{ in } Y, \\ \mathbf{K}^{kl1}(\mathbf{y}) \text{ is periodic in } \mathbf{y} \text{ with periodicity cell } Y. \end{cases} \tag{4.37}$$

The problem (4.36) coincides with the periodicity cell problem (3.27) for composite material with no initial stresses, thus,

$$\mathbf{K}^{kl0}(\mathbf{y}) = \mathbf{N}^{kl}(\mathbf{y}). \tag{4.38}$$

Formula (4.34) for computation of the homogenized coefficients $C_{ijkl}(\sigma)$ takes the form (we save only linear terms below and take into account formula (4.38))

$$\begin{aligned} C_{ijkl}(\sigma) &= \langle (c_{ijmn}(\mathbf{y}) + \mu b_{ijmn}(\mathbf{x}, \mathbf{y}))(N^{kl}_{m,ny}(\mathbf{y}) + \mu K^{kl1}_{m,ny}(\mathbf{y})) \\ &\quad + (c_{ijkl}(\mathbf{y}) + \mu b_{ijkl}(\mathbf{x}, \mathbf{y}))\rangle \\ &= \langle c_{ijmn}(\mathbf{y})N^{kl}_{m,ny}(\mathbf{y}) + c_{ijkl}(\mathbf{y})\rangle \\ &\quad + \mu[\langle c_{ijmn}(\mathbf{y})K^{kl1}_{m,ny}(\mathbf{y}) + b_{ijmn}(\mathbf{x}, \mathbf{y})N^{kl}_{m,ny}(\mathbf{y}) + b_{ijkl}(\mathbf{x}, \mathbf{y})\rangle]. \end{aligned} \tag{4.39}$$

The first term on the right-hand part (4.39) coincides with the formula (3.27) for computation of the homogenized constant $C_{ijkl} = C_{ijkl}(0)$ of composite with no initial stresses, as a result, we arrive at the following formula:

$$\begin{aligned} C_{ijkl}(\sigma) &= C_{ijkl}(0) \\ &\quad + \mu[\langle c_{ijmn}(\mathbf{y})K^{kl1}_{m,ny}(\mathbf{y}) + b_{ijmn}(\mathbf{x}, \mathbf{y})N^{kl}_{m,ny}(\mathbf{y}) + b_{ijkl}(\mathbf{x}, \mathbf{y})\rangle]. \end{aligned} \tag{4.40}$$

Taking into account definition of b_{ijkl} (4.32), we obtain

$$\begin{aligned} C_{ijkl}(\sigma) &= C_{ijkl}(0) \\ &\quad + \mu[\langle c_{ijmn}(\mathbf{y})K^{kl1}_{m,ny}(\mathbf{y}) + \sigma^*_{jn}(\mathbf{x}, \mathbf{y})N^{kl}_{i,ny}(\mathbf{y}) + \delta_{ik}\sigma^*_{jl}(\mathbf{x}, \mathbf{y})\rangle], \end{aligned} \tag{4.41}$$

where $\mathbf{K}^{kl1}(\mathbf{y})$ is solution to the problem (4.37).

4.6 G-Limit of the Operators and Functionals of the Elasticity Theory with Initial Stresses

Although the G-limit is an abstract mathematical concept, its use may to provide us with some important conclusions about the mechanics of composite materials. A specific property of G–limit is that the usual rule of calculus (Rudin 1967): "The limit of a sum is equal to the sum of the limits" is not true for the G-limit.

The linear elasticity theory problem for an elastic body with initial stresses σ^*_{ij} has the following form, see Sect. 4.2:

$$L_\varepsilon \mathbf{u}^\varepsilon + M_\varepsilon \mathbf{u}^\varepsilon + \mathbf{f}^\varepsilon(\mathbf{x}) = 0 \text{ in } G, \tag{4.42}$$

$$\mathbf{u}^\varepsilon(\mathbf{x}) = 0 \text{ on } \partial G, \tag{4.43}$$

where

$$L_\varepsilon \mathbf{u} = \frac{\partial}{\partial x_j}\left(c_{ijkl}(\mathbf{x}/\varepsilon)\frac{\partial u_k}{\partial x_l}\right), \tag{4.44}$$

$$M_\varepsilon \mathbf{u} = \frac{\partial}{\partial x_j}\left(\delta_{ik}\sigma^*_{jl}(\mathbf{x}, \mathbf{x}/\varepsilon)\frac{\partial u_k}{\partial x_l}\right). \tag{4.45}$$

The coefficients of the both elasticity theory operator L_ε (4.44) and operator M_ε (4.45), which account the initial stresses, are strongly oscillating functions, depending on the variable $\mathbf{x}/\varepsilon$. The differential operators L_ε and M_ε have the same (second) order. Thus, when considering a composite body with initial stresses, we encounter a situation where the limit of the sum is not equal to the sum of the limits (we are talking about G–limits).

The problem (4.42) and (4.43) admits the following equivalent formulation in the form of minimization problem: find $\mathbf{u}^\varepsilon$ from the solution of the minimization problem

$$F_\varepsilon(\mathbf{u}) - \langle \mathbf{f}^\varepsilon, \mathbf{u}\rangle \to \min, \mathbf{u} \in H^{01}(G); \tag{4.46}$$

where

$$F_\varepsilon(\mathbf{u}) = J_\varepsilon(\mathbf{u}) + I_\varepsilon(\mathbf{u}) \tag{4.47}$$

and functionals $J_\varepsilon(\mathbf{u})$ and $I_\varepsilon(\mathbf{u})$ are determined as follows:

$$J_\varepsilon(\mathbf{u}) = \frac{1}{2}\int_G c^\varepsilon_{ijkl}(\mathbf{x})e_{ij}(\mathbf{u})e_{kl}(\mathbf{u})d\mathbf{x}, \tag{4.48}$$

$$I_\varepsilon(\mathbf{u}) = \frac{1}{2}\int_G \delta_{ik}\sigma^{*\varepsilon}_{jl}(\mathbf{x})e_{ij}(\mathbf{u})e_{kl}(\mathbf{u})d\mathbf{x}. \tag{4.49}$$

In accordance to G-limit approach, one has to compute G-limit of the sum $J_\varepsilon + I_\varepsilon$ in order to carry out the homogenization procedure for a composite with initial stresses.

Note. The first thing that comes to mind for calculating the G-limit of the sum of the functionals (4.48) and (4.49) is to follow the procedure used for a homogeneous body with initial stresses Washizu (1992). Namely, determine (theoretically or experimentally) the homogenized elastic constants $C_{ijkl}(0) = C_{ijkl}$ of the body without initial stresses and the homogenized stresses $\langle\sigma^{*\varepsilon}_{jl}\rangle$ which would arise in this body.

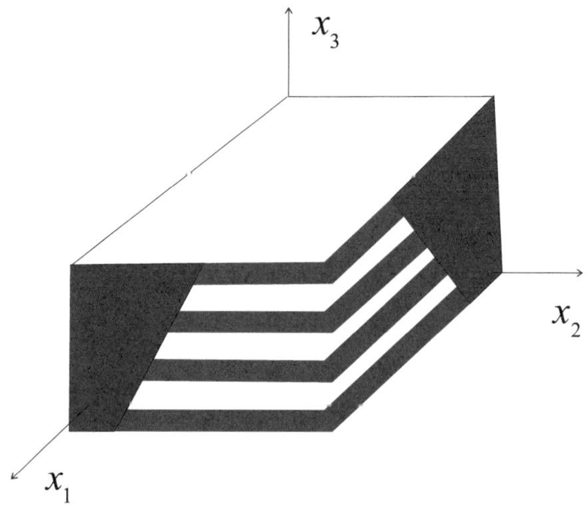

Fig. 4.2 Laminated media

Then compose the combination $C_{ijkl}(0) + \langle \sigma_{jl}^{*\varepsilon} \rangle \delta_{ik}$, which arises if we consider a real homogeneous elastic body with elastic constants $C_{ijkl}(0)$ and initial stresses $\langle \sigma_{jl}^{*\varepsilon} \rangle$ (Washizu 1992). The following example shows that such an approach is, generally speaking, incorrect.

4.7 An Example. Layered Body with Initial Stresses

We consider the layered structure shown in Fig. 4.2. Such kind layered structures with initial stresses were often analyzed in connections with problems in geophysics and rock mechanics, see, e.g., Backus (1962), Bai and Pollard (2000a, b, c) and design of materials possessing the required mechanical characteristics see, e.g., Kolpakov (2007), Kolpakov and Kolpakov (2005), Xu et al. (2018), Liu et al. (2023), Gurdal et al. (1999). For layered structures formed of layers of homogeneous isotropic materials, the solution to the elasticity theory problem can be obtained in explicit form.

Let the layers be parallel to the Ox_1x_2-plane and the Ox_3-axis be directed perpendicular to the layers. In this case, the solution to the elasticity theory problem depends on unique variable x_3, which will be denoted by x below in this section.

We assume that the structure is subjected to the initial stress $\sigma_{33}^* = \sigma^* = const$ and force $F^\varepsilon(x)$ acts in the direction Ox_3-axis (perpendicular to the layers). The displacements are assumed to have the form $(u, 0, 0)$. The elasticity theory problem (4.15) about the deformation of this structure takes the following form (prime means the derivative with respect to the variable $x = x_3$):

$$((E(x/\varepsilon) + \sigma^*)u')' = F^\varepsilon(x), \tag{4.50}$$

where function $(E(y)$ is periodic with period m.

The homogenized problem corresponding to (4.50) is (details may be found in Kalamkarov and Kolpakov 1997)

$$\left(\frac{1}{\left\langle \dfrac{1}{E(y)+\sigma^*} \right\rangle} u'\right)' = \langle F^{\varepsilon}(x) \rangle, \tag{4.51}$$

where

$$\langle\rangle = \frac{1}{m}\int_0^m dy$$

is the average value over the period m cell of the structure.

For a body with no initial stresses, the homogenized elastic constant is $C(0) = \dfrac{1}{\left\langle \dfrac{1}{E(y)} \right\rangle}$. For the body with initial stresses, the homogenized elastic constant is $C(\sigma) = \dfrac{1}{\left\langle \dfrac{1}{E(y)+\sigma^*} \right\rangle}$, see (4.51). Investigate the elastic constants $C(\sigma)$. We consider the case when $\sigma^* \ll E(y)$ and write the following approximate equalities

$$\begin{aligned}
\frac{1}{\left\langle \dfrac{1}{E(y)+\sigma^*} \right\rangle} &= \frac{1}{\left\langle \dfrac{1}{E(y)\left(1+\dfrac{\sigma^*}{E(y)}\right)} \right\rangle} = \frac{1}{\left\langle \dfrac{1}{E(y)}\left(1-\dfrac{\sigma^*}{E(y)}\right) \right\rangle} \\
&= \frac{1}{\left\langle \dfrac{1}{E(y)} \right\rangle - \left\langle \dfrac{\sigma^*}{E^2(y)} \right\rangle} = \frac{1}{\left\langle \dfrac{1}{E(y)} \right\rangle \left(1 - \dfrac{1}{\left\langle \dfrac{1}{E(y)} \right\rangle}\left\langle \dfrac{\sigma^*}{E^2(y)} \right\rangle\right)} \\
&= \frac{1}{\left\langle \dfrac{1}{E(y)} \right\rangle}\left(1 + \frac{1}{\left\langle \dfrac{1}{E(y)} \right\rangle}\left\langle \dfrac{\sigma^*}{E^2(y)} \right\rangle\right) = \frac{1}{\left\langle \dfrac{1}{E(y)} \right\rangle} + \sigma^* \frac{\left\langle \dfrac{1}{E^2(y)} \right\rangle}{\left\langle \dfrac{1}{E(y)} \right\rangle^2} \\
&= C(0) + \sigma^* \frac{\left\langle \dfrac{1}{E^2(y)} \right\rangle}{\left\langle \dfrac{1}{E(y)} \right\rangle^2}.
\end{aligned} \tag{4.52}$$

Writing the final equalities in (4.52), we have taken into account that $\sigma^* = const$.

Write the right-hand side of (4.52) as

$$C(0) + \sigma^* L,$$

where

$$L = \frac{\left\langle \dfrac{1}{E^2(y)} \right\rangle}{\left\langle \dfrac{1}{E(y)} \right\rangle^2}. \tag{4.53}$$

For homogeneous materials, multiplier $L = 1$, and (4.53) has the form $C(0) + \sigma^*$. It would be interesting to know the possible values of L for inhomogeneous material. To find the possible values of L, it is sufficient to calculate the possible values of two functionals $\left\langle \dfrac{1}{E(y)} \right\rangle$ and $\left\langle \dfrac{1}{E^2(y)} \right\rangle$. The problem of possible values of these functionals can be completely solved using the method of convex combinations (Kalamkarov and Kolpakov 1997).

We present the solution to the problem for the simplest case of a two-component medium. Denote E_{min} and E_{max} the Young's moduli and λ_{min} and $\lambda_{max} = 1 - \lambda_{min}$ the volume fractions of the soft and stiff components, correspondingly. For $\lambda_{min} = 0$ and $\lambda_{min} = 1$, the material is homogeneous, thus $L = 1$ for $\lambda_{min} = 0$ and $\lambda_{min} = 1$. For the intermediate values $0 < \lambda_{min} < 1$, due to Cauchy inequality (Rudin 1967), $L \geq 1$. It is interesting to know the possible deviation of L from the unity. The plots of L as function of λ_{min} are drawn in Fig. 4.3 for $\dfrac{E_{max}}{E_{max}}$ equal to 10 and 20 (this is realistic ratio for rock materials Bourn 2003). The maximum values of L are 3 and 5.5, correspondingly. It is see in Fig. 4.3 that the maximum values of L occur at a small volume fraction of the soft component—$\lambda_{min} < 0.1$ in the case under consideration. Recall that at $\lambda_{min} = 0$ the coefficient $L = 1$. From a mechanical point of view, this means that thin soft layers in a composite material significantly increase the effect of initial stresses on the material as a whole.

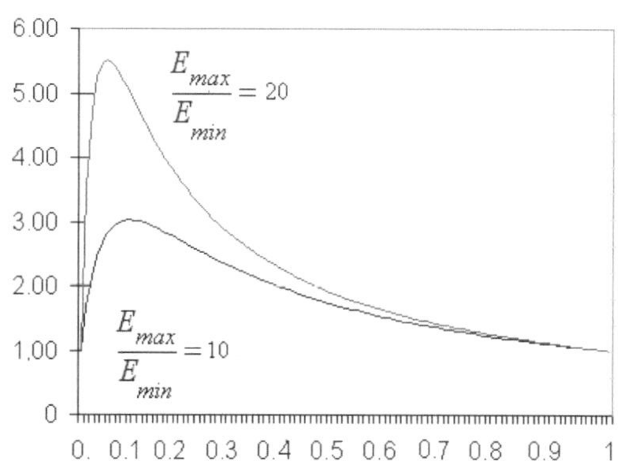

Fig. 4.3 Plot L (4.53) as function of λ_{min}

References

Andrianov IV, Awrejcewicz J, Manevich LI (2004) Asymptotic mechanics of thin-walled structures. Springer, Berlin

Ashby MF, Jones DRH (2005) Engineering materials. In two volumes. Butterworth-Heinemann, Boston

Backus GE (1962) Long-wave anisotropy produced by horizontal layering. J Geophys Res 67:4427–4440

Bai T, Pollard DD (2000) Explanation for fracture spacing in layered materials. Nature 403(6771):753–756

Bai T, Pollard DD (2000) Closely spaced fractures in layered rocks; initiation mechanism and propagation kinematics. J Struct Geol 22(10):1409–1425

Bai T, Pollard DD (2000) Fracture spacing in layered rocks: a new explanation based on the stress transition. J Struct Geol 22(1):43–57

Bensoussan A, Lions J-L, Papanicolaou G (1978) Asymptotic analysis for periodic structures. North-Holland Publ, Amsterdam

Bourn SJ (2003) Contrast of elastic properties between rock layers as a mechanism for the initiation and orientation of tensile failure under uniform remote compression. J Geophys Res 108(B8):2395

Brandes EA, Brook GB (1992) Smithells metals reference book, 7th edn. Butterworth-Heinemann, Boston

Courant RS, Hilbert D (1953) Methods of mathematical physics. Wiley, New York

Gurdal Z, Haftka RT, Hajela P (1999) Design and optimization of laminated composite materials. Wiley-Interscience, New York

Kalamkarov AL, Kolpakov AG (1997) Analysis, design and optimization of composite structures. Wiley, Chichester

Knops RJ, Payne LE (1971) Uniqueness theorems in linear elasticity. Springer, Berlin

Kolpakov AA (2007) Design of laminated plates possessing the required stiffnesses using the minimal number of materials and layers. J Elast 86:245–261

Kolpakov AA, Kolpakov AG (2005) Solution of the laminated plate design problem: new problems and algorithms. Comput Struct 83(12–13):964–975

Lagerstrom PA (1988) Matched asymptotic expansions. Springer, New York

Lions J-L, Magenes E (1973) Non-homogeneous boundary value problems and applications, vol 1. Springer, Berlin

Liu X, Qin J, Zhao K, Featherston CA, Kennedy D, Jing Y, Yang G (2023) Design optimization of laminated composite structures using artificial neural network and genetic algorithm. Comput Struct 305:116500

Murray JD (1984) Asymptotic analysis. Springer, Berlin

Rudin W (1967) Principles of mathematical analysis, 3rd edn. McGraw-Hill, New York

Washizu K (1992) Variational methods in elasticity and plasticity. Pergamon Press, Oxford

Xu Y, Zhu J, Wu Z, Cao Y, Zhao Y, Zhang W (2018) A review on the design of laminated composite structures: constant and variable stiffness design and topology optimization. Adv Comput Hybrid Mater 1:460–477

Chapter 5
Uniform Plates with Initial Stresses

This book is devoted to thin-walled structural elements with initial stresses. Distinguish the following basic thin-walled structural elements: plates, shells, membranes, rods, beams, and strings. The behavior of plates, shells, rods, and beams is determined equally by initial stresses in the structural element and the elastic characteristics of the structural element. The behavior of membranes and strings is determined only by initial stresses in the structural element. In this book, we will consider plates and beams. Those interested in non-uniform membranes and strings can find some information in Kolpakov (2004).

We start our consideration with plates of constant thickness made of homogeneous material (hereinafter referred to as uniform plates). This is the case considered in classical plate theories, including the Kirchhoff-Love plate theory. The asymptotic analysis of uniform plates allows us to clearly trace the emergence of the terms corresponding to the "additional normal force" in the classical theories from the asymmetry of coefficients in the elasticity theory problem with initial stresses.

The asymptotic analysis presented in this section may be considered an alternative deriving of the classical plate model based exclusively on the mathematical analysis of three-dimensional elasticity problem.

5.1 Engineering Approaches to the Initial Stresses in Plates

It is interesting to compare the classical (engineering) and asymptotic approaches. For this reason, before proceeding to the presentation of the asymptotic theory, we present in a condensed form the classical engineering approaches to the construction of the model of plates with initial stresses. Detailed presentation of the classical engineering approaches can be found in Timoshenko (1961, 1971), Timoshenko and Woinowsky-Krieger (1959), Washizu (1992).

A. Kolpakov, *Thin-Walled Structures with Initial Stresses*, Mathematical Engineering,
https://doi.org/10.1007/978-3-032-07247-4_5

The earliest engineering analysis of stability of thin plates was based on the analysis of deformation of an element of stressed plate and computation of additional normal forces arising as a result of additional deformation of an element of a structure, see historical reviews in Rhodes (2002), Timoshenko (1983). Later, there was developed an alternative approach to the stability of thin plates based on the variational principles, in particular, computation of work of initial stresses on the additional displacement (Washizu 1992).

5.1.1 Deriving the Model of Uniform Plate with Initial Stresses by Using Geometrical Analysis

We consider two-dimensional plate made of homogeneous material, which have constant thickness and is placed in Ox_1x_2-plane (Ox_3-axis is perpendicular to the plate). It is assumed that the plate was loaded and, as a result, stresses σ_x^*, σ_y^*, and σ_{xy}^* have arisen in the plate. After that, additional loads are applied to the plate. Let's consider normal deflection which may to arise in the plate and analyze the corresponding stress-strain state in the plate.

Here we use for coordinates notations (x, y, z) instead (x_1, x_2, x_3) as is customary in technical literature.

We consider a plate element of sides dx and dy subjected to the resultant in-plane initial stresses $N_x^*(x, y)$, $N_y^*(x, y)$, and $N_{xy}^*(x, y) = N_{yx}^*(x, y)$. Top and side views of such element are shown in Fig. 5.1 (left and right figures, respectively).

The resultant in-plane initial stresses are introduced as

$$N_x^* = \int_{-H/2}^{H/2} \sigma_x^* dz, \quad N_y^* = \int_{-H/2}^{H/2} \sigma_y^* dz, \quad N_{xy}^* = N_{yx}^* = \int_{-H/2}^{H/2} \sigma_{xy}^* dz.$$

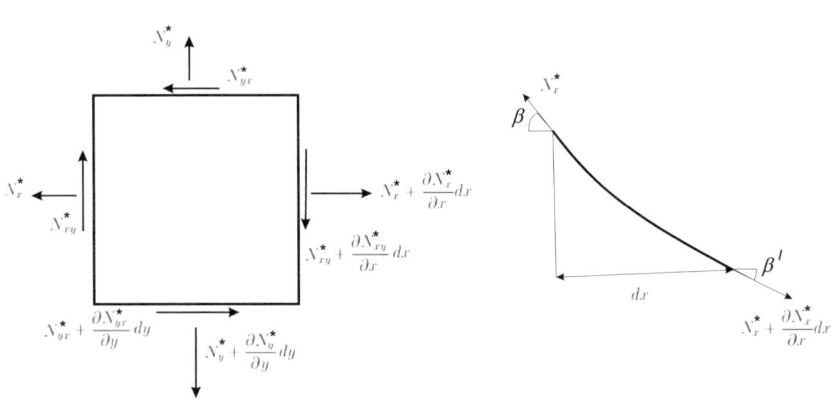

Fig. 5.1 An element of plate

It is assumed that the resultant in-plane initial stresses, which are functions of x and y, satisfy the equilibrium equations

$$\frac{\partial N_x^*}{\partial x} + \frac{\partial N_{xy}^*}{\partial y} = F_x, \quad \frac{\partial N_{yx}^*}{\partial x} + \frac{\partial N_y^*}{\partial y} = F_y. \tag{5.1}$$

Consider an element $dxdy$ of the plate, see Fig. 5.1 (left). Let the plate gets additional displacements, namely, normal deflection $w(x, y)$. Let's compute the projection of all the forces, which arise in the result of this additional normal deflection, on the Oz-axis. Due to the bending of the plate in the Oxz plane, the z component of N_x^* is equal to (see Fig. 5.1 right)

$$N_x^* dy \sin\beta + \left(N_x^* - \frac{\partial N_x^*}{\partial x} dx\right) dy \sin\beta'. \tag{5.2}$$

Since β and β' are small, we can write the following approximate equalities: $\sin\beta \approx \beta \approx \frac{\partial w}{\partial x}$ and $\sin\beta' \approx \beta'$, where $\beta' = \beta + d\beta$. The latter can be represented as

$$\beta' \approx \beta + \frac{\partial \beta}{\partial x} dx = \frac{\partial w}{\partial x} + \frac{\partial^2 w}{\partial x^2} dx.$$

Substituting this equality into (5.2), we have

$$N_x^* dy \sin\beta + \left(N_x^* - \frac{\partial N_x^*}{\partial x} dx\right)\left(\frac{\partial w}{\partial x} + \frac{\partial^2 w}{\partial x^2} dx\right) dy. \tag{5.3}$$

Making computations and neglecting high-order terms in Eq. (5.3), we obtain

$$N_x^* \frac{\partial^2 w}{\partial x^2} dxdy - \frac{\partial N_x^*}{\partial x} \frac{\partial w}{\partial x} dxdy.$$

The z components of the resultant in-plane shear forces N_{xy}^* on the x edges of the element are determined as follows. The slopes of the deflection surface in the y direction on the x edges are equal to $\frac{\partial w}{\partial y}$ and $\frac{\partial w}{\partial y} + \frac{\partial^2 w}{\partial x \partial y} dx$, respectively. The z directed component of N_{xy}^* is

$$N_{xy}^* \frac{\partial^2 w}{\partial x \partial y} dxdy - \frac{\partial N_{xy}^*}{\partial x} \frac{\partial w}{\partial y} dxdy.$$

In similar way, one obtains an expression for the z projection of the resultant in-plane shear forces N_{yx}^* acting on the y edges:

$$N_{yx}^{*}\frac{\partial^2 w}{\partial x \partial y}dxdy - \frac{\partial N_{yx}^{*}}{\partial y}\frac{\partial w}{\partial x}dxdy,$$

and the z component of N_y^*:

$$-N_y^{*}\frac{\partial^2 w}{\partial y^2}dxdy + \frac{\partial N_y^{*}}{\partial y}\frac{\partial w}{\partial y}dxdy.$$

The sum of the z projections of the resultant in-plane forces without the multiplier $dxdy$ is the additional normal force, which appears as a result of additional deformation of the element $dxdy$. This sum is

$$N_x^{*}\frac{\partial^2 w}{\partial x^2} + N_y^{*}\frac{\partial^2 w}{\partial y^2} + N_{xy}^{*}\frac{\partial^2 w}{\partial x \partial y} -$$
$$+\left(\frac{\partial N_x^{*}}{\partial x} + \frac{\partial N_{yx}^{*}}{\partial y} - F_x\right)\frac{\partial w}{\partial x} + \left(\frac{\partial N_{xy}^{*}}{\partial x} + \frac{\partial N_y^{*}}{\partial y} - F_y\right)\frac{\partial w}{\partial y}.$$

The terms within the parentheses in (5.4) vanish by virtue of (5.1) and we finally obtain the following formula for the computation of additional normal force, arising as a result of additional normal deflection w of the plate:

$$N_x^{*}\frac{\partial^2 w}{\partial x^2} + N_y^{*}\frac{\partial^2 w}{\partial y^2} + 2N_{xy}^{*}\frac{\partial^2 w}{\partial x \partial y}. \tag{5.4}$$

To obtain the equation of stability, we write the equilibrium equation for the plate

$$\Delta\Delta w = f$$

with the normal force f given by (5.4). As a result, we obtain the engineering equation of buckling of plate subjected to in-plane resultant initial forces N_x^*, N_y^* and N_{xy}^*:

$$\Delta\Delta w = N_x^{*}\frac{\partial^2 w}{\partial x^2} + N_y^{*}\frac{\partial^2 w}{\partial y^2} + 2N_{xy}^{*}\frac{\partial^2 w}{\partial x \partial y}, \tag{5.5}$$

where Δ is Laplace operator $\frac{\partial^2}{\partial x^2} + \frac{\partial^2}{\partial y^2}$.

5.1.2 Deriving the Model of Uniform Plate with Initial Stresses by Using Variational Principle

Another widely used method deriving the model of uniform plate with initial stresses is based on the application of variational principles. This approach also assumes the using of kinematic hypothesis. Following Washizu (1992), we accept that the local displacement $\mathbf{u}^{\varepsilon}$ in plate are given by formula

$$\mathbf{u}^{\varepsilon} = w(x_1, x_2)\mathbf{e}_3 - x_3 w_{3,\beta x}(x_1, x_2)\mathbf{e}_{\beta} \ \ (\beta = 1, 2), \tag{5.6}$$

and additional stresses $\sigma_{ij}^{\varepsilon}$ are related to the initial stresses σ_{ij}^{*} and additional strains e_{mn} by the formula

$$\sigma_{ij}^{\varepsilon} = (c_{ijmn} + \delta_{im}\sigma_{jn}^{*})e_{mn}.$$

Substituting these expressions into the variational equation (4.8), we obtain the variational equation

$$\begin{aligned}
&-\int_D \left(M_x \frac{\partial^2 \delta w}{\partial x^2} + M_y \frac{\partial^2 \delta w}{\partial x^2} + 2M_{xy} \frac{\partial^2 \delta w}{\partial x \partial y} \right) dxdy+ \\
&+\lambda \int_D \left[\left(N_x^* \frac{\partial w}{\partial x} + N_{xy}^* \frac{\partial w}{\partial y} \right) \frac{\partial \delta w}{\partial x} + \left(N_{xy}^* \frac{\partial w}{\partial x} + N_y^* \frac{\partial w}{\partial y} \right) \frac{\partial \delta w}{\partial y} \right] dxdy.
\end{aligned} \tag{5.7}$$

Here D is a domain on the Oxy-plane occupied by the plane considered as a two-dimensional object. The resultant moments M_x, M_y and $M_{xy} = M_{yx}$ are defined in accordance with the engineering theory as

$$M_x = \int_{-H/2}^{H/2} \sigma_x z dz, \quad M_y = \int_{-H/2}^{H/2} \sigma_y z dz, \tag{5.8}$$

$$M_{xy} = \int_{-H/2}^{H/2} \sigma_{xy} z dz, \quad M_{yx} = \int_{-H/2}^{H/2} \sigma_{yx} z dz.$$

Integrating (5.7) by parts and substituting the expressions (5.8) for the moments in accordance with the governing equations

$$M_x = \frac{Eh^3}{12(1-\nu^2)} \left(\frac{\partial^2 w}{\partial^2 x} + \nu \frac{\partial^2 w}{\partial^2 y} \right),$$

$$M_y = \frac{Eh^3}{12(1-\nu^2)} \left(\nu \frac{\partial^2 w}{\partial^2 x} + \frac{\partial^2 w}{\partial^2 y} \right),$$

$$M_y = \frac{Eh^3}{12(1-\nu^2)} \frac{\partial^2 w}{\partial x \partial y},$$

we arrive at the Eq. (5.5). Details exposition of this approach can be found in Washizu (1992).

From the brief presentation above, it is seen that the classical engineering approach to plates subjected to initial stresses essentially uses the hypothesis of Kirchhoff-Love, see formula (5.6), as well as classical definition of resultant in-plane forces and moments (5.8).

5.1.3 An Approach Based on the Asymptotic Integration Method

The mathematical justification of two-dimensional model of uniform plates was started in Morgenstern (1959a, b, c) and extended by the following researchers (a historical review may be found in Nazarov 2002). In the 1960 and 1970s, the so-called "asymptotic integration method" was developed to reduce the problem of elasticity theory in a thin region to a two-dimensional problem(s). The method clearly pointed out the asymptotic nature of plate and shell theories and used asymptotic series as a method for analyzing the problems. The main ideas of the asymptotic integration method can be found in Gol'denveizer (1962, 1963), Green (1962), Green and Naghdi (1965), Rutten (1969), John (1965), John (1969), Sencenmg (1968), Koiter (1960), Gol'denveizer (1968), Novotny (1970). To the best of the author's knowledge, the mentioned researches were devoted to the analysis of uniform plates and the problem with initial stresses has not been addressed in the mentioned papers.

5.2 Investigation of Plates with Initial Stresses on the Basis of the Asymptotic Method

In the section, we apply two-scale asymptotic method for deriving two-dimensional model of uniform plate directly from three-dimensional elasticity theory model.

Our analysis is based on the application of the formalism of two-scale expansions to the three-dimensional elasticity problem in a thin uniform elastic layer. A plate is called uniform if it has constant thickness and made of homogeneous material. Although the layer under consideration is uniform, we naturally distinguish two different spatial scales in the problem under consideration, see Fig. 5.2. Due to the small thickness of the layer, the stress-strain state in the transverse (across the plate) direction Ox_3 changes significantly faster than in the directions Ox_1 and Ox_2. As a result, "fast" coordinate naturally appears in the transversal direction Ox_3.

The fast change of the local stress-strain state in the plate in the direction across the plate thickness takes place not only in the bending/torsion mode but also in the in-plane deformation mode due to Poisson's effect. Taking into account the transverse displacements, the asymptotic model deviates from the Kirchhoff-Love model at the very beginning.

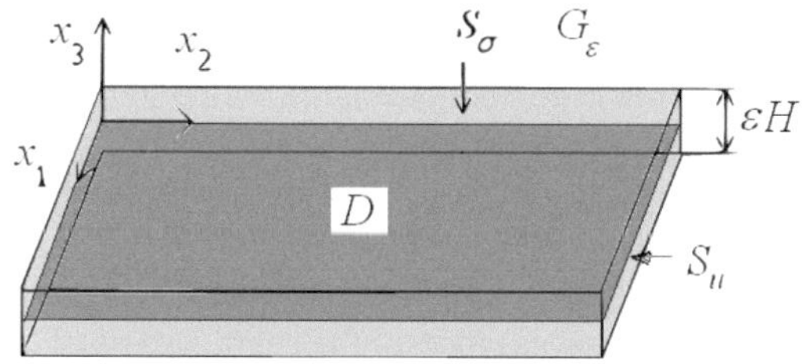

Fig. 5.2 Three-dimensional thin layer—a uniform plate

We will examine a body, which occupies a cylindrical domain G_ε of small thickness εH:

$$G_\varepsilon = D \times [-\varepsilon H/2, \varepsilon H/2],$$

where D is the middle plane of the cylinder G_ε belonging to Ox_1x_2-plane, Fig. 5.2 (it is also the middle plane of the plate).

The parameter ε denotes the characteristic thickness of the layer G_ε. It is assumed the thickness to be small: $\varepsilon \ll 1$. Since ε can be as small as required, it is appropriate to replace the condition $\varepsilon \ll 1$ by the condition $\varepsilon \to 0$. As $\varepsilon \to 0$, the three-dimensional layer (three-dimensional plate) G_ε tends to domain D on the Ox_1x_2-plane—two-dimensional plate. Our aim is to derive a model describing this two-dimensional plate.

The starting point of our investigation is the exact three-dimensional formulation of an elasticity problem for the body with initial stresses without any simplifying assumptions, presented in Sect. 4.1. In accordance with this model, the equilibrium equations of the plate as a three-dimensional body with initial stresses can be written in the form,

$$\frac{\partial \sigma_{ij}^\varepsilon}{\partial x_j} + f_i^\varepsilon(x_1, x_2) = 0 \text{ in } G_\varepsilon, \tag{5.9}$$

where

$$\sigma_{ij}^\varepsilon = h_{ijmn}^\varepsilon \frac{\partial u_m^\varepsilon}{\partial x_n}. \tag{5.10}$$

The boundary conditions can be written in the form,

$$\begin{gathered} \sigma_{i3}^\varepsilon = h_i^\varepsilon(x_1, x_2) \text{ for } x_3 = \pm\varepsilon H/2, \\ \mathbf{u}^\varepsilon(\mathbf{x}) = 0 \text{ on } S_u. \end{gathered} \tag{5.11}$$

Here:

$S_\varepsilon = \{x_3 = \pm\varepsilon H/2\}$ are the top and bottom surfaces of the body G_ε;

S_u are the lateral surface along which the body is clamped,

$\mathbf{u}$ are the displacements;

$\mathbf{f}^\varepsilon = (\varepsilon^{-1} f_1, \varepsilon^{-1} f_2, \varepsilon^{-3} f_3)(x_1, x_2)$ are the mass forces,

$\mathbf{h}^\varepsilon = (h_1, h_2, \varepsilon^{-3} h_3)(x_1, x_2)$ are the surface forces. The normal component of the mass and surface forces have order of ε^{-3}.

The most general form of the coefficients $h_{ijmn}(x_1, x_2, y_3)$ in (5.10) is the following:

$$h_{ijmn}(x_1, x_2, y_3) = \varepsilon^{-3} c_{ijmn} + \varepsilon^{-3} b^{(-3)}_{ijmn}(x_1, x_2, x_3/\varepsilon) \tag{5.12}$$

$$+\varepsilon^{-2} b^{(-2)}_{ijmn}(x_1, x_2, x_3/\varepsilon) + \varepsilon^{-1} b^{(-1)}_{ijmn}(x_1, x_2, x_3/\varepsilon),$$

where

$$b^{(p)}_{ijmn}(x_1, x_2) = \sigma^{*\varepsilon(p)}_{jn}(x_1, x_2, x_3/\varepsilon)\delta_{im} \ (p = -3, -2, -1). \tag{5.13}$$

Using the two-scale method in versions of plate, a function of argument $(x_1, x_2, x_3/\varepsilon)$ is considered as a function of "slow" variables x_1, x_2 and "fast" variable $y_3 = x_3/\varepsilon$.

The plate is assumed to be made of a homogeneous material. Thus, elastic characteristics c_{ijmn} are constants which do not depend not "slow" nor "fast" variables. The initial stresses $\sigma^{*\varepsilon}$ significantly change at the small distance of the order of ε in the direction Ox_3. In the directions Ox_1 and Ox_2, the initial stresses change significantly at the distance of the order of unity. In accordance with this note, the functions in (5.12) have the the form c_{ijmn}, and $h_{ijmn}(x_1, x_2, y_3)$ (x_1, x_2 are the "slow" variables and $y_3 = x_3/\varepsilon$ is the "fast" variable).

Note on the orders of the terms in (5.12) **and coupling of material and geometrical characteristics in the model of plate**. In (5.12) and (5.13), one can see the product of the powers of ε (the geometrical characteristic ε has the meaning of the thickness of the plate), material constants c_{ijmn}, and initial stresses $\sigma^{*(-3)}_{jn}$, $\sigma^{*(-2)}_{jn}$, and $\sigma^{*(-1)}_{jn}$.

Such a relationship between geometric and material properties is typical for thin-walled structures, and one can hear "The elastic constants are of the order of ε (raised to proper power)" or "The initial stresses are of the order of ε (raised to proper power)". Such formulations sometimes initiate the question: "How can material properties and/or initial stresses be proportional to geometric properties?" It seems that the best answer is the classical formulas for calculating the stiffnesses of a homogeneous plate: $\dfrac{EH}{1-\nu^2}$—tensile in-plane stiffness of a plane or $\dfrac{EH^3}{12(1-\nu^2)}$—cylindrical bending stiffness of a plate. These formulas do not establish any relationship between Young's modulus E, Poisson's ratio ν and the thickness H of the plate, but they show that the plate stiffnesses are independent characteristics of the plate, which are determined by combinations of characteristics of different nature.

Now, discuss the degrees of ε in (5.12). The multiplier ε^{-3} at c_{ijmn} guarantees that the bending stiffnesses of the plate will be nonzero, as $\varepsilon \to 0$.

The term $\varepsilon^{-1}b^{(-1)}_{ijmn} = \varepsilon^{-1}\sigma^{*\varepsilon(-1)}_{jn}\delta_{im}$ corresponds to the resultant initial in-plane forces, which are independent of ε, as $\varepsilon \to 0$.

If the resultant in-plane forces are zero,

$$\langle\sigma^{*\varepsilon(-1)}_{\alpha\beta}\rangle = 0 \ (\alpha, \beta = 1, 2), \tag{5.14}$$

the local initial stresses in the three-dimensional plate may nevertheless be non-zero. In order to analyze this situation, we introduce the term $\varepsilon^{-2}b^{(-2)}_{ijmn} = \varepsilon^{-2}\sigma^{*\varepsilon(-2)}_{jn}\delta_{im}$, corresponding to the resultant moments of the initial stresses, which are independent of ε, as $\varepsilon \to 0$.

The term $\varepsilon^{-3}b^{(-3)}_{ijmn} = \varepsilon^{-3}\sigma^{*\varepsilon(-1)}_{jn}\delta_{im}$ corresponds to the initial stresses which have the order of the elastic constants.

When ε is small, we have the following hierarchy of the powers: $\varepsilon^{-1} \ll \varepsilon^{-2} \ll \varepsilon^{-3}$. This means that the minimum initial stresses which affect the plate have an order of ε^{-1}. This stresses correspond to in-plane resultant forces and lead to the classical loss of stability of the plate. Only if these resultant in-pane forces are zero, the moments of initial stresses can manifest themselves. Else, the plate loss stability before the moments of the initial stresses manifest themselves. The resultant moments are produced by the initial stresses, which have order of ε^{-2}. An action of the initial stresses of order of ε^{-3} (these are the maximum possible initial stresses) may be observed only if the resultant initial forces and the resultant moments of the initial stresses are zero.

5.3 Resultant in-Plane Initial Forces

First, we consider the case when the resultant initial in-plane forces are non-zero. As was noted above, this case is described by the non-zero term $\varepsilon^{-1}b^{(-1)}_{ijmn} = \varepsilon^{-1}\sigma^{*\varepsilon(-1)}_{jn}\delta_{im}$ in (5.12).

5.3.1 Asymptotic Expansions and Limit Equilibrium Equations

We will study global deformations of the plate, as $\varepsilon \to 0$. To do this, we use the following asymptotic expansions:

asymptotic expansion for displacements

$$\mathbf{u}^{\varepsilon} = \mathbf{u}^{(0)}(x_1, x_2) + \sum_{k=1}^{\infty} \varepsilon^k \mathbf{u}^{(k)}(x_1, x_2, y_3); \tag{5.15}$$

asymptotic expansion for stresses

$$\sigma_{ij}^{\varepsilon} = \sum_{p=-3}^{\infty} \varepsilon^p \sigma_{ij}^{(p)}(x_1, x_2, y_3). \tag{5.16}$$

Note the difference between the series (5.15) and (5.16). The leading terms in the series (5.15) depends only on “slow” variables, thus, the deflection of plate has limit value, as $\varepsilon \to 0$. The leading term in the series (5.16) does not have limit value independent on $y_3 = x_3/\varepsilon$. But average values of stresses, as well as moments of stresses, have limit values.

When using the two-scale expansion, the differential operators are represented in the form of the sum of differential operators in the “fast” and “slow” variables (Sanchez-Palencia 2000). For the function $Z(x_1, x_2, y_3)$ of the arguments x_1, x_2 and y_3, like on the right-hand sides of (5.15) and (5.16), the differential operators are represented in the following form:

$$\frac{\partial Z}{\partial x_\alpha} \to Z_{,\alpha x}, \ \frac{\partial Z}{\partial x_3} \to \varepsilon^{-1} Z_{,3y}. \tag{5.17}$$

Here and below in this chapter, the Latin indexes take the values 1, 2, 3 and Greek indexes take the values 1, 2, index, αx means derivative $\partial/\partial x_\alpha$ and index , $3y$ means derivative $\partial/\partial y_3$.

The analysis of problem (5.9)–(5.13) breaks down into two stages. The first entails obtaining the equations of equilibrium for the plate considered as a two-dimensional object. This stage is not involved with local governing equations and it is the same stage for any governing equations.

To obtain the homogenized equilibrium equations, consider a weak form of problem (5.9)

$$-\int_{G_\varepsilon} \sigma_{ij}^{\varepsilon} \frac{\partial v_i}{\partial x_j} d\mathbf{x} + \int_{S_\varepsilon} \mathbf{h}^{\varepsilon} \mathbf{v} d\mathbf{x} + \int_{G_\varepsilon} \mathbf{f}^{\varepsilon} \mathbf{v} d\mathbf{x} = 0 \text{ for any } \mathbf{v} \in V_\varepsilon, \tag{5.18}$$

where

$$V_\varepsilon = \{\mathbf{v} \in \{H^1(G_\varepsilon)\}^3 : \mathbf{v}(\mathbf{x}) = 0 \text{ on } S_u\}$$

is the set of admissible displacements of the plate considered as three-dimensional body. In the terms of the differential equations theory, $\mathbf{v}$ is trial function (Lions and Magenes 1973; Mizohata 1973).

Substituting (5.16) in (5.18), we obtain, with allowance for the differentiation rule (5.17)

$$
\begin{aligned}
-\sum_{p=-3}^{\infty}\int_{G_\varepsilon}\varepsilon^p\sigma_{ij}^{(p)}(v_{i,\alpha x}\delta_{j\alpha}+\varepsilon^{-1}\delta_{i3}v_{i,3y})d\mathbf{x}+\int_{S_\varepsilon}\mathbf{h}^\varepsilon\mathbf{v}d\mathbf{x}\\
+\int_{G_\varepsilon}\mathbf{f}^\varepsilon\mathbf{v}d\mathbf{x}=0 \text{ for any } \mathbf{v}\in V_\varepsilon.
\end{aligned}
\tag{5.19}
$$

Let's define the functional space

$$
U=\{H^{01}(D)\}^3=\{\mathbf{v}(x_1,x_2)\in\{H^1(D)\}^3:\mathbf{v}(x_1,x_2)=0 \text{ on } \partial D\},
$$

which is the set of admissible displacements of the plate considered as two-dimensional object. It is evident that $U\subset V_\varepsilon$.

We take $\mathbf{v}=\mathbf{w}(x_1,x_2)\in U$. For this choice of trial function, $w_{i,3y}=0$, and (5.19) takes the following form:

$$
-\sum_{p=-3}^{\infty}\int_{G_\varepsilon}\varepsilon^p\sigma_{i\alpha}^{(p)}w_{i,\alpha x}d\mathbf{x}+\int_{S_\varepsilon}\mathbf{h}^\varepsilon\mathbf{w}d\mathbf{x}+\int_{G_\varepsilon}\mathbf{f}^\varepsilon\mathbf{w}d\mathbf{x}=0 \tag{5.20}
$$

for any $\mathbf{w}\in U$.

Denote the average value over the plate thickness

$$
\langle Z(y_3)\rangle=\int_{-H/2}^{H/2}Z(y_3)dy_3, \tag{5.21}
$$

and the sum value over the top and bottom surfaces S_σ of the plate

$$
\langle Z(y_3)\rangle_S=Z(H/2)+Z(-H/2). \tag{5.22}
$$

Proposition 5.1 *Let function $Z(x_1,x_2,y_3)$ continuous. Then,*

$$
\varepsilon^{-1}\int_{G_\varepsilon}Z(x_1,x_2,x_3/\varepsilon)d\mathbf{x}=\int_D\langle Z\rangle(x_1,x_2)dx_1dx_2;
$$

$$
\int_{S_\varepsilon}Z(\mathbf{x},\mathbf{x}/\varepsilon)d\mathbf{x}=\int_D\langle Z\rangle_S(\mathbf{x})dx_1dx_2.
$$

The first equation follows from Fubini theorem (Rudin 1967). The second one is evident.

Taking into account Proposition 5.1, we obtain form (5.20) the following equality:

$$
\begin{aligned}
-\sum_{p=-3}^{\infty}\int_D\varepsilon^{p+1}\langle\sigma_{i\alpha}^{(p)}\rangle w_{i,\alpha x}dx_1dx_2+\int_D\langle\mathbf{h}^\varepsilon\rangle_S\mathbf{w}dx_1dx_2=\\
+\int_D\langle\mathbf{f}^\varepsilon\rangle\mathbf{w}dx_1dx_2=0 \text{ for any } \mathbf{w}\in U.
\end{aligned}
\tag{5.23}
$$

The Eq. (5.23) indicates that in the present analysis, the resultant in-plane stresses must be introduced by the formula

$$N_{ij}^{(p)} = \langle \sigma_{ij}^{(p)} \rangle = \int_{-H/2}^{H/2} \sigma_{ij}^{(p)}(x_1, x_2, y_3) dy_3 \tag{5.24}$$

as in the engineering theory.

Substituting $N_{ij}^{(p)}$ instead $\langle \sigma_{ij}^{(p)} \rangle$ in (5.23), we obtain

$$\begin{aligned} -\sum_{p=-3}^{\infty} \int_D \varepsilon^{p+1} N_{i\alpha}^{(p)} w_{i,\alpha x} dx_1 dx_2 + \int_D \langle \mathbf{h}^\varepsilon \rangle_S \mathbf{w} dx_1 dx_2 \\ + \int_D \langle \mathbf{f}^\varepsilon \rangle \mathbf{w} dx_1 dx_2 = 0 \text{ for any } \mathbf{w} \in U. \end{aligned} \tag{5.25}$$

Integrating (5.25) by parts, we obtain an infinite sequence of equilibrium equations for the resultant in-plane forces ($i, j = 1, 2, 3$ and $\alpha, \beta = 1, 2$):

$$\begin{aligned} N_{\alpha\beta,\beta x}^{(-3)} &= 0; \\ N_{\alpha\beta,\beta x}^{(-2)} + \langle f_\alpha^\varepsilon \rangle + \langle h_\alpha^\varepsilon \rangle_S &= 0; \\ N_{3\beta,\beta x}^{(-1)} + \langle f_3^\varepsilon \rangle + \langle h_3^\varepsilon \rangle_S &= 0. \end{aligned} \tag{5.26}$$

In (5.26), we save only the leading equations corresponding to $p = -3, -2, -1$.

Now let's choose a trial function as $\mathbf{v} = \mathbf{w}(x_1, x_2) y_3$, where $\mathbf{v}(x_1, x_2) \in U$. For this function (5.19) takes the following form:

$$\begin{aligned} -\sum_{p=-3}^{\infty} \int_{G_\varepsilon} \varepsilon^p \sigma_{ij}^{(p)} (w_{i,\alpha x} \delta_{j\alpha} y_3 + \varepsilon^{-1} w_i \delta_{j3}) d\mathbf{x} + \int_{S_\varepsilon} \mathbf{h}^\varepsilon \mathbf{w}(\mathbf{x}) y_3 d\mathbf{x} \\ + \int_{G_\varepsilon} \mathbf{f}^\varepsilon \mathbf{w}(\mathbf{x}) y_3 d\mathbf{x} = 0 \text{ for any } \mathbf{w} \in U. \end{aligned}$$

With regard to Proposition 5.1 we obtain from this equation

$$\begin{aligned} -\sum_{p=-3}^{\infty} \int_D \varepsilon^{p+1} (\langle \sigma_{ij}^{(p)} y_3 \rangle w_{i,\alpha x} \delta_{j\alpha} + \varepsilon^{-1} \langle \sigma_{ij}^{(p)} \rangle w_i \delta_{j3}) dx_1 dx_2 \\ + \int_D \langle \mathbf{h}^\varepsilon y_3 \rangle_S \mathbf{w} dx_1 dx_2 + \int_D \langle \mathbf{f}^\varepsilon y_3 \rangle \mathbf{w} dx_1 dx_2 = 0 \text{ for any } \mathbf{w} \in U. \end{aligned} \tag{5.27}$$

The Eq. (5.27) indicates that in the present analysis, the resultant moments must be introduced as

$$M_{\alpha\beta}^{(p)} = -\langle \sigma_{\alpha\beta}^{(p)} y_3 \rangle = -\int_{-H/2}^{H/2} \sigma_{ij}^{(p)}(x_1, x_2, y_3) y_3 dy_3. \tag{5.28}$$

As for the resultant forces, this definition also coincide with the definition from the engineering theory.

Changing $\langle \sigma_{\alpha\beta}^{(p)} y_3 \rangle$ and $\langle \sigma_{ij}^{(p)} \rangle$ in (5.27) in accordance with (5.28) and (5.24), correspondingly, we obtain

$$\begin{gathered} -\sum_{p=-3}^{\infty} \int_D \varepsilon^{p+1} (M_{ij}^{(p)} w_{i,\alpha x} + \varepsilon^p N_{i3}^{(p)} w_i) dx_1 dx_2 \\ + \int_D \langle \mathbf{h}^\varepsilon y_3 \rangle_S \mathbf{w} dx_1 dx_2 + \int_D \langle \mathbf{f}^\varepsilon y_3 \rangle w_3 dx_1 dx_2 = 0 \text{ for any } \mathbf{w} \in U. \end{gathered} \tag{5.29}$$

Integrating in (5.29) by parts, we obtain the next equilibrium equations ($i, j = 1, 2, 3$ and $\alpha, \beta = 1, 2$):

$$M_{\alpha\beta,\beta x}^{(-3)} - N_{\beta 3}^{(-2)} = 0; \tag{5.30}$$

$$M_{\alpha\beta,\beta x}^{(-2)} - N_{\beta 3}^{(-1)} + \langle f_\alpha^\varepsilon y_3 \rangle + \langle h_\alpha^\varepsilon y_3 \rangle_S = 0; \tag{5.31}$$

$$N_{i3}^{(-3)} = 0. \tag{5.32}$$

In (5.30), we save only the leading equations corresponding to $p = -3, -2$.

The Eqs. (5.26) and (5.30), (5.31) are similar to the equilibrium equations of the classical plate theory Timoshenko and Woinowsky-Krieger (1959). But there exists a fundamental difference related to the terms $N_{3\beta}^{(-1)}$ and $N_{\beta 3}^{(-1)}$ in (5.26) and (5.31), which will be discussed below, in Sect. 5.3.3.

Substituting (5.16) in (5.9), we obtain, with allowance for the differentiation rule (5.17):

$$\sum_{p=-3}^{\infty} \varepsilon^p \left(\sigma_{i\alpha,\alpha x}^{(p)} + \varepsilon^{-1} \frac{d\sigma_{i3}^{(p)}}{dy_3} \right) + f_i^\varepsilon = 0. \tag{5.33}$$

We change here $\dfrac{\partial}{\partial y_3}$ for $\dfrac{d}{dy_3}$ because the variables x_1, x_2 are considered as parameters and differentiation is carried out with respect to the unique variable y_3. It is important for the following that the boundary-value problem in partial derivatives transforms into boundary-value problem for ordinary differential equations.

Equating the terms with identical power of ε in (5.33), we obtain the following sequence of local equations:

$$\begin{aligned}
&\frac{d\sigma_{i3}^{(-3)}}{dy_3} = 0 \text{ in } Y, (p = -3)\\
&\frac{d\sigma_{i3}^{(p)}}{dy_3} + \sigma_{i\beta,\beta x}^{(p-1)} = 0 \text{ in } Y(p = -2, -1, ...),\\
&\sigma_{ij}^{(p)} = 0 \text{ for } y_3 = \pm H/2(p = -3, -2, -1, ...).
\end{aligned} \tag{5.34}$$

We will not use the equations corresponding to $p =, 0, 1, \ldots$.

5.3.2 The Limit Governing Equations

The second stage of analysis of the problem consists of obtaining the governing equations for the plate. If the previous stage did not involve the local governing equations, then this stage essentially involves the local governing equations.

We consider initial stresses corresponding to in-plane stresses. In this case (5.12) takes the form

$$\sigma_{ij}^{\varepsilon} = \left(\varepsilon^{-3}c_{ijmn} + \varepsilon^{-1}b_{ijmn}(x_1, x_2, y_3)\right)\frac{\partial u_m}{\partial x_n}. \tag{5.35}$$

Substituting (5.15) and (5.16) in (5.35), we obtain, with allowance for the differentiation rule (5.17),

$$\begin{aligned}
\sum_{p=-3}^{\infty} \varepsilon^p \sigma_{ij}^{(p)} &= \sum_{k=0}^{\infty} \varepsilon^k (\varepsilon^{-3}c_{ijmn} + \varepsilon^{-1}b_{ijmn}(x_1, x_2, y_3))\\
&\times \left(u_{m,\beta x}^{(k)}\delta_{m\beta} + \varepsilon^{-1}\delta_{n3}\frac{du^{(k)}}{dy_3}\right).
\end{aligned} \tag{5.36}$$

Equating in (5.36) the terms with identical power of ε, we obtain in particular,

$$\sigma_{ij}^{(p)} = c_{ijm\beta}u_{m,\beta x}^{(p+3)} + c_{ijm3}\frac{du_m^{(p+2)}}{dy_3} \quad (p = -3, -2), \tag{5.37}$$

$$\begin{aligned}
\sigma_{ij}^{(-1)} &= c_{ijm\beta}u_{m,\beta x}^{(2)} + c_{ijm3}\frac{du_m^{(3)}}{dy_3}\\
&+b_{ijm3}(x_1, x_2, y_3)\frac{du_m^{(1)}}{dy_3} + b_{ijm\beta}(x_1, x_2, y_3)u_{m,\beta x}^{(0)}.
\end{aligned} \tag{5.38}$$

Let us consider the problem (5.34), (5.37) with $p = -3$ in both equations. It can be written in the form

$$\begin{cases} \dfrac{d}{dy_3}\left(c_{i3m3}\dfrac{du_m^{(1)}}{dy_3} + c_{i3m\alpha}u_{m,\alpha x}^{(0)}(x_1, x_2)\right) = 0 \text{ in } [-H/2, H/2], \\ \\ c_{i3m3}\dfrac{du_m^{(1)}}{dy_3} + c_{i3m\alpha}u_{m,\alpha x}^{(0)}(x_1, x_2) = 0 \text{ for } y_3 = \pm H/2. \end{cases} \tag{5.39}$$

Solution to the boundary-value problem (5.39) is not zero due to the boundary condition at the top and bottom surfaces $y_3 = \pm H/2$ of the plate. From the mechanical point of view, the non-zero solution is the result of transversal deformation of the plate due to Poisson effect. Allowing for the fact that the function $\mathbf{u}^{(0)}(x_1, x_2)$ plays the role of a parameter in the problems in the "fast" variable y_3, solution to the problem (5.39) can be found in the form,

$$\mathbf{u}^{(1)} = \mathbf{N}^{0m\alpha}(y_3)u_{m,\alpha x}^{(0)}(x_1, x_2) + \mathbf{V}(x_1, x_2), \tag{5.40}$$

where function $\mathbf{N}^{0m\alpha}(y_3)$ in (5.40) is the solution to the following problem ($\alpha, \beta = 1, 2$; $m = 1, 2, 3$):

$$\begin{cases} \dfrac{d}{dy_3}\left(c_{i3k3}\dfrac{dN_k^{0m\alpha}}{dy_3} + c_{i3m\alpha}\right) = 0 \text{ in } [-H/2, H/2], \\ \\ c_{i3k3}\dfrac{dN_k^{0m\alpha}}{dy_3} + c_{i3m\alpha} = 0 \text{ for } y_3 = \pm H/2, \end{cases} \tag{5.41}$$

and $\mathbf{V}(x_1, x_2)$ in (5.40) is an arbitrary function, which will be determined below.

Proposition 5.2 *For $m = 3$, the solution to problem (5.41) is given by the following formula:*

$$N^{03\alpha}(y_3) = -y_3\mathbf{e}_\alpha, \tag{5.42}$$

where $\{\mathbf{e}_i\}$ are basis vectors of the coordinate system.

To verify Proposition 5.2, we substitute $N^{03\alpha}(y_3)$ (5.42) into the expression $c_{i3k3}\dfrac{dN_k^{03\alpha}}{dy_3} + c_{ij3\alpha}$. As a result of the substitution, this expression takes the form $-c_{i3\alpha3} + c_{i33\alpha}$. Due to the symmetry of the elasticity constants, $-c_{i3\alpha3} + c_{i33\alpha} = 0$.

Since solution to the problem (5.41) for $m = 3$ is known, the following problems ($\alpha, \beta = 1, 2$) remain from (5.41):

$$\begin{cases} \dfrac{d}{dy_3}\left(c_{i3k3}\dfrac{dN_k^{0\alpha\beta}}{dy_3} + c_{i3\alpha\beta}\right) = 0 \text{ in } [-H/2, H/2], \\ \\ c_{i3k3}\dfrac{dN_k^{0\alpha\beta}}{dy_3} + c_{i3\alpha\beta} = 0 \text{ for } y_3 = \pm H/2. \end{cases} \tag{5.43}$$

These problems will be solved below. Now, we use the exact solution (5.42), only and rewrite (5.40) as

$$\mathbf{u}^{(1)} = \mathbf{N}^{0\beta\alpha}(y_3)u^{(0)}_{\beta,\alpha x}(x_1, x_2) - y_3\mathbf{e}_\alpha u^{(0)}_{3,\alpha x}(x_1, x_2) + \mathbf{V}(x_1, x_2). \qquad (5.44)$$

Substituting (5.44) in (5.37, $p = -3$) gives

$$\sigma^{(-3)}_{ij} = \left(c_{ij\alpha\beta} + c_{ijk3}\frac{dN^{0\alpha\beta}_k}{dy_3}(y_3)\right)u^{(0)}_{\alpha,\beta x}.$$

Averaging this equation over the interval $[-H/2, H/2]$ for $ij = \gamma\delta$, we obtain in accordance with the definition (5.24)

$$N^{(-3)}_{\gamma\delta} = A_{\gamma\delta\alpha\beta}u^{(0)}_{\alpha,\beta x}, \qquad (5.45)$$

where

$$A^0_{\gamma\delta\alpha\beta} = \left\langle c_{\gamma\delta\alpha\beta}(y_3) + c_{\gamma\delta k3}\frac{dN^{0\alpha\beta}_k}{dy_3}(y_3)\right\rangle \qquad (5.46)$$

are the in-plane stiffnesses (often called rigidities) of the plate considered as a two-dimensional object.

In accordance with the equilibrium equation (5.26, $p = -3$), we obtain from (5.45) the following equation:

$$(A^0_{\gamma\delta\alpha\beta}u^{(0)}_{\alpha,\beta x})_{,\delta x} = 0 \text{ in } D. \qquad (5.47)$$

The boundary condition for $u^{(0)}_\alpha$ ($\alpha = 1, 2$) follows from (5.11) and (5.15) and has the form

$$u^{(0)}_\alpha(x_1, x_2) = 0 \text{ on } \partial D. \qquad (5.48)$$

The solution to problem (5.47), (5.48) is $u^{(0)}_\alpha = 0$ ($\alpha = 1, 2$). Then (5.44) takes a form

$$\mathbf{u}^{(1)} = -y_3\mathbf{e}_\alpha u^{(0)}_{3,\alpha x}(x_1, x_2) + \mathbf{V}(x_1, x_2), \qquad (5.49)$$

The right-hand side (5.49) corresponds to the Kirchhoff-Love hypothesis, see, e.g., Timoshenko and Woinowsky-Krieger (1959), Love (1929). Note that formula (5.49) is derived here directly from the three-dimensional problem of elasticity theory without additional hypotheses, only using the method of asymptotic analysis.

Substituting (5.49) into equalities (5.37, $p = -3, -2$), we obtain

$$\sigma_{ij}^{(-3)} = 0, \tag{5.50}$$

$$\sigma_{ij}^{(-2)} = c_{ijmn} u_{m,ny}^{(2)} - c_{ij\alpha\beta} y_3 u_{3,\alpha x\beta x}^{(0)}(x_1, x_2) + c_{ijm\beta} V_{m,\beta x}(x_1, x_2). \tag{5.51}$$

Note that due to (5.50) the equality $N_{i3}^{(-3)} = 0$ from (5.30) is satisfied.

Let us examine problem (5.34, $p = -2$), (5.51). It can be written in the form

$$\begin{cases} \dfrac{d}{dy_3}\left(c_{i3m3}\dfrac{du_m^{(2)}}{dy_3} - c_{ij\alpha\beta} y_3 u_{3,\alpha x\beta x}^{(0)}(x_1, x_2) + c_{i3m\beta} V_{m,\beta x}(x_1, x_2)\right) = 0 \\ \text{in } [-H/2, H/2]; \\ c_{i3m3}\dfrac{du_m^{(2)}}{dy_3} - c_{ij\alpha\beta} y_3 u_{3,\alpha x\beta x}^{(0)}(x_1, x_2) + c_{i3m\beta} V_{m,\beta x}(x_1, x_2) = 0 \\ \text{for } y_3 = \pm H/2. \end{cases} \tag{5.52}$$

In order to solve problems of this kind, we introduce the following problem, corresponding to bending of a plate

$$\begin{cases} \dfrac{d}{dy_3}\left(c_{i3m3}\dfrac{dN_m^{1\alpha\beta}}{dy_3} - y_3 c_{ij\alpha\beta}\right) = 0 \text{ in } [-H/2, H/2], \\ c_{i3m3}\dfrac{dN_m^{1\alpha\beta}}{dy_3} - y_3 c_{ij\alpha\beta} = 0 \text{ for } y_3 = \pm H/2. \end{cases} \tag{5.53}$$

The solution to problem (5.52) can be expressed through the functions $\mathbf{N}^{\nu\alpha\beta}(y_3)$ ($\nu = 0, 1$)—solutions to problems (5.41) and (5.53). Allowing for the fact that the function $\mathbf{u}^{(0)}(x_1, x_2)$ plays the role of a parameter in the problems in the "fast" variable y_3, solution to the problem (5.53) can be written in the form,

$$\mathbf{u}^{(2)} = \mathbf{N}^{0\alpha\beta}(y_3) u_{3,\alpha x\beta x}^{(0)}(x_1, x_2) - y_3 V_{3,\beta x}(x_1, x_2)\mathbf{e}_\beta + \mathbf{N}^{1\alpha\beta}(y_3) V_{\alpha,\beta x}(x_1, x_2). \tag{5.54}$$

Substituting (5.54) in (5.51), we obtain

$$\sigma_{ij}^{(-2)} = \left(c_{ijm3}\frac{dN_m^{0\alpha\beta}}{dy_3} + c_{ij\alpha\beta}\right) V_{\alpha,\beta x}(x_1, x_2) + \left(c_{ijm3}\frac{dN_m^{1\alpha\beta}}{dy_3} - y_3 c_{ij\alpha\beta}\right) u_{3,\alpha x\beta x}^{(0)}(x_1, x_2). \tag{5.55}$$

Note that although V_3 appears in equality (5.54), it does not appear in equality (5.55). This is because V_3 is present in equality (5.54) with the multiplier y_3, which plays a special role in the problem under consideration due to Proposition 5.2.

Averaging (5.55) with $ij = \gamma\delta$ $(\gamma, \delta = 1, 2)$ over the interval $[-H/2, H/2]$ (thickness of the plate), we obtain the governing equations for the in-plane resultant forces

$$N_{\gamma\delta}^{(-2)} = A_{\gamma\delta\alpha\beta}^{0} V_{\alpha,\beta x} + A_{\gamma\delta\alpha\beta}^{1} u_{3,\alpha x\beta x}^{(0)}. \tag{5.56}$$

Multiplying (5.55) with $ij = \gamma\delta$ ($\gamma, \delta = 1, 2$) by $-y_3$ and averaging over the interval $[-H/2, H/2]$, we obtain the following governing equation for the resultant moments:

$$M_{\gamma\delta}^{(-2)} = A_{\gamma\delta\alpha\beta}^{1} V_{\alpha,\beta x} + A_{\gamma\delta\alpha\beta}^{2} u_{3,\alpha x\beta x}^{(0)}. \tag{5.57}$$

The coefficients in (5.56) and (5.57) are

$$A_{\gamma\delta\alpha\beta}^{\mu+\nu} = \left\langle \left((-1)^{\nu} c_{\gamma\delta\alpha\beta} y_3^{\nu} + c_{\gamma\delta m3} \frac{dN_m^{\nu\alpha\beta}}{dy_3} \right) (-1)^{\mu} y_3^{\mu} \right\rangle \tag{5.58}$$

$(\nu, \mu = 0, 1; \alpha, \beta, \gamma, \delta = 1, 2)$.

The obtained Eqs. (5.56) and (5.57) are the limit governing equations of plate considered as a two-dimensional structure. The coefficients $A_{\gamma\delta\alpha\beta}^{\mu+\nu}$ are the stiffnesses of the plate considered as two-dimensional object: $A_{\gamma\delta\alpha\beta}^{0}$ are in-plane stiffnesses and $A_{\gamma\delta\alpha\beta}^{2}$ are bending stiffnesses. As far as the asymmetrical (out-of-plane)stiffnesses, $A_{\gamma\delta\alpha\beta}^{1}$, they are zero for the uniform plates. All the stiffnesses are expressed through solutions to the problems (5.41) and (5.53).

5.3.3 *Elimination of Shear Forces in the Presence of Initial Stresses*

In this section, we discuss the specificity of the limit equilibrium equations (5.26) and (5.30) in the case of the presence of initial stresses. The specificity does indeed exist. The equilibrium equations (5.26) and (5.30) include the quantities $N_{3\gamma}^{(-1)}$ and the resultant shear forces $N_{\gamma 3}^{(-1)}$, both of which are not yet defined. What to do with them?

What is the situation with the equilibrium equations (5.26) and (5.30) and the resultant forces $N_{3\gamma}^{(-1)}$ and $N_{\gamma 3}^{(-1)}$ in the case of absence of initial stresses?

In the absence of initial stresses, σ_{ij} are symmetric in i and j: $\sigma_{ij}^{(-1)} = \sigma_{ji}^{(-1)}$ due to $c_{ijkl} = c_{jikl}$, see (5.38). Then the corresponding resultant forces are also symmetric in i and j: $N_{ij}^{(-1)} = N_{ji}^{(-1)}$, and we can eliminate $N_{\gamma 3}^{(-1)}$ and $N_{3\gamma}^{(-1)}$ from Eqs. (5.26), (5.31) using this symmetry, only, as done in both the classical theory of plates (Love 1929; Timoshenko and Woinowsky-Krieger 1959) and in the asymptotic theory for plates with no initial stresses (Caillerie 1982, 1984).

Let us insert $\mathbf{u}^{(1)}$ into (5.38) in accordance with (5.49). We obtain

$$\begin{aligned}\sigma_{ij}^{(-1)} &= c_{ijm\beta} u_{m,\beta x}^{(2)} + c_{ijm3}\frac{du_m^{(3)}}{dy_3} \\ -b_{ij\beta 3}(x_1,x_2,y_3)&u_{3,\beta x}^{(0)}(x_1,x_2) + b_{ij3\beta}(x_1,x_2,y_3)u_{3,\beta x}^{(0)}(x_1,x_2).\end{aligned} \tag{5.59}$$

The first and second terms on the right-hand side of equation (5.59) are symmetric with respect to the indices i and j by virtue of the symmetry of the elastic constants c_{ijkl} in i and j. But other terms on the right-hand side of equation (5.59) are not symmetric in i and j because b_{ijmn} are not symmetric in i and j, see (5.13). Thus, $\sigma_{ij}^{(-1)}$ *is not symmetric* in i and j. As a result, the corresponding resultant forces $N_{ij}^{(-1)}$ are not symmetric with respect to i and j, and the classical method of eliminating of shear forces turns out to be unsuitable for plates with initial stresses. To overcome this problem, we consider the asymmetric part of the tensor $\sigma_{ij}^{(-1)}$, which is equal to

$$\begin{aligned}&\sigma_{ij}^{(-1)} - \sigma_{ji}^{(-1)} = \\ &= -b_{ij\beta 3}(x_1,x_2,y_3)u_{3,\beta x}^{(0)}(x_1,x_2) + b_{ij3\beta}(x_1,x_2,y_3)u_{3,\beta x}^{(0)}(x_1,x_2) \\ &+b_{ji\beta 3}(x_1,x_2,y_3)u_{3,\beta x}^{(0)}(x_1,x_2) - b_{ji\beta 3}(x_1,x_2,y_3)u_{3,\beta x}^{(0)}(x_1,x_2).\end{aligned} \tag{5.60}$$

Averaging (5.60) over the interval $[-H/2, H/2]$, we obtain

$$\begin{aligned}&N_{ij}^{(-1)} - N_{ji}^{(-1)} \\ &= -\langle b_{ij\beta 3}(x_1,x_2,y_3)\rangle u_{3,\beta x}^{(0)}(x_1,x_2) + \langle b_{ij3\beta}(x_1,x_2,y_3)\rangle u_{3,\beta x}^{(0)}(x_1,x_2) \\ &+\langle b_{ji\beta 3}(x_1,x_2,y_3)\rangle u_{3,\beta x}^{(0)}(x_1,x_2) - \langle b_{ji3\beta}(x_1,x_2,y_3)\rangle u_{3,\beta x}^{(0)}(x_1,x_2)\end{aligned} \tag{5.61}$$

Taking into account definition of b_{ijmn} (5.13), we obtain from (5.61)

$$\begin{aligned}&N_{ij}^{(-1)} - N_{ji}^{(-1)} \\ &= -\delta_{i\beta}\langle\sigma_{j3}^{*\varepsilon}(x_1,x_2,y_3)\rangle u_{3,\beta x}^{(0)}(x_1,x_2) + \delta_{i3}\langle\sigma_{j\beta}^{*\varepsilon}(x_1,x_2,y_3)\rangle u_{3,\beta x}^{(0)}(x_1,x_2) \\ &+\delta_{j\beta}\langle\sigma_{i3}^{*\varepsilon}(x_1,x_2,y_3)\rangle u_{3,\beta x}^{(0)}(x_1,x_2) - \delta_{j3}\langle\sigma_{i\beta}^{*\varepsilon}(x_1,x_2,y_3)\rangle u_{3,\beta x}^{(0)}(x_1,x_2).\end{aligned} \tag{5.62}$$

For $i = 3$ and $j = \gamma$, we obtain from (5.62) and definition of b_{ijmn} (5.62) the equality

$$N_{3\gamma} - N_{\gamma 3} = K_\gamma, \tag{5.63}$$

where K_γ is defined by the follows formula:

$$
\begin{aligned}
K_\gamma &= -\delta_{3\beta}\langle\sigma^{*\varepsilon}_{\gamma 3}(x_1,x_2,y_3)\rangle u^{(0)}_{3,\beta x}(x_1,x_2) + \delta_{33}\langle\sigma^{*\varepsilon}_{\gamma\beta}(x_1,x_2,y_3)\rangle u^{(0)}_{3,\beta x}(x_1,x_2)\\
&+\delta_{\gamma\beta}\langle\sigma^{*\varepsilon}_{33}(x_1,x_2,y_3)\rangle u^{(0)}_{3,\beta x}(x_1,x_2) - \delta_{\gamma 3}\langle\sigma^{*\varepsilon}_{3\beta}(x_1,x_2,y_3)\rangle u^{(0)}_{3,\beta x}(x_1,x_2)\\
&= N^*_{33}u^{(0)}_{3,\gamma x}(x_1,x_2) - N^*_{\gamma\beta}u^{(0)}_{3,\beta x}(x_1,x_2)
\end{aligned}
\tag{5.64}
$$

(we use here that $\delta_{33} = 1$ and $\delta_{\gamma 3} = 0$ for $\gamma = 1, 2$.)

It is seen that the asymmetric part K_γ of shear forces is expressed through the normal deflection $u_3^{(0)}$. In other words, (5.64) can be treated as a governing equation which relates K_γ and $u_3^{(0)}$. There is no analog of such kind governing equation in the theory of plates with no initial stresses.

Proposition 5.3 *Let the initial stresses $\sigma^{*\varepsilon}_{ij}$ satisfy the equilibrium equations,*

$$
\frac{\partial\sigma^{*\varepsilon}_{ij}}{\partial x_j} = \varepsilon^A F_i \ in\ Q_\varepsilon,\quad \sigma^{*\varepsilon}_{ij}n_j = H_i\ on\ S_\varepsilon. \tag{5.65}
$$

If $A \neq 1$ and $\mathbf{H} = 0$, then $\left\langle \sigma^{*\varepsilon}_{i3}\frac{dZ}{dy_3}(y_3)\right\rangle = 0$ for any differentiable function $Z(y_3)$. In particular, $\langle\sigma^{*\varepsilon}_{i3}\rangle = 0$, $\langle\sigma^{*\varepsilon}_{i3}y_3\rangle = 0$.

From (5.65) taking into account (5.17), we obtain

$$
\begin{cases}
\dfrac{d\sigma^{*\varepsilon}}{dy_3} = \mathbb{F}_i \text{ in } [-a,a]\ (\mathbb{F}_i = 0 \text{ if } A \neq 1, \mathbb{F}_i = F_i \text{ if } A = -1),;\\[2ex]
\sigma^{*\varepsilon}_{i3}(x_1,x_2,y_3) = 0 \text{ for } y_3 = \pm H/2.
\end{cases}
\tag{5.66}
$$

Multiplying equation in (5.66) by $Z(y_3)$ and integrating by parts, we obtain

$$
\int_{-a}^{a}\frac{\sigma^{*\varepsilon}_{j3}}{dy_3}Z(y_3)dy_3 = -\int_{-a}^{a}\sigma^{*\varepsilon}_{j3}Z(y_3)dy_3
$$

$$
+\int_{-a}^{a}\mathbb{F}_i\frac{dZ}{dy_3}(y_3)dy_3 + \sigma^{*\varepsilon}_{j3}Z(y_3)\Big|_{y_3=-a}^{y_3=a}.
$$

Taking $Z(y_3) = y_3$ and $Z(y_3) = y_3^2/2$, we obtain the required results.

5.3.4 Limit Model of Plate with Initial Stresses

Now, we can exclude the quantities $N_{3\gamma}$ and $N_{\gamma 3}$ from the equilibrium equations (5.26), (5.30) and obtain the closed two-dimensional model of plate.

Let conditions of Proposition 5.3 be satisfied. Then $N^*_{33} = 0$ and (5.64) takes the form

$$K_\gamma = -N^*_{\gamma\beta} u^{(0)}_{3,\beta x}(x_1, x_2). \tag{5.67}$$

Differentiating (5.31), we have

$$M^{(-2)}_{\alpha\beta,\beta x\alpha x} - N^{(-1)}_{\beta 3,\alpha x} + (\langle f^\varepsilon_\alpha y_3 \rangle + \langle h^\varepsilon_\alpha y_3 \rangle_S)_{,\alpha x} = 0. \tag{5.68}$$

Using (5.26) and (5.63), we obtain

$$\begin{gathered}(A^1_{\delta\gamma\alpha\beta} V_{\alpha,\beta x} + A^2_{\delta\gamma\alpha\beta} u^{(0)}_{3,\alpha x\beta x})_{,\gamma x\delta x} = K_{\gamma,\gamma x} \\ + \langle f^\varepsilon_3 \rangle + \langle h^\varepsilon_3 \rangle_S + (\langle f^\varepsilon_\gamma y_3 \rangle + \langle h^\varepsilon_\gamma y_3 \rangle_S)_{,\gamma x}.\end{gathered} \tag{5.69}$$

Substituting K_δ in accordance with (5.67) we obtain

$$\begin{gathered}(A^1_{\delta\gamma\alpha\beta} V_{\alpha,\beta x} + A^2_{\delta\gamma\alpha\beta} u^{(0)}_{3,\alpha x\beta x})_{,,\gamma x\delta x} \\ + \langle f_3 \rangle + \langle h_3 \rangle_S + (\langle f^\varepsilon_\delta y_3 \rangle + \langle h^\varepsilon_\delta y_3 \rangle_S)_{,\delta x}. = (N^*_{\delta\beta} u^{(0)}_{3,\gamma x})_{,\gamma x}\end{gathered} \tag{5.70}$$

From (5.26, $p = -2$) and (5.56) we obtain

$$(A^0_{\gamma\delta\alpha\beta} V_{\alpha,\beta x} + A^1_{\gamma\delta\alpha\beta} u^{(0)}_{3,\alpha x\beta x})_{,\delta x} + \langle f^\varepsilon_\gamma \rangle + \langle g^\varepsilon_\gamma \rangle_S = 0. \tag{5.71}$$

Boundary conditions follow from the original boundary conditions (5.11) and formula (5.49) for the displacements. The boundary conditions have the

$$V_\alpha(x_1, x_2) = 0 \quad (\alpha = 1, 2), \tag{5.72}$$

$$u^{(0)}_3(x_1, x_2) = \frac{\partial u^{(0)}_3}{\partial \mathbf{n}}(x_1, x_2) = 0 \text{ on } \partial D. \tag{5.73}$$

5.3.5 Solution to the Problems (5.43) and (5.53)

The problems (5.43) and (5.53), which appear above and play an important role in our consideration, may be solved in explicit form.

The problem (5.43), corresponding to the in-plane deformation of the plate in the plane, and the problem (5.53), corresponding to the bending of the plate, can be written in the following uniform form:

$$\begin{cases} \dfrac{d}{dy_3}\left(c_{i3p3}\dfrac{dN_p^{\nu\alpha\beta}}{dy_3} + (-1)^\nu y_3^\nu c_{i3\alpha\beta}\right) = 0 \text{ on } [-H/2, H/2], \\ c_{i3p3}\dfrac{N_p^{\nu\alpha\beta}}{dy_3} + (-1)^\nu y_3^\nu c_{i3\alpha\beta} = 0 \text{ for } y_3 = -H/2, H/2. \end{cases} \tag{5.74}$$

Here $\nu = 0$ corresponds to the in-plane deformation of the plate (see problem (5.43)) and $\nu = 1$ corresponds to the bending of the plate (see problem (5.53)).

From (5.74), we have

$$c_{i3m3}\frac{dN_m^{\nu\alpha\beta}}{dy_3}(y_3) + (-1)\nu y_3^\nu c_{i3\alpha,\beta} = 0,$$

and then

$$\frac{dN_m^{\nu\alpha\beta}}{dy_3}(y_3) = -(-1)^\nu y_3^\nu c_{m3i3}^{-1} c_{i3\alpha\beta}, \tag{5.75}$$

where $\{c_{m3i3}^{-1}\}$ means the matrix inverse to 3×3 matrix $\{c_{m3i3}\}$.

The formulas (5.58) for calculating the plate stiffnesses can be written in a uniform form

$$A_{\gamma\delta\alpha\beta}^{\mu+\nu} = \left\langle \left((-1)^\nu c_{\gamma\delta\alpha\beta} y_3^\nu + c_{\gamma\delta m3}\frac{N_m^{\nu\alpha\beta}}{dy_3}(y_3)\right)(-1)^\mu y_3^\mu \right\rangle \tag{5.76}$$

$(\nu, \mu = 0, 1; \alpha, \beta, \gamma, \delta = 1, 2)$.

Integrating (5.75), one can obtain explicit formulas for the computation of the functions $\mathbf{N}^{0\alpha\beta}(y_3)$ and $\mathbf{N}^{1\alpha\beta}(y_3)$. However, to compute the stiffnesses of the plate, it is sufficient to know only the derivatives of these functions. The derivatives of the functions are given by (5.75).

Substituting (5.75) into (5.76), we obtain the following formula:

$$A_{\gamma\delta\alpha\beta}^{\mu+\nu} = (-1)^{\mu+\nu}\left\langle (c_{\gamma\delta\alpha\beta} - c_{\gamma\delta m3}c_{m3i3}^{-1}c_{i3\alpha\beta})y_3^{\mu+\nu}\right\rangle. \tag{5.77}$$

Formulas for computation of the stiffnesses of the in-plane stiffnesses $A_{\gamma\delta\alpha\beta}^0$ and the bending stiffnesses $A_{\gamma\delta\alpha\beta}^2$ of uniform plate are

$$A^0_{\gamma\delta\alpha\beta} = \left\langle c_{\gamma\delta\alpha\beta} + c_{\gamma\delta m3}\frac{N_m^{0\alpha\beta}}{dy_3}(y_3)\right\rangle = \langle c_{\gamma\delta\alpha\beta} - c_{\gamma\delta m3}c^{-1}_{m3i3}c_{i3\alpha\beta}\rangle$$
$$= (c_{\gamma\delta\alpha\beta} - c_{\gamma\delta m3}c^{-1}_{m3i3}c_{i3\alpha\beta})H,$$
$$A^2_{\gamma\delta\alpha\beta} = \left\langle c_{\gamma\delta\alpha\beta}y_3^2 + c_{\gamma\delta m3}\frac{N_m^{1\alpha\beta}}{dy_3}(y_3)y_3\right\rangle - \langle(c_{\gamma\delta\alpha\beta} - c_{\gamma\delta m3}c^{-1}_{m3i3}c_{i3\alpha\beta})y_3^2\rangle$$
$$= (c_{\gamma\delta\alpha\beta} - c_{\gamma\delta m3}c^{-1}_{m3i3}c_{i3\alpha\beta})\frac{H^3}{12}. \tag{5.78}$$

All the coupling stiffnesses $A^1_{\gamma\delta\alpha\beta}$ of uniform plate are equal to zero.

If the plate is made of an isotropic material with Young's modulus E and Poisson's ratio ν, the elastic constants have the form

$$c_{ijkl} = \frac{E\nu}{(1+\nu)(1-2\nu)}\delta_{ij}\delta_{kl} + \frac{E}{1+\nu}\delta_{ik}\delta_{jl}, \tag{5.79}$$

in particular,

$$c_{1111} = \frac{E(1-\nu)}{(1+\nu)(1-2\nu)},\ c_{1122} = \frac{E\nu}{(1+\nu)(1-2\nu)},\ c_{1212} = \frac{E}{1+\nu},$$

and $c^{-1}_{3333} = \dfrac{1}{c_{3333}}$.

As a result, formulas (5.78) take the form

$$A^0_{1111} = (c_{1111} - c_{1133}c^{-1}_{3333}c_{3311})H = \frac{EH}{1-\nu^2},$$
$$A^0_{1122} = (c_{1122} - c_{1133}c^{-1}_{3333}c_{3322})H = \frac{E\nu H}{1-\nu^2},$$
$$A^0_{1212} = c_{1212}H = \frac{EH}{1+\nu},$$
$$A^2_{1111} = (c_{1111} - c_{1133}c^{-1}_{3333}c_{3311})\frac{H^3}{12} = \frac{EH^3}{12(1-\nu^2)}, \tag{5.80}$$
$$A^2_{1122} = c_{1122} - c_{1133}c^{-1}_{3333}c_{3322}\frac{H^3}{12} = \frac{E\nu H^3}{12(1-\nu^2)},$$
$$A^2_{1212} = c_{1212}\frac{H^3}{12} = \frac{EH^3}{12(1+\nu)}.$$

The formulas (5.80) coincide with the formulas for stiffnesses of plate in Kirchhoff-Love theory.

It may be instructive to follow the calculations for stiffnesses in (5.80) by using formulas (5.78). Although these calculations are fairly simple mathematically, they are interesting from a mechanical point of view. For example, stiffness A^0_{1111} is calculated as follows:

$$
\begin{aligned}
c_{1111} - c_{1133} c_{3333}^{-1} c_{3311} &= \frac{E(1-\nu)}{(1+\nu)(1-2\nu)} - \frac{\left(\dfrac{E\nu}{(1+\nu)(1-2\nu)}\right)^2}{\dfrac{E(1-\nu)}{(1+\nu)(1-2\nu)}} \\
&= \frac{E}{(1+\nu)(1-2\nu)}\left(1-\nu-\frac{\nu^2}{1-\nu}\right) \\
&= \frac{E}{(1+\nu)(1-2\nu)}\left(\frac{1-2\nu}{1-\nu}\right) = \frac{E}{1-\nu^2}.
\end{aligned} \tag{5.81}
$$

Stiffness A^0_{1122} is calculated as follows:

$$
\begin{aligned}
c_{1122} - c_{1133} c_{3333}^{-1} c_{3322} &= \frac{E\nu}{(1+\nu)(1-2\nu)} - \frac{\left(\dfrac{E\nu}{(1+\nu)(1-2\nu)}\right)^2}{\dfrac{E(1-\nu)}{(1+\nu)(1-2\nu)}} \\
&= \frac{E}{(1+\nu)(1-2\nu)}\left(\nu-\frac{\nu^2}{1-\nu}\right) \\
&= \frac{E}{(1+\nu)(1-2\nu)}\left(\nu\frac{1-2\nu}{1-\nu}\right) = \frac{E\nu}{1-\nu^2}.
\end{aligned} \tag{5.82}
$$

The computations in (5.81) and (5.82) involve the terms accounting the across the plate deformations of the plate due to Poisson effect. These terms are ignored in Kirchhoff-Love theory. Nevertheless the final resulting formulas in (5.81) and (5.82) completely coincide with the engineering formulas.

5.4 Moments of Initial Stresses

Initial stresses can generate in the plate both the in-plane resultant initial forces and the resultant initial moments. However, the author did not find any references to the influence of the resultant moments of initial stress on the behavior of plates, in particular, there are no references to the loss of stability of plates under the action of moments. At the same time, the phenomenon of loss of stability of a beam under the action of a torsion is well known. Such differences arouse additional interest in the problem.

5.4.1 Statement of the Problem

Let's consider the case when the initial stresses in (5.12) satisfy the following conditions: $\{\sigma_{ij}^{*(-1)}\} = 0$, $\{\sigma_{ij}^{*(-2)}\} \neq 0$, and $\sigma_{ij}^{*(-2)}$ satisfy the condition (5.14)—the in-plane resultant initial forces produced by the local initial stresses $\sigma_{ij}^{*(-2)}$ are equal to zero.

As above, the starting point of our consideration is the three-dimensional problem of elasticity theory (9.1), (9.5). In the case under consideration, the coefficients h_{ijmn} (9.3) have the form

$$h_{ijmn}(x_1, x_2, \mathbf{x}/\varepsilon) = \varepsilon^{-3} c_{ijmn} + \varepsilon^{-2} b_{ijmn}(x_1, x_2, \mathbf{x}/\varepsilon), \tag{5.83}$$

where

$$b_{ijmn}(x_1, x_2, \mathbf{x}/\varepsilon) = \sigma_{jn}^{*}(x_1, x_{2,}\mathbf{x}/\varepsilon)\delta_{im}. \tag{5.84}$$

For $h_{ijmn}(x_1, x_2, \mathbf{y})$ defined by (5.83), the relationship between the additional stresses σ_{ij} and the displacements $\mathbf{u}^{\varepsilon}$ takes the form:

$$\sigma_{ij}^{\varepsilon} = \left(\varepsilon^{-3} c_{ijmn} + \varepsilon^{-2} b_{ijmn}(x_1, x_2, \mathbf{x}/\varepsilon)\right) \frac{\partial u_m^{\varepsilon}}{\partial x_n}. \tag{5.85}$$

The terms $\varepsilon^{-2} b_{ijmn}(x_1, x_2, \mathbf{x}/\varepsilon) = \varepsilon^{-2} \sigma_{jn}^{*}(x_1, x_2, \mathbf{x}/\varepsilon)\delta_{im}$ correspond to resultant moments of initial stresses that do not depend on ε, see Sect. 6.2.3. Really, the integral $\int_{-\varepsilon H/2}^{\varepsilon H/2} \sigma_{jn}^{*} x_3 dx_3$ has the order of ε^2. In order that the integral has order of unity, the stresses σ_{jn}^{*} must have the order of ε^{-2}.

5.4.2 Asymptotic Expansions and Limit Equilibrium Equations

As above, we use the asymptotic expansions (9.6) and (9.7). The analysis of the problem (5.9), (5.11), (5.85) is separated into two stages. The first stage consists in obtaining the limit equilibrium equations for the plate, considered as a two-dimensional structure. This stage does not involve local governing equations (in our case, equation (5.85)), and is the same for any governing equations. The limit equilibrium equations are (5.26), and (5.30)–(5.32).

5.4.3 Limit Governing Equations

The second step is to derive the governing equations for the plate as a two-dimensional structure. This step makes significant use of the governing equations and we will consider it in detail.

Substituting (9.6) and (9.7) into the local governing equation (5.85), and taking into account the differentiating rule (5.17), we obtain

$$\sum_{p=-3}^{\infty} \varepsilon^p \sigma_{ij}^{(p)} = \sum_{k=0}^{\infty} \varepsilon^k (\varepsilon^{-3} c_{ijmn} + \varepsilon^{-2} b_{ijmn}(x_1, x_2, \mathbf{y})) \times (u_{m,nx}^{(k)} \delta_{n\beta} + \varepsilon^{-1} u_{m,3y}^{(k)} \delta_{n3}). \tag{5.86}$$

Equating the terms with identical power of ε in (5.86), we obtain an infinite sequence of local equations: the first three of them are as follows:

$$\sigma_{ij}^{(-3)} = c_{ijm\beta} u_{m,\beta x}^{(0)} + c_{ijm3} u_{m,3y}^{(1)}; \tag{5.87}$$

$$\sigma_{ij}^{(-2)} = c_{ijm\beta} u_{m,\beta x}^{(1)} + c_{ijmn} u_{m,ny}^{(2)} + b_{ijm\beta}(x_1, x_2, \mathbf{y}) u_{m,\beta x}^{(0)} + b_{ijm3}(x_1, x_2, \mathbf{y}) u_{m,3y}^{(1)}; \tag{5.88}$$

$$\sigma_{ij}^{(-1)} = c_{ijm\beta} u_{m,\beta x}^{(2)} + c_{ijmn} u_{m,ny}^{(3)} + b_{ijm\beta}(x_1, x_2, y_3) u_{m,\beta x}^{(1)} + b_{ijm3}(x_1, x_2, y_3) u_{m,3y}^{(2)}. \tag{5.89}$$

Let us consider the problem (5.34, $p = -3$) with $\sigma_{ij}^{(-3)}$ given by (5.87). Taking into account that $\mathbf{u}^{(0)}$ depends only on the variables x_1, x_2, we can write this problem as

$$\begin{cases} \dfrac{d}{dy_3}\left(c_{i3m3} \dfrac{du_m^{(1)}}{dy_3} + c_{i3m\alpha} u_{m,\alpha x}^{(0)}(x_1, x_2) \right) = 0 \text{ on } [-H/2, H/2], \\ c_{i3m3} \dfrac{du_m^{(1)}}{dy_3} + c_{i3m\alpha} u_{m,\alpha x}^{(0)}(x_1, x_2) = 0 \text{ for } y_3 = -H/2, H/2. \end{cases} \tag{5.90}$$

The solution to problem (5.90) has the form (compare with (5.44)).

$$\mathbf{u}^{(1)} = -y_3 \mathbf{e}_\alpha u_\beta^{(0)}(x_1, x_2) + \mathbf{N}^{0\alpha\beta}(y_3) u_{\alpha,\beta}^{(0)}(x_1, x_2) + \mathbf{V}(x_1, x_2). \tag{5.91}$$

In (5.91) $\mathbf{V}(x_1, x_2)$ is an arbitrary function of the variables x_1, x_2 (it will be defined below), $\mathbf{N}^{0\alpha\beta}(y_3)$ is a solution to the problem (5.41).

Substituting (5.91) into (5.87) yields the following equation

$$\sigma_{ij}^{(-3)} = c_{ij\alpha\beta} u_{\alpha,\beta x}^{(0)} + c_{ijkl} N_{k,ly}^{0\alpha\beta}(y_3) u_{\alpha,\beta x}^{(0)}(x_1, x_2). \tag{5.92}$$

Averaging (5.92) over the thickness $[-H/2, H/2]$ of the plate with $ij = \gamma\delta$, we obtain

$$N_{\gamma\delta}^{(-3)} = \langle\sigma_{\gamma\delta}^{(-3)}\rangle = A_{\gamma\delta\alpha\beta}^{0} u_{\alpha,\beta x}^{(0)}, \tag{5.93}$$

where $A_{\gamma\delta\alpha\beta}^{0}$ are given by formula (5.58) with $\mu = \nu = 0$.

In the case under consideration, the limit equilibrium equations for $N_{\gamma\delta}^{(-3)}$ (5.26) remain the same (as was noted, their derivation did not depend on the governing relations). From (5.26, $p = -3$) and (5.93) we obtain

$$(A_{\gamma\delta\alpha\beta}^{0} u_{\alpha,\beta x}^{(0)})_{,\delta x} = 0 \text{ in } D. \tag{5.94}$$

The boundary conditions for $u_{\alpha}^{(0)}$ (α =1, 2) follow from (5.11) and (5.15) and are of the form

$$u_{\alpha}^{(0)}(x_1, x_2) = 0 \text{ on } \partial D \quad (\alpha = 1, 2). \tag{5.95}$$

The solution to problem (5.94) and (5.95) is $u_{\alpha}^{(0)} = 0 \quad (\alpha = 1, 2)$. Then (5.91) takes the form

$$\mathbf{u}^{(1)} = -y_3 \mathbf{e}_{\alpha} u_{\alpha}^{(0)}(x_1, x_2) + \mathbf{V}(x_1, x_2). \tag{5.96}$$

Substituting equality (5.96) into (5.87) and (5.88), we obtain

$$\sigma_{ij}^{(-3)} = 0, \tag{5.97}$$

$$\begin{aligned} \sigma_{ij}^{(-2)} &= c_{ijmn} u_{m,ny}^{(2)} - c_{ij\alpha\beta} y_3 u_{3,\alpha x\beta x}^{(0)}(x_1, x_2) \\ &+ c_{ij\alpha\beta} V_{\alpha,\beta x}(x_1, x_2) \\ &+ (b_{ij3\alpha}(x_1, x_2, y_3) - b_{ij\alpha 3}(x_1, x_2, y_3)) u_{3,\alpha x}^{(0)}(x_1, x_2). \end{aligned} \tag{5.98}$$

Let's consider the problem (5.34, $p = -2$) with (5.98). We can write this as follows:

$$\begin{cases} \dfrac{d}{dy_3}\left(c_{i3m3}\dfrac{du_m^{(2)}}{dy_3} - c_{i3\alpha\beta} y_3 u_{3,\alpha x\beta x}^{(0)}(x_1, x_2) + c_{i3\alpha\beta} V_{\alpha,\beta x}(x_1, x_2)\right. \\ \left. + (b_{i33\alpha}(x_1, x_2, y_3) - b_{i3\alpha 3}(x_1, x_2, y_3)) u_{3,\alpha x}^{(0)}(x_1, x_2)\right) = 0 \text{ in } [-H/2, H/2], \\ c_{i3m3}\dfrac{du_m^{(2)}}{dy_3} - c_{i3\alpha,\beta} y_3 u_{3,\alpha x\beta x}^{(0)}(x_1, x_2) + c_{i3\alpha\beta} V_{\alpha,\beta x}(x_1, x_2) \\ + (b_{i33\alpha}(x_1, x_2, y_3) - b_{i3\alpha 3}(x_1, x_2, y_3)) u_{3,\alpha x}^{(0)}(x_1, x_2) = 0 \text{ for } y_3 = \pm H/2, \end{cases} \tag{5.99}$$

Proposition 5.4 *If the conditions of Proposition 5.3 are satisfied than*

$$\frac{db_{ij\alpha 3}}{dy_3} = 0 \text{ in } [-H/2, H/2] \text{ and } b_{i3\alpha 3} = 0 \text{ on } S, \tag{5.100}$$

$$\frac{db_{i33\alpha}}{dy_3} = 0 \text{ in } [-H/2, H/2] \text{ and } b_{i33\alpha} = 0 \text{ on } S.$$

Indeed, $\dfrac{db_{i3\alpha 3}}{dy_3} = \delta_{i\alpha}\dfrac{d\sigma^*_{33}}{dy_3} = 0$ in $[-H/2, H/2]$ and $b_{i3\alpha 3} = \delta_{i\alpha}\sigma^*_{33} = 0$ due to the symmetry of the initial stresses in i and j and taking into account (9.56).

By virtue of (5.100), we can omit the terms containing $b_{ij\alpha 3}$ and $b_{ij3\alpha}$ in the problem (5.99). As a result, we arrive at the problem (5.52), the solution to which can be expressed through the functions $\mathbf{N}^{\nu\alpha\beta}(y_3)$ ($\nu = 0, 1$) defined by (5.43) and (5.53) in the form

$$\begin{aligned}\mathbf{u}^{(2)} &= \mathbf{N}^{1\alpha\beta}(y_3)u^{(0)}_{3,\alpha x\beta x}(x_1, x_2) - y_3 V_{3,\beta x}(x_1, x_2)\mathbf{e}_\beta \\ &\quad + \mathbf{N}^{0\alpha\beta}(y_3)V_{\alpha,\beta x}(x_1, x_2).\end{aligned} \tag{5.101}$$

Substituting (5.101) into (5.88), we have

$$\begin{aligned}\sigma^{(-2)}_{ij} &= \left(c_{ijm3}\frac{dN^{0\alpha\beta}_m}{dy_3} + c_{ij\alpha\beta}\right)V_{\alpha,\beta x}(x_1, x_2) \\ &+ \left(c_{ijm3}\frac{dN^{1\alpha\beta}_m}{dy_3} - y_3 c_{ij\alpha\beta}\right)u^{(0)}_{3,\alpha x\beta x}(x_1, x_2) \\ &+ (b_{ij3\alpha}(x_1, x_2, y_3) - b_{ij\alpha 3}(x_1, x_2, y_3))u^{(0)}_{3,\alpha x}(x_1, x_2).\end{aligned} \tag{5.102}$$

Averaging the equality (5.102) with $ij = \gamma\delta$ ($\gamma, \delta = 1, 2$) over $[-H/2, H/2]$, we obtain

$$N^{(-3)}_{\gamma\delta} = A^0_{\gamma\delta\alpha\beta}V_{\alpha,\beta x} + A^1_{\gamma\delta\alpha\beta}u^{(0)}_{3,\alpha x\beta x} + B^0_{\gamma\delta\alpha}u^{(0)}_{3,\alpha x}. \tag{5.103}$$

Multiplying (5.102) with $ij = \gamma\delta$ ($\gamma\delta$ =1, 2) by $-y_3$ and averaging over $[-H/2, H/2]$, we obtain

$$M^{(-3)}_{\gamma\delta} = A^1_{\gamma\delta\alpha\beta}V_{\alpha,\beta x} + A^2_{\gamma\delta\alpha\beta}u^{(0)}_{3,\alpha x\beta x} + B^1_{\gamma\delta\alpha}u^{(0)}_{3,\alpha x}. \tag{5.104}$$

The coefficients in the relations (5.103), (5.104) are

$$\begin{aligned}A^{\mu+\nu}_{\gamma\delta\alpha\beta} &= \left\langle\left((-1)^\nu c_{\gamma\delta\alpha\beta}y_3 + c_{\gamma\delta mn}\frac{dN^{\nu\alpha\beta}_m}{dy_3}\right)(-1)^\mu y_3^\mu\right\rangle, \\ B^\mu_{\gamma\delta\alpha} &= \langle(-1)^\mu b_{\gamma\delta\alpha 3}(x_1, x_2, y_3)y_3^\mu\rangle\end{aligned} \tag{5.105}$$

(ν, μ =0, 1; $\alpha, \beta, \gamma, \delta$ =1, 2).

Here it is taken into account that $b_{\gamma\delta 3\alpha} = 0$ by virtue of the definition of the quantities b_{ijkl} (5.84).

Proposition 5.5 *Under the conditions of Proposition 5.3* $B^{\mu}_{\gamma\delta\alpha}$ *=0* $(\mu = 0, 1)$.

Proposition follow directly from the definition $B^{\mu}_{\gamma\delta\alpha}$ (5.105), definition b_{ijmn} (5.84), and Proposition 5.3.

As follows from (5.103)–(5.105) and Proposition 5.5, in the case under consideration the initial stresses do not affect the limit governing equations relating the resultant in-plane forces and moments to the strains and curvatures of the plate.

5.4.4 Symmetry of Shear Forces

Consider the shear forces $N^{(-1)}_{\gamma 3}$ and $N^{(-1)}_{3\gamma}$. Substitute $\mathbf{u}^{(1)}$ into (5.89) in accordance with (5.96). We obtain

$$\begin{gathered}\sigma^{(-1)}_{ij} = c_{ijm\beta}u^{(2)}_{m,\beta x} + c_{ijm3}u^{(3)}_{m,3y}\\ -b_{ij\alpha\beta}(x_1,x_2,y_3)y_3u^{(0)}_{\alpha,\beta x}(x_1,x_2) + b_{ijk\beta}(x_1,x_2,y_3)V_{k,\beta x}(x_1,x_2)\\ +b_{ijmn}(x_1,x_2,y_3)u^{(2)}_{m,ny}.\end{gathered} \tag{5.106}$$

The first two terms on the right-hand side of equation (9.32) are symmetrical in i and j due to the symmetry of the elastic constants ($c_{ijmn} = c_{jimn}$, see condition C1 Sect. 3.1). Due to which we have

$$\begin{gathered}\sigma^{(-1)}_{ij} - \sigma^{(-1)}_{ji}\\ = (-b_{ij\alpha\beta}(x_1,x_2,y_3) + b_{ji\alpha\beta}(x_1,x_2,y_3))y_3u^{(0)}_{\alpha,\beta x}(x_1,x_2)\\ +(b_{ijk\beta}(x_1,x_2,y_3) - b_{jik\beta}(x_1,x_2,y_3))V_{k,\beta x}(x_1,x_2)\\ +(b_{ijmn}(x_1,x_2,y_3) - b_{jim3}(x_1,x_2,y_3))u^{(2)}_{m,3y}.\end{gathered} \tag{5.107}$$

Averaging (5.107) over $[-H/2, H/2]$, we obtain

$$\begin{gathered}N^{(-1)}_{ij} - N^{(-1)}_{ji}\\ = \langle(-b_{ij\alpha\beta}(x_1,x_2,y_3) + b_{ji\alpha\beta}(x_1,x_2,y_3))y_3\rangle u^{(0)}_{\alpha,x\beta x}\\ +\langle(b_{ijmn}(x_1,x_2,y_3) - b_{jim3}(x_1,x_2,y_3))u^{(2)}_{m,3y}\rangle.\end{gathered} \tag{5.108}$$

The average values of $b_{ijk\beta}$ and $b_{ijk\beta}$ in (5.107) are zero. Indeed, $\langle b_{ijk\beta}\rangle = \langle\sigma^*_{j\beta}\delta_{ik}\rangle$ and $\langle\sigma^*_{\alpha\beta}\rangle = 0$ by condition (5.14) and $\langle\sigma^*_{3\beta}\rangle = 0$ by Proposition 5.2. Also, $\langle b_{jik\beta}\rangle =$

$\langle \sigma^*_{i\beta}\delta_{jk}\rangle$ and $\langle \sigma^*_{i\beta}\rangle = 0$ by virtue of condition (5.14) and Proposition 5.2. As a result, there is no term with $V_{k,\beta x}$ in (5.108).

If the condition of Proposition 5.3 is satisfied,

$$\left\langle b_{ijm3}(x_1, x_2, y_3)\frac{du_m^{(2)}(y_3)}{dy_3}\right\rangle = 0,$$

which follows from Proposition 5.3 for $Z = u_m^{(2)}(y_3)$.

For i=3 and $j = \gamma$ we obtain from (5.108) taking into account the proposition 5.3:

$$N^{(-1)}_{3\gamma} - N^{(-1)}_{\gamma 3} = K_\gamma, \tag{5.109}$$

where K_γ is defined by the formula

$$\begin{aligned} K_\gamma &= (\langle b_{\gamma 3\alpha\beta}(x_1, x_2, y_3)y_3\rangle - \langle b_{3\gamma\alpha\beta}(x_1, x_2, y_3)y_3\rangle)u^{(0)}_{\alpha,\beta x}(x_1, x_2) \\ &= (\delta_{\gamma\alpha}\langle \sigma^*_{3\beta}(x_1, x_2, y_3)y_3\rangle - \delta_{3\alpha}\langle \sigma^*_{\gamma\beta}(x_1, x_2, y_3)y_3\rangle)u^{(0)}_{\alpha,\beta x}(x_1, x_2) \\ &= \langle \sigma^*_{3\beta}(x_1, x_2, y_3)y_3\rangle u^{(0)}_{\gamma,\beta x}(x_1, x_2). \end{aligned} \tag{5.110}$$

In the case under consideration, according to (5.109), (5.110) and the Proposition 5.3,

$$N^{(-1)}_{3\gamma} - N^{(-1)}_{\gamma 3} = 0.$$

It means that the resultant shear forces *are symmetrical*, despite the fact that the corresponding local stresses *are not symmetrical*. Then, in the case under consideration, we can use the classical procedure for eliminating the shear forces (Timoshenko and Woinowsky-Krieger 1959), leading to the classical model of plate bending without initial stresses (despite the fact that the initial stresses are present in the original problem (5.9)–(5.13)).

The purely formal calculations performed have significant mechanics meaning. The obtained result of computations means that the initial bending/torque moments do not affect the plate behavior. In particular, it means that bending/torsion moments cannot lead to loss of stability of plate. It is instructive to compare the influence of resultant forces and moments of the initial stresses on deformation of plates and beams. In-plane resultant forces in plane, as well as, resultant axial force in beam may lead to the loss of stability. In the contrast, resultant moments of initial stresses in beam may to lead to the loss of stability of beam, but resultant moments of initial stresses in plate cannot lead to the loss of stability of plate.

5.5 Local Stresses in Uniform Plate with Initial Stresses

The total stresses $\sigma_{ij}^{total}(\mathbf{x})$ in the body with initial stresses are the sum of the initial stresses $\sigma_{ij}^{*}(\mathbf{x})$ and additional stresses $\sigma_{ij}^{\varepsilon}(\mathbf{x})$ (Washizu 1992):

$$\sigma_{ij}^{total}(\mathbf{x}) = \sigma_{ij}^{*}(\mathbf{x}) + \sigma_{ij}^{\varepsilon}(\mathbf{x}). \tag{5.111}$$

5.5.1 *In-Plane Resultant Initial Stresses*

The in-plane resultant initial stresses are result of the action of the local initial stresses $\sigma_{ij}^{*} = \sigma_{ij}^{*(-1)}$, which characteristic value is inversely proportional to the thickness ε of the plate.

In accordance with (5.16), the additional stresses

$$\sigma_{ij}^{\varepsilon} = \varepsilon^{-3}\sigma_{ij}^{(-3)}(x_1, \mathbf{y}) + \varepsilon^{-2}\sigma_{ij}^{(-2)}(x_1, \mathbf{y}) + \cdots . \tag{5.112}$$

In accordance with (5.50), $\sigma_{ij}^{(-3)} = 0$, thus, the local stresses in the plate are determined by the term $\sigma_{ij}^{(-2)}$ given by (5.55), and we arrive at the following formula for the computation of the total stresses σ_{ij}^{total} in the uniform plate with initial stresses σ_{ij}^{*}:

$$\begin{aligned} \sigma_{ij}^{total} &= \sigma_{ij}^{*} + \left(c_{ijm3}\frac{dN_m^{0\alpha\beta}}{dy_3} + c_{ij\alpha\beta} \right) V_{\alpha,\beta x}(x_1, x_2) \\ &+ \left(c_{ijm3}\frac{dN_m^{1\alpha\beta}}{dy_3} - y_3 c_{ij\alpha\beta} \right) u_{3,\alpha x\beta x}^{(0)}(x_1, x_2). \end{aligned} \tag{5.113}$$

In accordance with (5.75)

$$\frac{dN_m^{0\alpha\beta}}{dy_3}(y_3) = -c_{m3i3}^{-1}c_{i3\alpha\beta}, \quad \frac{dN_m^{1\alpha\beta}}{dy_3}(y_3) = -y_3 c_{m3i3}^{-1}c_{i3\alpha\beta}. \tag{5.114}$$

Substituting $\dfrac{dN_m^{0\alpha\beta}}{dy_3}$ and $\dfrac{dN_m^{1\alpha\beta}}{dy_3}$ (5.114) into (5.113), we obtain that the actual total stresses σ_{ij}^{total} in the plate are the following:

$$\begin{aligned} \sigma_{ij}^{total} &= \sigma_{ij}^{*} + (-c_{ijm3}c_{m3i3}^{-1}c_{i3\alpha\beta} + c_{ij\alpha\beta})V_{\alpha,\beta x}(x_1, x_2) \\ &-(c_{ijm3}c_{m3i3}^{-1}c_{i3\alpha\beta} + c_{ij\alpha\beta}) y_3 u_{3,\alpha x\beta x}^{(0)}(x_1, x_2). \end{aligned} \tag{5.115}$$

5.5.2 Moments of Initial Stresses

The moments of initial stresses are result of the action of the local initial stresses $\sigma^*_{ij} = \sigma^{*(-2)}_{ij}$, which characteristic value is inversely proportional to the square of the thickness ε of the plate.

In accordance with (5.97), $\sigma^{(-3)}_{ij} = 0$, thus, the local stresses in the plate are determined by the term $\sigma^{(-2)}_{ij}$ given by (5.102), and we arrive at the following formula for the computation of the total stresses σ^{total}_{ij} in the uniform plate with initial stresses σ^*_{ij}:

$$\begin{aligned}\sigma^{total}_{ij} &= \sigma^*_{ij} + \left(c_{ijm3}\frac{dN^{0\alpha\beta}_m}{dy_3} + c_{ij\alpha\beta}\right)V_{\alpha,\beta x}(x_1,x_2)\\ &+\left(c_{ijm3}\frac{dN^{1\alpha\beta}_m}{dy_3} - y_3c_{ij\alpha\beta}\right)u^{(0)}_{3,\alpha x\beta x}(x_1,x_2)\\ &+(b_{ij3\alpha}(x_1,x_2,y_3) - b_{ij\alpha 3}(x_1,x_2,y_3))u^{(0)}_{3,\alpha x}(x_1,x_2).\end{aligned} \tag{5.116}$$

Substituting $\dfrac{dN^{0\alpha\beta}_m}{dy_3}$ and $\dfrac{dN^{\nu\alpha\beta}_m}{dy_3}$ (5.114) into into (5.116), and taking into account definition of $b_{ij\alpha 3}$ (5.84), we obtain that the actual total stresses σ^{total}_{ij} in the plate are the following:

$$\begin{aligned}\sigma^{total}_{ij} &= \sigma^*_{ij} + (-c_{ijm3}\overset{-1}{c}_{m3i3}c_{i3\alpha\beta} + c_{ij\alpha\beta})V_{\alpha,\beta x}(x_1,x_2)\\ &-(c_{ijm3}\overset{-1}{c}_{m3i3}c_{i3\alpha\beta} + c_{ij\alpha\beta})y_3u^{(0)}_{3,\alpha x\beta x}(x_1,x_2)\\ &+(\delta_{i3}\sigma^*_{j\alpha}(x_1,x_2,y_3) - \delta_{i\alpha}\sigma^*_{j3}(x_1,x_2,y_3))u^{(0)}_{3,\alpha x}(x_1,x_2).\end{aligned} \tag{5.117}$$

In an uniform plate, the transversal stress $\sigma^*_{j3} = 0$.

The initial stresses $\sigma^*_{ij} = \sigma^{*(-2)}_{ij}$ create moments that do not affect the limit boundary value problem. Nevertheless, as follows from (5.117), the initial stresses influence the local stresses. The initial stresses influence the total local stresses not only through the summand $\sigma^{*\varepsilon}_{ij}$, but also through the less obvious summand $\delta_{i3}\sigma^{*(-2)}_{j\alpha}(x_1,x_2,y_3)u^{(0)}_{3,\alpha x}(x_1,x_2)$.

References

Caillerie D (1982) Plaques elastiques minces a structute periodique de periode et depaisseur comparable. C R Acad Sci Ser 2 294(3):159–162

Caillerie D (1984) Thin elastic, periodic plates. Math Methods Appl Sci 6:159–191

Gol'denveizer AL (1962) Derivation of an approximate theory of bending of a plate by the method of asymptotic integration of the equations of the theory of elasticity (in Russian). Prikl Math Mekh 26:668–686

Gol'denveizer AL (1963) Derivation of an approximate theory of shells by means of asymptotic integration of the equations of the theory of elasticity (in Russian). Prikl Math Mekh 27:593–608

Gol'denveizer AL (1968) Problems of foundations and refinement of shell theory (in Russian). Prikl Marh Mekh 32:684–695

Green AE (1962) On the linear theory of thin elastic shells. Proc R Soc Ser A 266:143–160

Green AE, Naghdi PM (1965) Some remarks on the linear theory of shells. Q J Mech Appl Math 18:257–276

John F (1969) Refined interior shell equations. In: Proceedings of second IUTAM symposium theory of thin shells. Springer, Copenhagen, pp 1–14

John F (1965) Estimates for the derivatives of the stresses in a thin shell and interior shell equations. Commun Pure Appl Math 18:235–267

Koiter WT (1960) A consistent first approximation in the general theory of thin elastic shells. In: Proceedings of IUTAM symposium on the theory of thin shells. Elsevier, North-Holland, Delft, pp 12–33

Kolpakov AG (2004) Stressed composite structures: homogenized models for thin-walled nonhomogeneous structures with initial stresses. Springer, Berlin

Lions J-L, Magenes E (1973) Non-homogeneous boundary value problems and applications, vol 1. Springer, Berlin

Love AEH (1929) A treatise on the mathematical theory of elasticity. Oxford University Press, Oxford

Mizohata S (1973) The theory of partial differential equations. Cambridge University Press, Cambridge

Morgenstern D (1959) Bernoullische hypothesen bei balken und platten theorie. Z Angew Math Mech 39(9–11):420–422

Morgenstern D (1959) Herleitung der plattenteorie sus der dreidimensionalen elastitzitatsteorie. Arch Ration Mech Anal 4(2):145–152

Morgenstern D (1959) Maththematische begrundund der scheibentheorie. Arch Ration Mech Anal 3(3):91–96

Nazarov SA (2002) Asymptotic analysis of thin plates and rods. Nauchay Kniga, Novosibirsk (in Russian)

Novotny B (1970) On the asymptotic of the three-dimensional non-linear equations of thin elastic shells and plates. Int J Solids Struct 6:433–451

Rhodes J (2002) Buckling of thin plates and members—and early work on rectangular tubes. Thin-Walled Struct 40(2):87–108

Rudin W (1967) Principles of mathematical analysis, 3rd edn. McGraw-Hill, New York

Rutten HS (1969) Asymptotic approximation in the three-dimensional theory of thin and thick elastic shells. In: Proceedings of Second IUTAM symposium theory of thin shells. Springer, Copenhagen, pp 15–34

Sanchez-Palencia E (2000) Non-homogeneous media and vibration theory. Springer, Berlin

Sencenmg CB (1968) A shell theory compared with the exact three-dimensional theory of elasticity. Int J Eng Sci 6:435–464

Timoshenko S (1961) Theory of elastic stability, 2nd edn. McGraw-Hill, New York

Timoshenko S (1983) History of strength of materials (with a brief account of the history of elasticity and theory of structures). Dover, New York

Timoshenko S (1971) Stability of bars, plates, and shells. Soviet Appl Mech 7:1175–1176

Timoshenko S, Woinowsky-Krieger K (1959) Theory of plates and shells. McGraw-Hill, New York

Washizu K (1992) Variational methods in elasticity and plasticity. Pergamon Press, Oxford

Chapter 6
Non-homogeneous Plates with Initial Stresses

In the previous chapter, we discussed the deriving of two-dimensional model of uniform plate with initial stresses directly from three dimensional elasticity theory problem. The presence of initial stresses manifested themselves in the loss of symmetry in an analog of Hooke's low, only. All others basic notions related to plate model in Chap. 3 coincided with the classical ones.

In this chapter, we demonstrate that the homogenization method allows making transition from three-dimensional elasticity problem to two-dimensional problem for plates of inhomogeneous structure, including plates for which Kirchhoff-Love hypothesis obviously cannot be satisfied (for example, ribbed Kalamkarov and Kolpakov 1997, wafer Ye et al. 2014; Kolpakov and Kolpakov 2020 or perforated plates Ghergu et al. 2007; Andrianov et al. 1999; Mignot et al. 1980).

The main difference between plates of non-uniform and uniform plates is that for non-uniform plates the concepts of neutral surface and cross-section are absent, and the concept of resultant forces and resultant moments are significantly changed. Note that even for such plates the first terms of the asymptotic expansions demonstrate a "reminiscence" of the Kirchhoff-Love hypothesis, although realized not directly in the "average-macroscopic" sense. Thus, the Kirchhoff-Love hypothesis shows itself to be the most stable component of the theory of plates.

We start by discussing the problems that arise when trying to apply classical concepts to non-uniform plates (made of composite materials or having variable thickness). After this discussion, we present a theory of thin plate structures with initial stresses based on the ideas of the asymptotic homogenization method.

In this book only elastic plates are considered. The influence of thermal, piezoelectric, phase transformation effects, embedding electronic circuits, etc., on thin plates was studied in Andrianov et al. (2024), Hadjiloizi et al. (2014a, b), Pan and Heyliger (2002), Brischetto and Cesare (2024), Eremeyev and Pietraszkiewicz (2004), Canon and Lenczner (2000).

A. Kolpakov, *Thin-Walled Structures with Initial Stresses*, Mathematical Engineering,
https://doi.org/10.1007/978-3-032-07247-4_6

6.1 Discussion on the Basic Notions of Classical Theory as Applied to Plates of Non-homogeneous Structure

Comparing models of uniform and non-uniform plates, see, e.g., Timoshenko and Woinowsky-Krieger (1959), Caillerie (1984), Kalamkarov and Kolpakov (1997), one find that the difference between the model are not only in quantitative things, but in conceptual ones. Let us discuss some of these differences.

6.1.1 Notion of Neutral Plane

A plate has three kind of stiffnesses: in-plane stiffnesses A^0_{ijkl}, coupled stiffnesses (also referred to as asymmetrical stiffnesses) A^1_{ijkl} and bending stiffnesses A^2_{ijkl}. For uniform plate of constant thickness H made of homogeneous material with Young's modulus E and Poisson's ratio ν, the stiffnesses are computed in accordance with the formula (coordinate z is associated with Oy_3)

$$A^{\mu}_{\alpha\alpha\alpha\alpha} = \frac{E}{1-\nu^2}\int_{-H/2}^{H/2} z^{\mu}dz \ \ (\mu = 0, 1, 2). \tag{6.1}$$

In accordance with (6.1), the stiffnesses $A^1_{\alpha\alpha\alpha\alpha} = 0$ if the middle plane of the plate coincides with the coordinate plane $z = 0$. The middle plane in the uniform plate is also the neutral plane, i.e. the plane which has zero stresses and strains if the plate is subjected to pure bending. In the general case, if the middle plane has position $z = h$, formula (6.1) takes the form

$$A^{\mu}_{\alpha\alpha\alpha\alpha} = \frac{E}{1-\nu^2}\int_{-H/2+h}^{H/2+h} z^{\mu}dz \ \ (\mu = 0, 1, 2). \tag{6.2}$$

For uniform plate, there exists the following relationship:

$$h = -\frac{A^1_{\alpha\alpha\alpha\alpha}}{A^0_{\alpha\alpha\alpha\alpha}}. \tag{6.3}$$

The relationship (6.3) can be expanded to the inhomogeneous plates, if to use as A^0_{ijkl} and A^1_{ijkl} the homogenized stiffnesses of the plate.

Consider a waffle plate with a supporting plate parallel to the plane Ox_1x_2 and a rectangular reinforcement system, see Fig. 6.1. The homogenized in-plane and out-of-plane stiffnesses of wafer plate are the following Kalamkarov and Kolpakov (1997):

$$A^0_{1111} = E\left(\frac{1}{1-\nu^2} + F_1\right), \quad A^1_{1111} = ES_1,$$
$$A^0_{2222} = E\left(\frac{1}{1-\nu^2} + F_2\right), \quad A^1_{2222} = ES_2, \tag{6.4}$$

where

$$F_1 = \frac{Ht_2}{h_2}, \quad S_1 = \frac{(H^2+H)t_2}{2h_2},$$
$$F_2 = \frac{Ht_1}{h_1}, \quad S_2 = \frac{(H^2+H)t_1}{2h_1}, \tag{6.5}$$

and E and ν are Young's modulus and Poisson's ratio of the material from which the plate is made; the geometrical parameters H, h_1, h_2, t_1, and t_2 are shown in Fig. 6.1. Formulas (6.6) are derived in the coordinate system shown in Fig. 6.1 ($z = 0$ plane coincides with the middle plane of the carrier plate).

6.1.2 Notion of Cross-Section

The cross-section of the plate is a two-dimensional figure resulting from the intersection of the plate with vertical planes $x_1 = const$ or $x_2 = const$. The cross-sections S_1 and S_2 of wafer plate are displayed in Fig. 6.1 (right) in the dependence on the $const$. Evidently, these cross-sections cannot be associated with the notion of cross-section in the classical theory of plates.

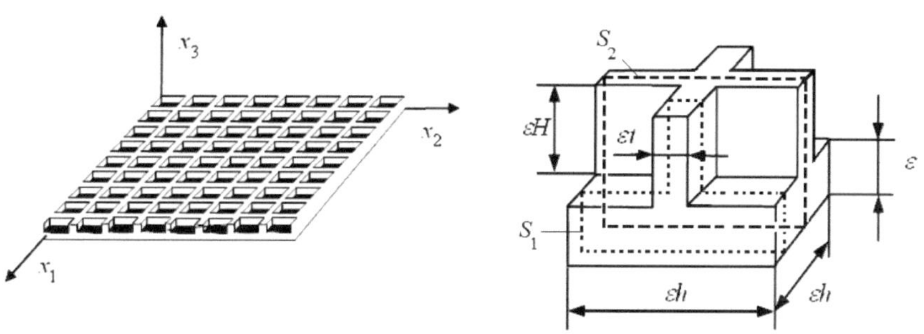

Fig. 6.1 Wafer plate and its periodicity cell

6.1.3 In-Plane Stresses in the "Neutral" Plane of Plate of Complex Structure

For a homogeneous plate, the defining property of the neutral plane is that the stresses in this plane are zero if the plate is subjected to pure bending. In Fig. 6.2, the numerically computed local stresses in corrugated plate are displayed. The plate in the figure is subjected to the pure bending. In the case under consideration, the corrugations have a symmetrical structure, and the plane $y_3 = 0$ can be taken as the neutral plane, see Fig. 6.2. It is seen that the local stresses on the neutral plane defined in this way are not equal to zero.

6.1.4 Is the "Neutral" Plane Unique for a Plate of Complex Structure?

The neutral plane can be introduced for plane of complex structure by formula (6.3). The position of the neutral plane given by the formula (6.3) depends on the index α, which is the number of the coordinate axis. Thus, theoretically, the answer the question is that a plate of complex structure may have several neutral planes. To clarify the situation we present an example of the plate which neutral planes in directions Ox_1 and Ox_2 do not coincide one with another. We consider the ribbed plate considered in the previous section.

The positions of the neutral plate corresponding to the bending in Ox_1x_3-plane is

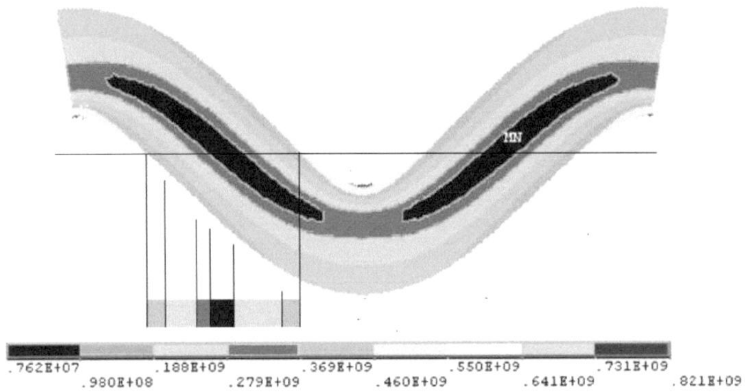

Fig. 6.2 von Mises stress in the "neutral" plane of corrugated plate

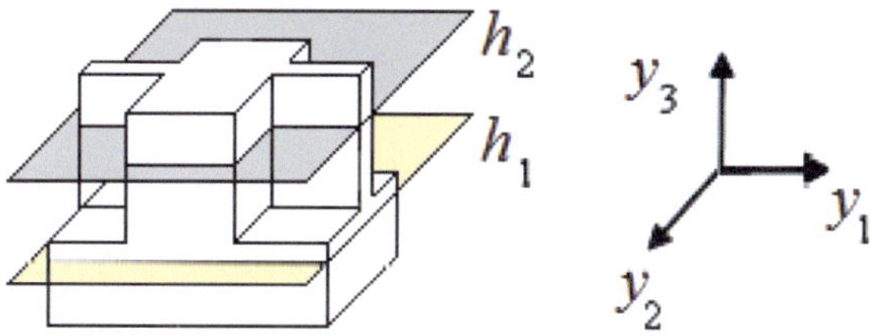

Fig. 6.3 Neutral planes of the wafer plate within one periodicity cell

$$A^0_{1111} = E\left(\frac{1}{1-\nu^2} + H\frac{t_2}{h_2}\right), \quad A^1_{1111} = EH\frac{(H+1)t_2}{2h_2},$$
$$A^0_{2222} = E\left(\frac{1}{1-\nu^2} + H\frac{t_1}{h_1}\right), \quad A^1_{2222} = EH\frac{(H+1)t_1}{2h_1}. \tag{6.6}$$

The positions of the neutral plate corresponding to the bending in Ox_1x_3-plane is

$$h_1 = -\frac{A^1_{1111}}{A^0_{1111}} = -\frac{H}{2} \cdot \frac{(H+1)X_2}{\frac{1}{1-\nu^2} + HX_2}. \tag{6.7}$$

The positions of the neutral plate corresponding to the bending in Ox_2x_3-plane is

$$h_2 = -\frac{A^1_{2222}}{A^0_{2222}} = -\frac{H}{2} \cdot \frac{(H+1)X_1}{\frac{1}{1-\nu^2} + HX_1}, \tag{6.8}$$

where $X_i = t_i/h_i$ $(i = 1, 2)$. Note that $0 \leq X_i \leq 1$.

In the computations in Kalamkarov and Kolpakov (1997), the coordinate plane $y_3 = 0$ coincides with the middle plane of the carrier plate. In the formulas (6.7) and (6.8), H and ν are the same (we assume that the carrier plate and wafer reinforcement are made of the same homogeneous material). The difference is X_1 in (6.7) and X_2 in 6.8). We can write $h_1 = h(X_2)$ and $h_2 = h(X_1)$, where the function $h(X)$ is the following:

$$h(X) = -\frac{H}{2} \cdot \frac{(H+1)X}{\frac{1}{1-\nu^2} + HX}. \tag{6.9}$$

In Fig. 6.4, the graph of the function $h(X)$ (6.9) is displayed for various H. The ratios $X_i = t_i/h_i$ $(i = 1, 2)$ in Fig. 6.4 take values from 0 to 0.7. In figure, the height H of the reinforcement takes values 1, 2, 3, and 4. It is seen that possible values of the function $h(X)$, thus, positions of the neutral planes h_1 and h_2, can vary in a wide

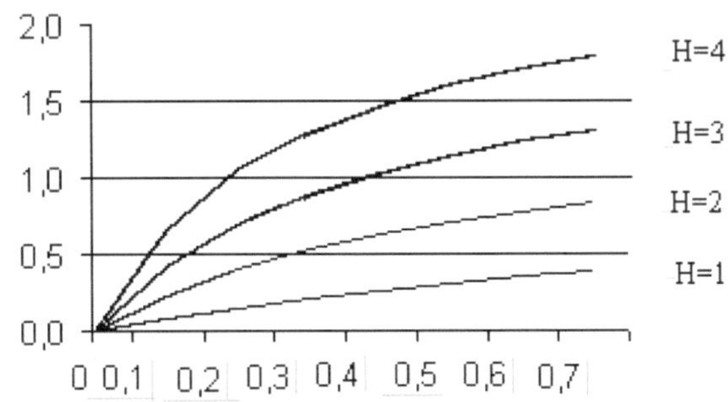

Fig. 6.4 Position of the neutral plate in wafer plate

interval in dependence on the geometry of the reinforcement—the values of H, and $X_i = t_i / h_i$ $(i = 1, 2)$.

We conclude that a plate of complex structure may have several neutral planes.

6.2 An Approach to the Plates with Initial Stresses Based on Two-Scale Expansions

The homogenized models of plates whose thickness is significantly less than the length of the period were obtained in Duvaut (1976), Kolpakov (1982). In particular, problems of stability and spectral problems for such kind plates were considered in Mignot et al. (1980), Villaggio (1979). Two-dimensional plate theory problem was taken as the initial model in the mentioned works.

The homogenization problem for periodic inhomogeneous plates, in the case where the thickness of the plate and the length of the period are compatible in size, was examined by Caillerie (1982). In the mentioned paper, it was proposed a modification of the asymptotic homogenization method, earlier developed for volumetric composite materials (Sanchez-Palencia 2000; Bensoussan et al. 1978; Cioranescu and Donato 2000; Jikov et al. 1994). In the papers by Caillerie (1982, 1984) and the following works (Sab 2020; Sab and Lebee 2015; Panasenko and Reztsov 1987; Kohn and Vogelius 1984) there were developed homogenization techniques for plates made of several materials (in particular, multilayered and reinforced plates) (Nazarova et al. 2011; Buannic et al. 2003), plates of varying thickness (Challagulla et al. 2010; Kalamkarov et al. 2022; Levinski 1995), frame-like plates (Lebee and Sab 2013) (in connection with the frame-like plates, mention a pioneering paper by Hrennikoff 1941), variational principles and estimates for the homogenized stiffnesses of plates (Kolpakov 1999), etc. The corresponding numerical techniques were developed in Muller et al. (2022), Lee and Yu (2011). In the mentioned papers, plates with no initial stresses were investigated. The analysis of the homogenization problem for thin elastic three-dimensional plates with initial was started in Kolpakov (1995), see also

Kolpakov (2000). The homogenization problem for a plate with matrix pre-strain was considered in Chakrabortty et al. (2024).

With the homogenization method, transition from three-dimensional elasticity problem to two-dimensional problem can be done for plates of inhomogeneous structure, including plates for which Kirchhoff-Love hypothesis obviously cannot be satisfied.

6.2.1 Statement of the Problem

We will examine a body G_ε of periodic structure obtained by repeating a certain small periodicity cell εY over the Ox_1x_2 plane, Fig. 6.5. Here ε is characteristic dimension of the periodicity cell, which is assumed to be small (that is formalized in the form $\varepsilon \to 0$). For $\varepsilon \to 0$, the three-dimensional plate G_ε tends to the domain D on the Ox_1x_2-plane—two-dimensional plate. Our aim is to derive a model describing the two-dimensional plate.

The starting point of our investigation is the three-dimensional elasticity problem for a body with initial stresses (without any simplifying assumptions). In accordance with this model, the equilibrium equations of the plate as a three-dimensional body with initial stresses are (see Chap. 4)

$$\frac{\partial \sigma_{ij}^{\varepsilon}}{\partial x_j} + f_i^{\varepsilon}(x_1, x_2, \mathbf{x}/\varepsilon) = 0 \text{ in } G_\varepsilon, \tag{6.10}$$

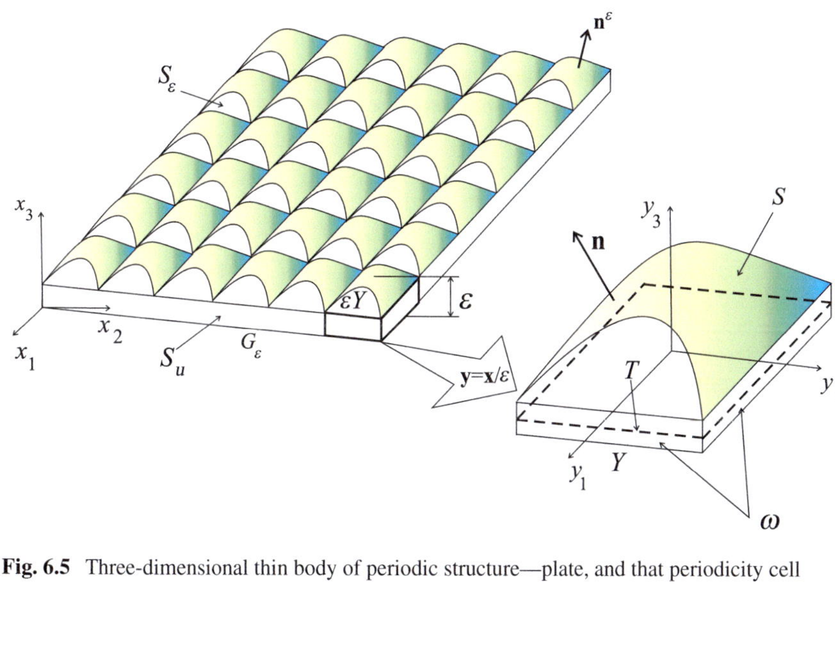

Fig. 6.5 Three-dimensional thin body of periodic structure—plate, and that periodicity cell

where the additional stresses $\sigma_{ij}^{\varepsilon}$ are related to the additional displacements $\mathbf{u}^{\varepsilon}$ by the equality

$$\sigma_{ij}^{\varepsilon} = h_{ijmn}^{\varepsilon}(x_1, x_2, \mathbf{x}/\varepsilon)\frac{\partial u_m^{\varepsilon}}{\partial x_n}. \tag{6.11}$$

The most general form of the coefficients h_{ijmn} in (6.11) is

$$\begin{aligned} h_{ijmn}(x_1, x_2, \mathbf{x}/\varepsilon) &= \varepsilon^{-3}c_{ijmn}(\mathbf{x}/\varepsilon) + \varepsilon^{-3}b_{ijmn}^{(-3)}(x_1, x_2, \mathbf{x}/\varepsilon) \\ &+\varepsilon^{-2}b_{ijmn}^{(-2)}(x_1, x_2, \mathbf{x}/\varepsilon) + \varepsilon^{-1}b_{ijmn}^{(-1)}(x_1, x_2, \mathbf{x}/\varepsilon), \end{aligned} \tag{6.12}$$

where

$$b_{ijmn}^{(p)}(x_1, x_2, \mathbf{x}/\varepsilon) = \sigma_{jn}^{*(p)}(x_1, x_{2,}\mathbf{x}/\varepsilon)\delta_{im} \quad (p = -3, -2, -1). \tag{6.13}$$

The boundary conditions can be written in the form,

$$\begin{aligned} \sigma_{ij}^{\varepsilon}n_j &= h_i^{\varepsilon}(x_1, x_2, \mathbf{x}/\varepsilon) \text{ on } S_{\varepsilon}, \\ \mathbf{u}^{\varepsilon}(\mathbf{x}) &= 0 \text{ on } S_u. \end{aligned} \tag{6.14}$$

Here:

S_{ε} is the lateral surface of the body G_{ε};

$\mathbf{n}^{\varepsilon}$ is a normal to S_{ε};

S_u is the lateral surface along which the body is clamped $\mathbf{u}^{\varepsilon}$ are the displacements;

$\mathbf{f}^{\varepsilon} = (f_1, f_2, \varepsilon^{-1}f_3)(\mathbf{x}/\varepsilon)$ are the mass forces (normal component of the mass force normal has order of ε^{-1}),

$\mathbf{h}^{\varepsilon} = (h_1, h_2, h_3)(\mathbf{x}/\varepsilon)$ are the surface forces.

Using the two-scale method as applied to thin plates of arbitrary periodic structure, we consider functions of the argument $x_1, x_2, \mathbf{x}/\varepsilon$. It differ the plates of periodic structure from the uniform plates considered in Sect. 5.3.1 (in Sect. 5.3.1, we considered functions of the argument $x_1, x_2, x_3/\varepsilon$).

We introduce “slow” variables x_1, x_2 and “fast” variables $\mathbf{y} = \mathbf{x}/\varepsilon$. Correspondingly, the functions above take the form $c_{ijmn}(\mathbf{y})$ and $\sigma_{jn}^{*(p)}(x_1, x_2, \mathbf{y}), h_{ijmn}(x_1, x_2, \mathbf{y})$. The functions are periodic in y_1, y_2 with periodicity cell T, where T is the projection of periodicity cell Y in the “fast” variables $\mathbf{y}$ on the Oy_1y_2-plane, see Fig. 6.5.

Note In (6.12), the multiplier ε^{-3} at c_{ijmn} guarantees that the bending stiffnesses of the plate will be nonzero, as $\varepsilon \to 0$. The term $\varepsilon^{-1}b_{ijmn}^{(-1)} = \varepsilon^{-1}\sigma_{jn}^{*(-1)}\delta_{im}$ in (6.12) corresponds to the resultant in-plane initial forces independent of ε. Really, the thickness of plate is of order of ε and the product of $\varepsilon^{-1}\sigma_{jn}^{*(-1)}\delta_{im}$ and ε is of order of unity.

If the resultant in-plane initial forces are zero,

$$\langle\sigma_{\alpha\beta}^{*(-2)}\rangle = 0(\alpha, \beta = 1, 2), \tag{6.15}$$

the local initial stresses in the three-dimensional plate can be nonzero. To take into consideration this case, we introduce the term $\varepsilon^{-2}b^{(-2)}_{ijmn} = \varepsilon^{-2}\sigma^{*(-2)}_{jn}\delta_{im}$, corresponding to the resultant moments of the initial stresses.

6.2.2 *Two-Scale Expansions and Homogenized Equilibrium Equations*

We will study global deformations of the plate, as $\varepsilon \to 0$. To do this, we use the following asymptotic expansion:

asymptotic expansion for displacements

$$\mathbf{u}^{\varepsilon} = \mathbf{u}^{(0)}(x_1, x_2) + \sum_{k=1}^{\infty} \varepsilon^k \mathbf{u}^{(k)}(x_1, x_2, \mathbf{y}), \tag{6.16}$$

asymptotic expansion for stresses

$$\sigma^{\varepsilon}_{ij} = \sum_{p=-3}^{\infty} \varepsilon^p \sigma^{(p)}_{ij}(x_1, x_2, \mathbf{y}). \tag{6.17}$$

The functions on the right-hand side of equalities (6.16) and (6.17) are assumed to be periodic in y_1, y_2 with periodicity cell T. Note that the leading terms in the series (6.16) depends only on "slow" variable, thus, the deflection of plate have limit value. The leading term in the series (6.17) is highly oscillating, as the result, the stresses do not have limit values (as $\varepsilon \to 0$, fast variables takes the impossible form "$\frac{x}{0}$"). The average values of stresses $\sigma^{\varepsilon}_{ij}$, as well as moments of these stresses, have limit values, as $\varepsilon \to 0$.

The differential operators are represented in the form of the sum of operators in the "slow" and "fast" variables. For function $Z(x_1, x_2, \mathbf{y})$ of the arguments x_1, x_2 and $\mathbf{y} = (y_1, y_2, y_3)$, like on the right-hand sides of (6.16) and (6.17), the differential operators are represented in the form (compare with the differentiation rule (6.18) for uniform plate)

$$\frac{\partial Z}{\partial x_\alpha} \to Z_{,\alpha x} + \varepsilon^{-1} Z_{,\alpha y} \quad (\alpha = 1, 2), \quad \frac{\partial Z}{\partial x_3} \to \varepsilon^{-1} Z_{,3y}. \tag{6.18}$$

Here and below in this chapter, the Latin indexes take the values 1, 2, 3 and Greek indexes take the values 1, 2; αx means $\partial/\partial x_\alpha$ and , $3y$ means $\partial/\partial y_3$.

The analysis of problem (6.10)–(6.13) breaks down into two stages. The first entails obtaining the equations of equilibrium for the plate considered as a two-dimensional structure. This stage does not involve local governing equations and it is the same stage for any governing equations.

Proposition 6.1 *Let function $Z(x_1, x_2, \mathbf{y})$ have continuous derivatives with respect to variable y_1, y_2 and it is periodic with respect to the variable y_1, y_2 with periodicity cell T. Then,*

$$\varepsilon^{-1}\int_{G_\varepsilon} Z(x_1, x_2, \mathbf{x}/\varepsilon)d\mathbf{x} \to \int_D \langle Z\rangle(x_1, x_2)dx_1dx_2,$$

$$\int_{S_\varepsilon} Z(x_1, x_2, \mathbf{x}/\varepsilon)d\mathbf{x} \to \int_D \langle Z\rangle_S(x_1, x_2)dx_1dx_2, \ \textit{as}\ \varepsilon \to 0.$$

It is denoted

$$\langle\rangle = \frac{1}{|T|}\int_Y d\mathbf{y}, \quad \langle\rangle_S = \frac{1}{|T|}\int_S d\mathbf{y}$$

the average values of function over periodic cell Y and its lateral surface S in "fast" variables $\mathbf{y}$, correspondingly, see Fig. 6.5 (right).

The proof of Proposition 6.1 is similar to the proof of Proposition 3.1 (see note after Proposition 3.1). Another proof can be found in Caillerie (1984).

To obtain the homogenized equilibrium equations, consider a weak form of problem (6.10)–(6.13)

$$-\int_{G_\varepsilon} \sigma_{ij}^{\varepsilon}\frac{\partial v_i}{\partial x_j}d\mathbf{x} + \int_{S_\varepsilon} \mathbf{h}^{\varepsilon}\mathbf{v}d\mathbf{x} + \int_{G_\varepsilon} \mathbf{f}^{\varepsilon}\mathbf{v}d\mathbf{x} = 0 \text{ for any } \mathbf{v} \in V_\varepsilon, \tag{6.19}$$

where

$$V_\varepsilon = \{\mathbf{v} \in H^1(G_\varepsilon)^3 : \mathbf{v}(\mathbf{x}) = 0 \text{ on } S_u\}$$

is the set of admissible displacements of the plate considered as a three-dimensional elastic body.

Substituting (6.17) in (6.19), we obtain, with allowance for (6.18)

$$\begin{gathered}-\sum_{p=-3}^{\infty}\int_{G_\varepsilon} \varepsilon^p\sigma_{ij}^{(p)}(v_{i,\alpha x}\delta_{j\alpha} + \varepsilon^{-1}v_{i,jy})d\mathbf{x} + \int_{S_\varepsilon} \mathbf{h}^{\varepsilon}\mathbf{v}d\mathbf{x} \\ + \int_{G_\varepsilon} \mathbf{f}^{\varepsilon}\mathbf{v}d\mathbf{x} = 0 \text{ for any } \mathbf{v} \in V_\varepsilon.\end{gathered} \tag{6.20}$$

We use functional space

$$U = \{H^{01}(D)\}^3 = \{\mathbf{v}(x_1, x_2) \in H^1(D)^3 : \mathbf{v}(x_1, x_2) = 0 \text{ on } \partial D\},$$

which is the set of admissible displacements of the plate considered as two-dimensional object. The functional space is defined as closure of finite $C^\infty(D)$ functions in the norm (3.6). It is evident that $U \subset V_\varepsilon$.

We take $\mathbf{v} = \mathbf{v}(x_1, x_2) \in U$. For this choice (6.20) takes the following form:

$$-\sum_{p=-3}^{\infty}\int_{G_\varepsilon}\varepsilon^p\sigma_{i,\alpha}^{(p)}v_{i,\alpha x}d\mathbf{x}+\int_{S_\varepsilon}\mathbf{h}^\varepsilon\mathbf{v}d\mathbf{x}+\int_{G_\varepsilon}\mathbf{f}^\varepsilon\mathbf{v}d\mathbf{x}=0 \tag{6.21}$$

for any $\mathbf{v}\in U$.

Taking into account Proposition 6.1, we obtain form (6.21) the following equality:

$$\begin{aligned}-\sum_{p=-3}^{\infty}\int_{D}\varepsilon^{p+1}\langle\sigma_{i\alpha}^{(p)}\rangle v_{i,\alpha x}dx_1dx_2+\int_{D}\langle\mathbf{h}^\varepsilon\rangle_S\mathbf{v}dx_1dx_2\\+\int_{D}\langle\mathbf{f}^\varepsilon\rangle\mathbf{v}dx_1dx_2=0\text{ for any }\mathbf{v}\in U.\end{aligned} \tag{6.22}$$

By virtue of Proposition 6.1, the average values of $\langle\sigma_{ij}^{(p)}\rangle$ in Eq. (6.22) is computed over the periodicity cell Y, not over the thickness of the plate. This suggests that to continue the analysis of the problem, one has to determine the resultant forces in the plane using the formula

$$N_{ij}^{(p)}=\langle\sigma_{ij}^{(p)}\rangle. \tag{6.23}$$

Substituting $\langle\sigma_{ij}^{(p)}\rangle$ for $N_{ij}^{(p)}$ in (6.22), we obtain

$$\begin{aligned}-\sum_{p=-3}^{\infty}\int_{D}\varepsilon^{p+1}\langle N_{i\alpha}^{(p)}\rangle v_{i,\alpha x}dx_1dx_2+\int_{D}\langle\mathbf{h}^\varepsilon\rangle_S\mathbf{v}dx_1dx_2\\+\int_{D}\langle\mathbf{f}^\varepsilon\rangle\mathbf{v}dx_1dx_2=0\text{ for any }\mathbf{v}\in U.\end{aligned} \tag{6.24}$$

Integrating (6.24) by parts and equating the terms corresponding to the identical powers of ε, we obtain an infinite sequence of equilibrium equations for the resultant in-plane forces ($i, j = 1, 2, 3$ and $\alpha, \beta = 1, 2$):

$$N_{\alpha\beta,\beta x}^{(-3)}=0, \tag{6.25}$$

$$N_{\alpha\beta,\beta x}^{(-2)}+\langle f_\alpha^\varepsilon\rangle+\langle h_\alpha^\varepsilon\rangle_S=0, \tag{6.26}$$

$$N_{3\beta,\beta x}^{(-1)}+\langle f_3^\varepsilon\rangle+\langle h_3^\varepsilon\rangle_S=0. \tag{6.27}$$

Now take the trial function $\mathbf{v}=\mathbf{v}(x_1,x_2)y_3$, where $\mathbf{v}(x_1,x_2)\in U$. For this function, (6.20) takes the following form:

$$\begin{aligned}-\sum_{p=-3}^{\infty}\int_{G_\varepsilon}\varepsilon^p\sigma_{ij}^{(p)}(v_{i,\alpha x}\delta_{j\alpha}y_3+\varepsilon^p v_i\delta_{j3})d\mathbf{x}+\int_{S_\varepsilon}\mathbf{h}^\varepsilon\mathbf{v}y_3d\mathbf{x}\\+\int_{G_\varepsilon}\mathbf{f}^\varepsilon\mathbf{v}y_3d\mathbf{x}=0\text{ for any }\mathbf{v}\in U.\end{aligned} \tag{6.28}$$

With regard to Proposition 6.1, we obtain from (6.28) the following equality:

$$-\sum_{p=-3}^{\infty}\int_D(\varepsilon^{p+1}\langle\sigma_{i\alpha}^{(p)}y_3\rangle v_{i,\alpha x}+\varepsilon^p\langle\sigma_{i3}^{(p)}\rangle v_i)dx_1dx_2$$
$$+\int_D\langle\mathbf{h}^\varepsilon y_3\rangle_S\mathbf{v}dx_1dx_2+\int_D\langle\mathbf{f}^\varepsilon y_3\rangle\mathbf{v}dx_1dx_2=0\text{ for any }\mathbf{v}\in U. \tag{6.29}$$

Equality (6.29) indicates that the resultant moments must be defined as

$$M_{\alpha\beta}^{(p)}=-\langle\sigma_{\alpha\beta}^{(p)}y_3\rangle. \tag{6.30}$$

Changing in (6.29) $\langle\sigma_{\alpha\beta}^{(p)}\rangle$ and $\langle\sigma_{\alpha\beta}^{(p)}y_3\rangle$ in accordance with (6.23) and (6.30), we obtain

$$-\sum_{p=-3}^{\infty}\int_D(\varepsilon^{p+1}M_{i\alpha}^{(p)}v_{i,\alpha x}+\varepsilon^pN_{i3}^{(p)}v_i)dx_1dx_2+$$
$$+\int_D\langle\mathbf{h}^\varepsilon y_3\rangle_S\mathbf{v}dx_1dx_2+\int_D\langle\mathbf{f}^\varepsilon y_3\rangle\mathbf{f}dx_1dx_2=0\text{ for any }\mathbf{v}\in U. \tag{6.31}$$

Integrating by parts in (6.31) and equating the terms corresponding to the identical power of ε, we obtain the next equations ($i, j = 1, 2, 3$ and $\alpha, \beta = 1, 2$):

$$M_{\alpha\beta,\beta x}^{(-3)}-N_{\beta3}^{(-2)}=0; \tag{6.32}$$

$$M_{\alpha\beta,\beta x}^{(-2)}-N_{\beta3}^{(-1)}+\langle f_\alpha^\varepsilon y_3\rangle+\langle h_\alpha^\varepsilon y_3\rangle_S=0; \tag{6.33}$$

$$N_{i3}^{(-3)}=0. \tag{6.34}$$

Although Eqs. (6.25), (6.26) and (6.32), (6.33) coincide in form with the classical equilibrium equations of the plate theory (Love AEH 1929; Timoshenko and Woinowsky-Krieger 1959). The resultant in-plane forces (6.23) and the resultant moments (6.30) in the equilibrium equations (6.25) and (6.32)–(6.34) are determined in not classical way.

Substituting (6.17) in (6.10), we obtain, with allowance (6.18):

$$\sum_{p=-3}^{\infty}\varepsilon^p(\sigma_{i\alpha,\alpha x}^{(p)}+\varepsilon^{-1}\sigma_{ij,jy}^{(p-1)})+f_i^\varepsilon=0. \tag{6.35}$$

Equating the terms with identical power in (6.35), we obtain the following sequence of local equations

$$\sigma^{(-3)}_{ij,jy} = 0 \text{ in } Y \quad (p = -3);$$
$$\sigma^{(-2)}_{ij,jy} + \sigma^{(-3)}_{i\beta,\beta x} = 0 \text{ in } Y \quad (p = -2); \tag{6.36}$$
$$\sigma^{(p)}_{ij} n_j = 0 \text{ on } S \ (p = -3, -2, ...).$$

We will not use the equations corresponding to $p = -1, 0, \ldots$.

6.2.3 The Homogenized Governing Equations

The second stage of analysis of the problem consists of obtaining the homogenized (limit) governing equations for the plate. If the previous stage did not involve the local governing equations, this stage essentially involves the local governing equations. We consider initial stresses corresponding to in-plane stresses. In this case, (6.14) takes the form

$$\sigma^{\varepsilon}_{ij} = \left(\varepsilon^{-3} c_{ijmn}(\mathbf{y}) + \varepsilon^{-1} b_{ijmn}(x_1, x_2, \mathbf{y}) \right) \frac{\partial u^{\varepsilon}_m}{\partial x_n}, \tag{6.37}$$

with

$$b_{ijmn}(x_1, x_2, \mathbf{x}/\varepsilon) = \sigma^{*(-1)}_{jn}(x_1, x_2, \mathbf{x}/\varepsilon)\delta_{im}. \tag{6.38}$$

Substituting (6.16) and (6.17) in (6.37), we obtain, with allowance for the differentiation rule (6.18),

$$\sum_{p=-3}^{\infty} \varepsilon^p \sigma^{(p)}_{ij} = \sum_{k=0}^{\infty} \varepsilon^k (\varepsilon^{-3} c_{ijmn}(\mathbf{y}) + \varepsilon^{-1} b_{ijmn}(x_1, x_2, \mathbf{y})) \times (u^{(k)}_{m,\beta x} \delta_{m\beta} + \varepsilon^{-1} u^{(k)}_{m,ny}). \tag{6.39}$$

Equating in (6.39) the terms with identical power of ε, we obtain in particular,

$$\sigma^{(p)}_{ij} = c_{ijm\beta}(\mathbf{y}) u^{(p+3)}_{m,\beta x} + c_{ijmn}(\mathbf{y}) u^{(p+2)}_{m,ny} \ (p = -3, -2); \tag{6.40}$$

$$\sigma^{(-1)}_{ij} = c_{ijm\beta}(\mathbf{y}) u^{(2)}_{m,\beta x} + c_{ijmn}(\mathbf{y}) u^{(3)}_{m,ny} + b_{ijmn}(x_1, x_2, \mathbf{y}) u^{(1)}_{m,ny} + b_{ijm\beta}(x_1, x_2, \mathbf{y}) u^{(0)}_{m,\beta x}. \tag{6.41}$$

As was assumed above,

$$\mathbf{u}^{(k)}(\mathbf{y}) \text{ is periodic in } y_1, y_2 \text{ with periodicity cell } T \ (k = 1, 2, ...). \tag{6.42}$$

Let us consider the problem (6.36, $p = -3$) with (6.40, $p = -3$). It can be written in the form

$$\begin{cases} (c_{ijmn}(\mathbf{y})u^{(1)}_{m,ny} + c_{ijm\alpha}(\mathbf{y})u^{(0)}_{m,\alpha x}(x_1, x_2))_{,jy} = 0 \text{ in } G_\varepsilon; \\ (c_{ijmn}(\mathbf{y})u^{(1)}_{m,ny} + c_{ijm\alpha}(\mathbf{y})u^{(0)}_{m,\alpha x}(x_1, x_2))n_j = 0 \text{ on } S_\varepsilon; \end{cases} \tag{6.43}$$

n means a normal to S, see Fig. 6.5 ($\alpha, \beta = 1, 2$; $m = 1, 2, 3$).

Allowing for the fact that the function $\mathbf{u}^{(0)}(x_1, x_2)$ plays the role of a parameter in the problems in the "fast" variable **y**, solution to the problem (6.43) can be found in the form,

$$\mathbf{u}^{(1)} = \mathbf{N}^{0m\alpha}(\mathbf{y})u^{(0)}_{m,\alpha x}(x_1, x_2) + \mathbf{V}(x_1, x_2). \tag{6.44}$$

Function $\mathbf{V}(x_1, x_2)$ from (6.44) will be determined below, and $\mathbf{N}^{0m\alpha}(\mathbf{y})$ is the solution to the following periodicity cell problem:

$$\begin{cases} (c_{ijmn}(\mathbf{y})N^{0m\alpha}_{m,ny} + c_{ijm\alpha}(\mathbf{y}))_{,jy} = 0 \text{ in } Y; \\ (c_{ijmn}(\mathbf{y})N^{0m\alpha}_{m,ny} + c_{ijm\alpha}(\mathbf{y}))n_j = 0 \text{ on } S; \\ \mathbf{N}^{0m\alpha}(\mathbf{y}) \text{ is periodic in } y_1, y_2 \text{ with periodicity cell } T. \end{cases} \tag{6.45}$$

Proposition 6.2 *For $m = 3$, the solution to the periodicity cell problem (6.45) is given by the following formula:*

$$\mathbf{N}^{03\alpha}(\mathbf{y}) = -y_3\mathbf{e}_\alpha, \tag{6.46}$$

where $\{\mathbf{e}_i\}$ are basis vectors of the coordinate system.

To prove the proposition, it is sufficient to substitute $\mathbf{N}^{03\alpha}(\mathbf{y})$ (6.46) into (6.45) and use the symmetry of the elastic constants c_{ijmn}.

Due to Proposition 6.2, the following non-trivial periodicity cell problems remain from the problems (6.45) ($\alpha, \beta = 1, 2$):

$$\begin{cases} (c_{ijmn}(\mathbf{y})N^{0\beta\alpha}_{m,ny} + c_{ijm\alpha}(\mathbf{y}))_{,jy} = 0 \text{ in } Y; \\ (c_{ijmn}(\mathbf{y})N^{0\beta\alpha}_{m,ny} + c_{ijm\alpha}(\mathbf{y}))n_j = 0 \text{ on } S; \\ \mathbf{N}^{0\beta\alpha}(\mathbf{y}) \text{ is periodic in } y_1, y_2 \text{ with periodicity cell } T. \end{cases} \tag{6.47}$$

Taking into account equality (6.46), we can write (6.44) as

$$\mathbf{u}^{(1)} = \mathbf{N}^{0\beta\alpha}(\mathbf{y})u^{(0)}_{\beta,\alpha x}(x_1, x_2) - y_3\mathbf{e}_\alpha u^{(0)}_{3,\alpha x}(x_1, x_2) + \mathbf{V}(x_1, x_2). \tag{6.48}$$

Substituting (6.48) in (6.40, $p = -3$) gives

$$\sigma^{(-3)}_{ij} = (c_{ij\alpha\beta}(\mathbf{y}) + c_{ijkl}(\mathbf{y})N^{0\alpha\beta}_{k,ly}(\mathbf{y}))u^{(0)}_{\alpha,\beta x}.$$

Averaging this equality over periodicity cell Y for $ij = \gamma\delta$, we obtain in accordance with Definition (6.23)

$$N_{\gamma\delta}^{(-3)} = A_{\gamma\delta\alpha\beta} u_{\alpha,\beta x}^{(0)}, \tag{6.49}$$

where

$$A_{\gamma\delta\alpha\beta}^{0} = \langle c_{\gamma\delta\alpha\beta}(\mathbf{y}) + c_{\gamma\delta kl}(\mathbf{y}) N_{k,ly}^{0\alpha\beta}(\mathbf{y})\rangle. \tag{6.50}$$

In accordance with the homogenized equilibrium equation (6.25, $p = -3$) from (6.49), we obtain

$$(A_{\gamma\delta\alpha\beta}^{0} u_{\alpha,\beta x}^{(0)})_{,\delta x} = 0 \text{ in } D. \tag{6.51}$$

The boundary condition for $u_{\alpha}^{(0)}$ ($\alpha = 1, 2$) follows from (6.14) and (6.16) and has the form

$$u_{\alpha}^{(0)}(x_1, x_2) = 0 \text{ on } \partial D. \tag{6.52}$$

The solution to problem (6.51), (6.52) is $u_{\alpha}^{(0)} = 0$ ($\alpha = 1, 2$). Then (6.48) takes a form

$$\mathbf{u}^{(1)} = -y_3 \mathbf{e}_{\alpha} u_{3,\alpha x}^{(0)}(x_1, x_2) + \mathbf{V}(x_1, x_2), \tag{6.53}$$

and equalities (6.40, $p = -3, -2$) take the form

$$\sigma_{ij}^{(-3)} = 0; \tag{6.54}$$

$$\begin{aligned} \sigma_{ij}^{(-2)} = {} & c_{ijmn}(\mathbf{y}) u_{m,ny}^{(2)} - c_{ij\alpha\beta}(\mathbf{y}) y_3 u_{3,\alpha x\beta x}^{(0)}(x_1, x_2) \\ & + c_{ijm\beta}(\mathbf{y}) V_{m,\beta x}(x_1, x_2). \end{aligned} \tag{6.55}$$

Note that due to (6.54) the equality $N_{i3}^{(-3)} = 0$ in (6.34) is satisfied identically.

Let us examine problem (6.36, $p = -2$) with $\sigma_{ij}^{(-2)}$ given by (6.55). It can be written in the form

$$\begin{cases} (c_{ijmn}(\mathbf{y}) u_{m,ny}^{(2)} - c_{ij\alpha\beta}(\mathbf{y}) y_3 u_{3,\alpha x\beta x}^{(0)}(x_1, x_2) \\ + c_{ijm\beta}(\mathbf{y}) V_{m,\beta x}(x_1, x_2))_{,jy} = 0 \text{ in } G_{\varepsilon}; \\ (c_{ijmn}(\mathbf{y}) u_{m,ny}^{(2)} - c_{ij\alpha\beta}(\mathbf{y}) y_3 u_{3,\alpha x\beta x}^{(0)}(x_1, x_2) \\ + c_{ijm\beta}(\mathbf{y}) V_{m,\beta x}(x_1, x_2)) n_j = 0 \text{ on } S_{\varepsilon}. \end{cases} \tag{6.56}$$

In order to solve problems of this kind, we introduce the periodicity cell problem, corresponding to bending of a plate

$$\begin{cases} (c_{ijmn}(\mathbf{y})N^{1\alpha\beta}_{m,ny} - y_3 c_{ij\alpha\beta}(\mathbf{y}))_{,jy} = 0 \text{ in } Y; \\ (c_{ijmn}(\mathbf{y})N^{1\alpha\beta}_{m,ny} - y_3 c_{ij\alpha\beta}(\mathbf{y}))n_j = 0 \text{ on } S; \\ \mathbf{N}^{1\alpha\beta}(\mathbf{y}) \text{ is periodic in } y_1, y_2 \text{ with periodicity cell } T. \end{cases} \tag{6.57}$$

The solution to problem (6.56) can be expressed through the functions $\mathbf{N}^{\nu\alpha\beta}(\mathbf{y})$ $(\nu = 0, 1)$ – solutions to the periodicity cell problems (6.47) and (6.57), as follows:

$$\begin{aligned} \mathbf{u}^{(2)} &= \mathbf{N}^{0\alpha\beta}(\mathbf{y})u^{(0)}_{3,\alpha x\beta x}(x_1, x_2) \\ -y_3 V_{3,\beta x}(x_1, x_2)\mathbf{e}_\beta &+ \mathbf{N}^{1\alpha\beta}(\mathbf{y})V_{\alpha,\beta x}(x_1, x_2). \end{aligned} \tag{6.58}$$

Here we use that $N^{03\alpha}(\mathbf{y}) = -y_3\mathbf{e}_\alpha$, see Proposition 6.2.

Substituting (6.58) in (6.55), we obtain

$$\begin{aligned} \sigma^{(-2)}_{ij} &= (c_{ijmn}(\mathbf{y})N^{0\alpha\beta}_{m,ny} + c_{ij\alpha\beta}(\mathbf{y}))V_{\alpha,\beta x}(x_1, x_2) \\ &+(c_{ijmn}(\mathbf{y})N^{1\alpha\beta}_{m,ny} - y_3 c_{ij\alpha\beta}(\mathbf{y}))u^{(0)}_{3,\alpha x\beta x}(x_1, x_2). \end{aligned} \tag{6.59}$$

Note that V_3 appears in (6.58), but it is absent in (6.59). This is because V_3 appears in (6.58) multiplied by the function $y_3\mathbf{e}_\beta$, which plays a special role in the asymptotic theory of plates.

Averaging (6.59) with $ij = \gamma\delta$ $(\gamma, \delta = 1, 2)$ over periodicity cell Y, we obtain the homogenized governing equations for the resultant in-plane forces

$$N^{(-2)}_{\gamma\delta} = A^0_{\gamma\delta\alpha\beta}V_{\alpha,\beta x} + A^1_{\gamma\delta\alpha\beta}u^{(0)}_{3,\alpha x\beta x}. \tag{6.60}$$

Multiplying (6.59) with $ij = \gamma\delta$ ($\gamma, \delta = 1, 2$) by $-y_3$ and averaging over periodicity cell Y, we obtain the following homogenization governing equation for the resultant moments

$$M^{(-2)}_{\gamma\delta} = A^1_{\gamma\delta\alpha\beta}V_{\alpha,\beta x} + A^2_{\gamma\delta\alpha\beta}u^{(0)}_{3,\alpha x\beta x}. \tag{6.61}$$

The coefficients are

$$A^{\mu+\nu}_{\gamma\delta\alpha\beta} = \langle((-1)^\nu c_{\gamma\delta\alpha\beta}y_3^\nu + c_{\gamma\delta mn}(\mathbf{y})N^{\nu\alpha\beta}_{m,ny})(-1)^\mu y_3^\mu\rangle \tag{6.62}$$

$(\nu, \mu = 0, 1; \alpha, \beta, \gamma, \delta = 1, 2)$.

The obtained Eqs. (6.60) and (6.61) are governing equations of plate considered as a two-dimensional structure. The coefficients $A^{\mu+\nu}_{\gamma\delta\alpha\beta}$ are the homogenized stiffnesses of the plate: $A^0_{\gamma\delta\alpha\beta}$ are the in-plane homogenized stiffnesses; $A^1_{\gamma\delta\alpha\beta}$ are the out-of-plane (asymmetric) homogenized stiffnesses; $A^2_{\gamma\delta\alpha\beta}$ are the bending homogenized stiffnesses. The homogenized stiffnesses are expressed through solutions to periodicity cell problems (6.47) and (6.57). As a result, for example, we cannon guaranty that the out-of-plane (asymmetric) homogenized stiffnesses $A^1_{\gamma\delta\alpha\beta}$ will be equal to zero.

6.2.4 Elimination of Shear Forces for Plate with Initial Stresses

In the general case of non-homogeneous plates considered here, we encounter the same problem of asymmetry of the resultant forces $N^{(-1)}_{ij}$ with respect to i and j that was discussed in Chap. 3 devoted to uniform plates, see Sect. 5.3.3.

In the plate with initial stresses, the resultant forces $N^{(-1)}_{ij}$ *are not symmetrical* in i and j due to (6.41) because b_{ijmn} are not symmetric in i and j, see (6.13). As a result, the classical method of elimination of shear forces does not work. To eliminate shear forces, we use the technique developed in Sect. 5.3.3.

Let us insert $\mathbf{u}^{(1)}$ into (6.41) in accordance with (6.53). Then we obtain

$$\begin{aligned}\sigma^{(-1)}_{ij} &= c_{ijm\beta}(\mathbf{y})u^{(2)}_{m,\beta x} + c_{ijmn}(\mathbf{y})u^{(3)}_{m,ny} \\ -b_{ij\beta 3}(x_1, x_2, \mathbf{y})&u^{(0)}_{3,\beta x}(x_1, x_2) + b_{ij3\beta}(x_1, x_2, \mathbf{y})u^{(0)}_{3,\beta x}(x_1, x_2).\end{aligned} \tag{6.63}$$

The first and second terms on the right-hand side of equation (6.63) are symmetrical with respect to i and j by virtue of the symmetry of the elastic constants. Then the asymmetrical part of the tensor $\sigma^{(-1)}_{ij}$ are

$$\begin{aligned}&\sigma^{(-1)}_{ij} - \sigma^{(-1)}_{ji} \\ &= -b_{ij\beta 3}(x_1, x_2, \mathbf{y})u^{(0)}_{3,\beta x}(x_1, x_2) + b_{ij3\beta}(x_1, x_2, \mathbf{y})u^{(0)}_{3,\beta x}(x_1, x_2) \\ &+b_{ji\beta 3}(x_1, x_2, \mathbf{y})u^{(0)}_{3,\beta x}(x_1, x_2) - b_{ji\beta 3}(x_1, x_2, \mathbf{y})u^{(0)}_{3,\beta x}(x_1, x_2).\end{aligned} \tag{6.64}$$

Averaging (6.64) over periodicity cell Y, we obtain

$$\begin{aligned}&N^{(-1)}_{ij} - N^{(-1)}_{ji} \\ &= -\langle b_{ij\beta 3}(x_1, x_2, \mathbf{y})\rangle u^{(0)}_{3,\beta x}(x_1, x_2) + \langle b_{ij3\beta}(x_1, x_2, \mathbf{y})\rangle u^{(0)}_{3,\beta x}(x_1, x_2) \\ &+\langle b_{ji\beta 3}(x_1, x_2, \mathbf{y})\rangle u^{(0)}_{3,\beta x}(x_1, x_2) - \langle b_{ji\beta 3}(x_1, x_2, \mathbf{y})\rangle u^{(0)}_{3,\beta x}(x_1, x_2).\end{aligned} \tag{6.65}$$

For $i = 3$ and $j = \alpha$, we obtain from (6.65) and definition of b_{ijmn}, see (6.65):

$$N^{(-1)}_{3\alpha} - N^{(-1)}_{\alpha 3} = K_\alpha, \tag{6.66}$$

where K_γ is defined by the follows formula:

$$K_\alpha = N_{33}^{*(-1)} u_{3,\alpha x}^{(0)}(x_1, x_2) - N_{\alpha\beta}^{*(-1)} u_{3,\beta x}^{(0)}(x_1, x_2). \tag{6.67}$$

It is seen that the symmetric part K_γ of shear forces is expressed through the normal deflection $u_3^{(0)}$. In other words, (6.67) can be treated as a governing equation which relates K_γ and $u_3^{(0)}$. There is no analog of this governing equation in the theory of plates with no initial stresses.

Proposition 6.3 *Let the initial stresses $\sigma_{ij}^{*(-1)}$ satisfy the equilibrium equations,*

$$\frac{\partial \sigma_{ij}^{*(-1)}}{\partial x_j} + \varepsilon^a F_i^\varepsilon = 0 \text{ in } Q_\varepsilon, \ \sigma_{ij}^{*(-1)} n_j = H_i^\varepsilon \text{ on } S_\varepsilon. \tag{6.68}$$

If $a \neq 1$ and $\mathbf{H}^\varepsilon = 0$, then $\langle b_{ijmn} Z_{,ny}\rangle = 0$ for every differentiable function $Z(\mathbf{y})$ periodic in y_1, y_2 with periodicity cell T. In particular, $\langle \sigma_{i3}^{*(-1)}\rangle = 0$, $\langle \sigma_{i3}^{*(-1)} y_3\rangle = 0$.

Taking into account the definition of b_{ijmn} (6.38), it sufficient to prove that $\langle \sigma_{jn}^{*(-1)} Z_{,ny}\rangle = 0$. From (6.68), taking into account the differentiation rule (6.18), we obtain

$$\begin{cases} \sigma_{ij,jy}^{*(-1)} + \mathbb{F}_i = 0 \text{ in } Y \ (\mathbb{F}_i = 0 \text{ if } a \neq 1, \mathbb{F}_i = F_i \text{ if } a = -1), \\ \sigma_{ij}^{*(-1)}(x_1, x_2, \mathbf{y}) \text{ periodic in } y_1, y_2 \text{ with periodicity cell } T. \end{cases} \tag{6.69}$$

Let us consider the value $\langle \sigma_{jn}^{*(-1)} Z_{,ny}\rangle$. Taking into account the definition of the average over the periodicity cell Y, and integrating by parts over Y, we find that this value is equal to

$$\begin{aligned} &\int_Y \sigma_{jn,ny}^{*(-1)} Z(\mathbf{y}) d\mathbf{y} + \int_S \sigma_{jn}^{*(-1)} n_n Z(\mathbf{y}) d\mathbf{y} + \int_\omega \sigma_{j\gamma}^{*(-1)} n_\gamma Z(\mathbf{y}) d\mathbf{y} \\ &\quad = \int_Y F_j Z(\mathbf{y}) d\mathbf{y} + \int_S h_j Z(\mathbf{y}) d\mathbf{y}. \end{aligned} \tag{6.70}$$

Here ω means the opposite sides of the periodicity cell Y, which (sides) are perpendicular to Oy_3-axis (see Fig. 6.5). The integrals over Y as well as the right-hand part of the Eq. (9.64) are zero due to (6.69) (if $a \neq -1$ and $\mathbf{H}^\varepsilon = 0$). Integral over ω is zero by virtue of the periodicity of the functions σ_{jn}^* and Z, and anti-periodicity of the normal-vector on ω (see Fig. 6.5). This proves the proposition.

6.2.5 *Homogenized Model of Plate with Initial Stresses*

Now, we can exclude the quantities $N_{3\gamma}^{(-1)}$ and $N_{\gamma 3}^{(-1)}$ from the equilibrium equations (6.27) and (6.33) and obtain the closed two-dimensional model of plate.

We consider the case when the conditions of Proposition 6.3 are satisfied. In this case,

$$K_\gamma = -N^*_{\gamma\beta} u^{(0)}_{3,\beta x}(x_1, x_2). \tag{6.71}$$

Write (6.27)

$$N^{(-1)}_{3\alpha,\alpha x} + \langle f^\varepsilon_3 \rangle + \langle h^\varepsilon_3 \rangle_S = 0. \tag{6.72}$$

Differentiating (6.33)

$$M^{(-2)}_{\alpha\beta,\beta x\alpha x} - N^{(-1)}_{\alpha 3,\alpha x} + (\langle f^\varepsilon_\alpha y_3 \rangle + \langle h^\varepsilon_\alpha y_3 \rangle_S)_{,\alpha x} = 0. \tag{6.73}$$

From (6.66) $N_{3\alpha,\alpha x} - N_{\alpha 3,\alpha x} = \langle K_{\alpha,\alpha x} \rangle$. Using this equality, we obtain

$$M^{(-2)}_{\alpha\beta,\beta x\alpha x} - N^{(-1)}_{3\alpha,\alpha x} + K_{\alpha,\alpha x} + (\langle f^\varepsilon_\alpha y_3 \rangle + \langle h^\varepsilon_\alpha y_3 \rangle_S)_{,\alpha x} = 0. \tag{6.74}$$

Using (6.72), we obtain from (6.74) the following equation:

$$M^{(-2)}_{\alpha\beta,\beta x\alpha x} + \langle f^\varepsilon_3 \rangle + \langle h^\varepsilon_3 \rangle_S + K_{\alpha,\alpha x} + (\langle f^\varepsilon_\alpha y_3 \rangle + \langle h^\varepsilon_\alpha y_3 \rangle_S)_{,\alpha x} = 0. \tag{6.75}$$

Using the homogenized governing equation (6.61), we can rewrite (6.75) as follows:

$$\begin{aligned} &(A^1_{\gamma\delta\alpha\beta} V_{\alpha,\beta x} + A^2_{\gamma\delta\alpha\beta} u^{(0)}_{3,\alpha x\beta x})_{,\delta x} \\ &+\langle f^\varepsilon_3 \rangle + \langle h^\varepsilon_3 \rangle_S + (\langle f^\varepsilon_\delta y_3 \rangle + \langle h^\varepsilon_\delta y_3 \rangle_S)_{,\delta x} = -K_{\delta,\delta x}. \end{aligned} \tag{6.76}$$

Substituting K_δ in accordance with (6.71) we obtain

$$\begin{aligned} &(A^1_{\gamma\delta\alpha\beta} V_{\alpha,\beta x} + A^2_{\gamma\delta\alpha\beta} u^{(0)}_{3,\alpha x\beta x})_{,\delta x} \\ &+\langle f^\varepsilon_3 \rangle + \langle h^\varepsilon_3 \rangle_S + (\langle f^\varepsilon_\delta y_3 \rangle + \langle h^\varepsilon_\delta y_3 \rangle_S)_{,\delta x} = (N^*_{\delta\beta} u^{(0)}_{3,\beta x})_{,\delta x}. \end{aligned} \tag{6.77}$$

From (6.25, $p = -2$) and (6.60) we obtain

$$(A^0_{\gamma\delta\alpha\beta} V_{\alpha,\beta x} + A^1_{\gamma\delta\alpha\beta} u^{(0)}_{3,\alpha x\beta x})_{,\delta x} + \langle f^\varepsilon_3 \rangle + \langle h^\varepsilon_3 \rangle_S = 0. \tag{6.78}$$

Boundary condition

$$V_\alpha(x_1, x_2) = 0 \ (\alpha = 1, 2);$$
$$u_3^{(0)}(x_1, x_2) = \frac{\partial u_3^{(0)}}{\partial \mathbf{n}}(x_1, x_2) = 0 \text{ on } \partial D. \tag{6.79}$$

Consider the case when the additional forces are absent: $\mathbf{f}^\varepsilon = 0, \mathbf{h}^\varepsilon = 0$, and initial stresses are proportional to parameter $\sigma_{ij}^* = \lambda\sigma_{ij}^0$ (λ is the parameter) and satisfy the conditions of Proposition 6.3. For this case, Eqs. (6.77) and (6.78) take the form

$$(A^0_{\gamma\delta\alpha\beta} V_{\alpha,\beta x} + A^1_{\gamma\delta\alpha\beta} u^{(0)}_{3,\alpha x\beta x})_{,\delta x} = 0, \tag{6.80}$$

$$(A^1_{\gamma\delta\alpha\beta} V_{\alpha,\beta x} + A^2_{\gamma\delta\alpha\beta} u^{(0)}_{3,\alpha x\beta x})_{,\delta x} = \lambda(N^*_{\delta\beta} u^{(0)}_{3,\beta x})_{,\delta x}. \tag{6.81}$$

Problems (6.77)–(6.79) and (6.79)–(6.81) coincides with the classical engineering model equations quantitatively (Timoshenko and Woinowsky-Krieger 1959). The methods of computation of the coefficients of the equations in the asymptotic theory and in the engineering theory are different.

The origin of the right-hand part in (6.81) in classical theories was discussed in Sect. 6.2. All the classical theories use Kirchhoff-Love hypothesis, some of them use geometric analysis in addition. In our method, term $(N^*_{\delta\beta} u^{(0)}_{3,\beta x})_{,\delta x}$ arises as a result of loss of symmetries by the quantities h_{ijkl} (6.11) as a consequence of presence of the initial stresses. Our approach does not need any hypothesis or any analysis outside the scope of the asymptotic analysis.

6.3 Local Stresses in Plate with Initial Stresses

In accordance with the general theory (Washizu 1992), the total stresses $\sigma_{ij}^{total}(\mathbf{x})$ in the body with initial stresses are the sum of the initial stresses $\sigma_{ij}^*(\mathbf{x})$ and additional stresses $\sigma_{ij}^\varepsilon(\mathbf{x})$ (Washizu 1992):

$$\sigma_{ij}^{total}(\mathbf{x}) = \sigma_{ij}^*(\mathbf{x}) + \sigma_{ij}^\varepsilon(\mathbf{x}). \tag{6.82}$$

In this chapter, we considered a non-homogeneous plate of periodic structure subjected to resultant in-plane initial forces, which are result of the action of the local initial stresses $\sigma_{ij}^* = \sigma_{ij}^{*(-1)}$, which characteristic value is inversely proportional to the thickness ε of the plate.

In the case under consideration, the additional stresses

$$\sigma_{ij}^\varepsilon = \varepsilon^{-3}\sigma_{ij}^{(-3)}(x_1, \mathbf{y}) + \varepsilon^{-2}\sigma_{ij}^{(-2)}(x_1, \mathbf{y}) + \dots. \tag{6.83}$$

In accordance with (6.54), $\sigma_{ij}^{(-3)} = 0$, thus, the local stresses in the plate are determined by the term $\sigma_{ij}^{(-2)}$ given by (6.59), and we arrive at the following formula for the computation of the total stresses σ_{ij}^{total} in the uniform plate with initial stresses σ_{ij}^{*}:

$$\sigma_{ij}^{total} = \sigma_{ij}^{*(-1)} + (c_{ljmn}(\mathbf{y})N_{m,ny}^{0\alpha\beta} + c_{ij\alpha\beta}(\mathbf{y}))V_{\alpha,\beta x}(x_1, x_2)$$
$$+(c_{ijmn}(\mathbf{y})N_{m,ny}^{1\alpha\beta} - y_3 c_{ij\alpha\beta}(\mathbf{y}))u_{3,\alpha x\beta x}^{(0)}(x_1, x_2). \quad (6.84)$$

We do not discuss the problem of a non-uniform plate subjected to resultant moments of initial stresses. In general, the situation for non-uniform plates with resultant moments of initial stresses is the same as for homogeneous plates subjected to resultant moments of initial stresses, see Sect. 5.4. Namely, resultant moments of initial stresses do not affect the homogenized problem. The corresponding initial stresses do affect the local stresses. Further details can be found in Kolpakov (2004).

6.4 Solution to the Periodicity Cell Problems

For uniform plates, analogs of the periodicity cell problems (6.47) and (6.57) are problems (5.43) and (5.53). The problems (5.43) and (5.53) represent themselves the ordinary differential equations, which may be solved in explicit form. The periodicity cell problems (6.47) and (6.57) are boundary-value problems for differential equations in partial derivatives, which cannot be solved in explicit forms, except few partial problems. When a periodicity cell problem is two-dimensional, or when a periodicity cell problem may be reduced to a two-dimensional problem, the methods of theory of complex variables may by used. The applications of the methods of theory of complex variables to the problems in the theory of composite materials and homogenization may be found in Mityushev and Rogozin (2000), Mityushev (1997), Mityushev and Adler (2002). Note that the approach based on the applications of the methods of theory of complex variables usually requires massive and non-trivial computations, see the papers cited above. When considering the three-dimensional periodicity cell problems or even two-dimensional periodicity cell problems corresponding to bending op plate, one cannot solve the problem in explicit form (except few partial problems). The only way to solve the problem, is the application of the numerical methods. There exist numerous literature on the numerical solution to the periodicity cell problems of the homogenization theory. Mention the papers (Lions 1978; Panasenko 1988; Gloria 2012; Kanit et al. 2003; Panasenko 1990) and books (Mishnaevsky 2007; Barbero 2013; Yvonnet 2019).

We present an example of solution to the periodicity cell problems for plate. We consider a plate reinforced by a periodic system of of unidirectional fibers, see Fig. 6.6a. We present solutions to the periodicity cell problems corresponding to in-plane tension of the plate and the bending of the plate, see Fig. 6.6b, c.

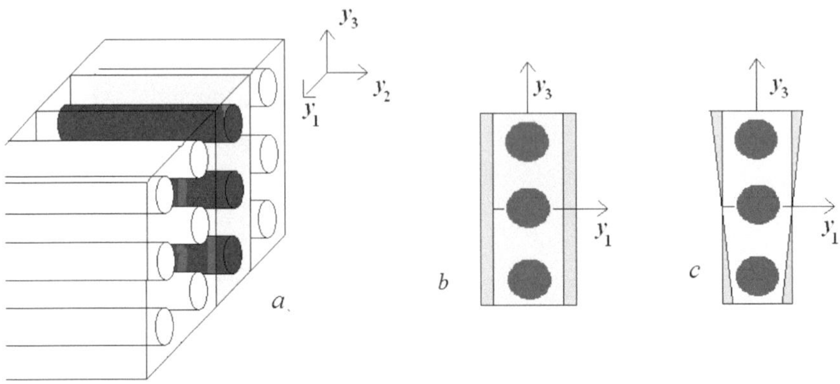

Fig. 6.6 Plate reinforced by a periodic system of of unidirectional fibers

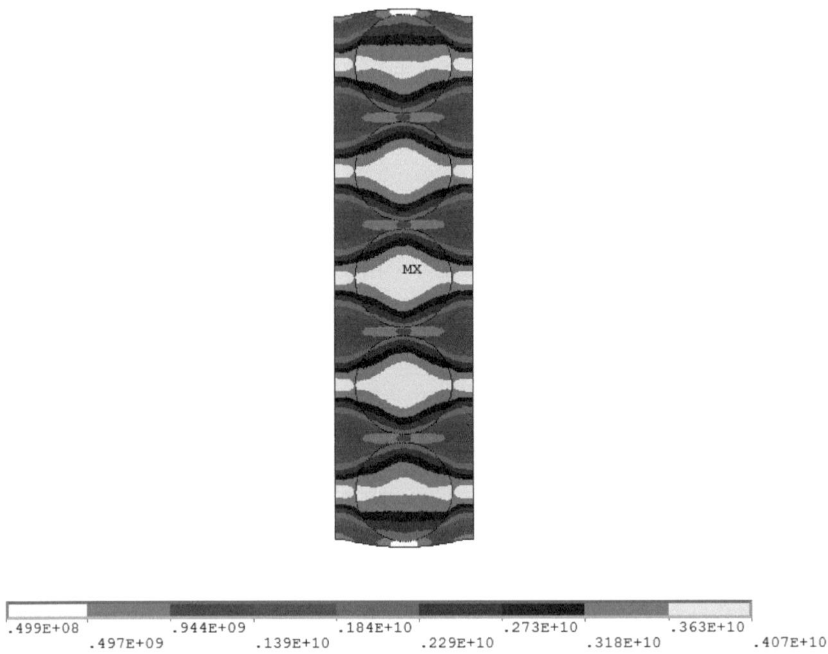

Fig. 6.7 Solutions to the periodicity cell problem corresponding to the in-pane tension

The numerical solution was done by using ANSYS finite-element software. In Figs. 6.7 and 6.8 numerical solutions to the periodicity cell problems for plate are presented. In Fig. 6.7, solutions to the periodicity cell problem corresponding to the in-pane tension is displayed. In Figs. 6.7 and 6.8, solutions to the periodicity cell corresponding to the bending of the plate is displayed. In Figs. 6.7 and 6.8, the stress intensity and deformed periodicity cells are displayed.

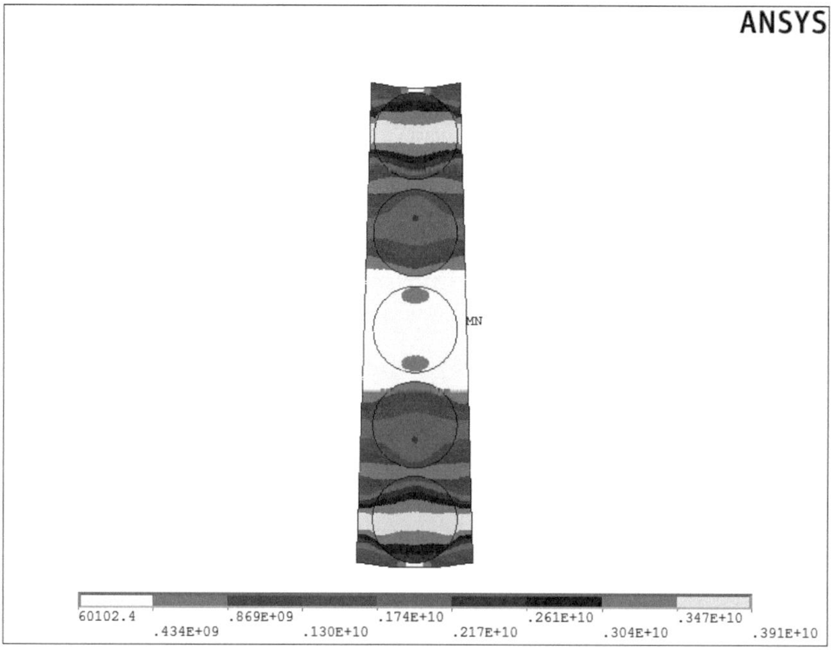

Fig. 6.8 Solutions to the periodicity cell problem corresponding to the bending

References

Andrianov IV, Starushenko GA, Danishevskiy VV, Tokarzewski S (1999) Homogenization procedure and Pade approximants for effective heat conductivity of composite material with cylindrical inclusions having square cross-section. Proc R Soc Ser A 455:3401–3413

Andrianov IV, Kolpakov AA, Faella L (2024) Asymptotic model of a piezoelectric composite beam. J Appl Mech Tech Phys 65(2):345–353

Barbero EJ (2013) Finite element analysis of composite materials using ANSYS. CRC Press, Boca Raton, FL

Bensoussan A, Lions J-L, Papanicolaou G (1978) Asymptotic analysis for periodic structures. North-Holland, Amsterdam

Brischetto S, Cesare D (2024) 3D electro-elastic static analysis of advanced plates and shells. Int J Mech Sci 280:109620

Buannic N, Cartraud P, Quesnel T (2003) Homogenization of corrugated core sandwich panel. Comput Struct 194(59):299–312

Caillerie D (1982) Plaques elastiques minces a structute periodique de periode et depaisseur comparable. C R Acad Sci Ser 2 294(3):159–162

Caillerie D (1984) Thin elastic, periodic plates. Math Methods Appl Sci 6:159–191

Canon E, Lenczner M (2000) Modelling of thin elastic plates with small piezoelectric inclusions and distributed electronic circuits. Models for inclusions that are small with respect to the thickness of the plate. J Elast 55:111–141

Chakrabortty A, Griso G, Orlik J (2024) Dimension reduction and homogenization of composite plate with matrix pre-strain. Asympt Anal 138(4):255–310

Challagulla KS, Georgiades AV, Kalamkarov AL (2010) Asymptotic homogenization modeling of smart composite generally orthotropic grid-reinforced shells: Part I—theory. Eur J Mech—A/Solids 29(4):530–540
Cioranescu D, Donato P (2000) An introduction to homogenization. Oxford University Press, Oxford
Duvaut G (1976) Analyse fonctionnelle et mecanique des milieux continus—application a l'etude des materiaux composites elastiques a structure periodique—homogeneisation. In: Koiter W (ed) Proceedings 14th IUTAM congress. North Holland, Amsterdam, pp 119–132
Eremeyev VA, Pietraszkiewicz W (2004) The nonlinear theory of elastic shells with phase transitions. J Elast 74:67–86
Ghergu M, Griso G, Mechkour H, Miara B (2007) Asymptotic methods in the statics and dynamics of perforated plates and shells with periodic structures. Math Model Numer Anal (ESAIM: M2AN) 41(5):875–895
Gloria A (2012) Numerical homogenization: survey, new results, and perspectives. ESAIM: Proc 37:50–116
Hadjiloizi DA, Kalamkarov AL, Metti Ch, Georgiades AV (2014) Analysis of smart piezo-magneto-thermo-elastic composite and reinforced plates: Part I—model development. Curved Layer Struct 1:11–31
Hadjiloizi DA, Kalamkarov AL, Metti Ch, Georgiades AV (2014) Analysis of smart piezo-magneto-thermo-elastic composite and reinforced plates: Part II—applications. Curved Layer Struct 1:32–58
Hrennikoff A (1941) Solution of problem of elasticity by framework method. J Appl Mech A169–A175
Jikov VV, Kozlov SM, Oleinik OA (1994) Homogenization of differential operators and integral functionals. Springer, Berlin
Kalamkarov AL, Hadjiloizi DA, Weaver PM, Georgiades AV (2022) Asymptotic homogenization of flexoelectric composite plates with periodically varying thickness. Math Mech Solids 29(12)
Kalamkarov AL, Kolpakov AG (1997) Analysis, design and optimization of composite structures. Wiley, Chichester
Kanit T, Forest S, Galliet I, Mounoury V, Jeulin D (2003) Determination of the size of the representative volume element for random composites: statistical and numerical approach. Int J Solids Struct 40(13–14):3647–3679
Kohn RV, Vogelius M (1984) A new model for thin plates with rapidly varying thickness. Int J Solids Struct 20:333–350
Kolpakov AG (1995) On the problem of plate with initial stresses. Izvestiay: Mech Solids 3:179–1871
Kolpakov AG (2004) Stressed composite structures: homogenized models for thin-walled nonhomogeneous structures with initial stresses. Springer, Berlin
Kolpakov AA, Kolpakov AG (2020) On the effective rigidities of corrugated plates of various geometries. Int J Eng Sci 154:103327
Kolpakov AG (1982) Effective rigidities of composite plates. J Appl Math Mech 46(4):529–535
Kolpakov AG (1999) Variational principles for stiffnesses of a non-homogeneous plate. J Mech Phys Solids 47:2075–2092
Kolpakov AG (2000) Homogenized model for plate periodic structure with initial stresses. Int J Eng Sci 38(18):2079–2094
Lebee A, Sab K (2013) Homogenization of a space frame as a thick plate. Comput Struct 127(C):88–101
Lee CY, Yu W (2011) Homogenization and dimensional reduction of composite plates with in-plane heterogeneity. Int J Solids Struct 48:1474–1484
Levinski T (1995) Effective stiffness of transversally non-homogeneous plates with unidirectional periodic structure. Int J Solids Struct 32(19):3261–3287

Lions J-L (1978) Notes on some computational aspects of the homogenization method in composite materials. In: Marchuk GI (ed) Computational methods in mathematics, geophysics and optimal control. Nauka, Novosibirsk
Love AEH (1929) A treatise on the mathematical theory of elasticity. Oxford University Press, Oxford
Mignot F, Puel J-P, Suquet P-M (1980) Homogenization and bifurcation of perforated plates. Int J Eng Sci 18:409–414
Mishnaevsky Jr LL (2007) Computational mesomechanics of composites: numerical analysis of the effect of microstructures of composites of strength and damage resistance. Wiley, Chichester
Mityushev V (1997) A functional equation in a class of analytic functions and composite materials. Demost Math 30:63–70
Mityushev V, Adler PM (2002) Longitudinal permeability of a doubly periodic rectangular array of cylinders. I Z Angew Math Mech 82:335–345
Mityushev V, Adler PM (2002) Longitudinal permeability of a doubly periodic rectangular array of cylinders. II. An arbitrary distribution of cylinders inside the unit cell. Z Angew Math Phys 53:486–517
Mityushev V, Rogozin SV (2000) Constructive methods for linear and nonlinear boundary value problems of the analytic functions theory. Chapman&Hall/CRC, Boca Raton, FL
Muller M, Klarmann S, Gruttmann F (2022) A new homogenization scheme for beam and plate structures without a priori requirements on boundary conditions. Comput Mech 70:1167–1187
Nazarova SA, Sweersbc GH, Slutskija AS (2011) Homogenization of a thin plate reinforced with periodic families of rigid rods. Russian Acad Sci Sb Math 202(8):1127–1168
Pan E, Heyliger RP (2002) Free vibrations of simply supported and multilayered magneto-electro-elastic plates. J Sound Vib 252(3):429–442
Panasenko GP (1988) Numerical solution of cell problems in averaging theory. USSR Comput Math Math Phys 28(1):183–186
Panasenko GP (1990) Numerical-asymptotic method of multicomponent homogenization of equations with contrasting values of coefficients. Comput Math Math Phys 30(2):134–142
Panasenko GP, Reztsov MV (1987) Homogenization of the three-dimensional elasticity problem for an inhomogeneous plate. Dokl Akad Nauk SSSR 294:1061–1065
Sab K (2020) Homogenization of thin periodic plates. In: Ochsner A, Altenbach H (eds) Encyclopedia of continuum mechanics. Springer, Berlin, pp 1231–1240
Sab K, Lebee A (2015) Homogenization of heterogeneous thin and thick plates. Wiley-ISTE, London
Sanchez-Palencia E (2000) Non-homogeneous media and vibration theory. Springer, Berlin
Timoshenko S, Woinowsky-Krieger K (1959) Theory of plates and shells. McGraw-Hill, New York
Villaggio P (1979) Buckling under unilateral constraints. Recent Dev Math Probl Eng 3(15):193–201
Washizu K (1992) Variational methods in elasticity and plasticity. Pergamon Press, Oxford
Ye Z, Berdichevsky VL, Yu W (2014) An equivalent classical plate model of corrugated structures. Int J Solids Struct 51(11–12):2073962083
Yvonnet J (2019) Computational homogenization of heterogeneous materials with finite elements. Springer Nature, Cham

Chapter 7
Initial Stresses Comparable to the Elastic Constants of the Material of the Plate

In this chapter, we discuss the transition from the three-dimensional problem of elasticity theory with initial stresses in a thin layer to two-dimensional problem of plate theory in the case when the initial stresses are of the same order as the elastic constants of the material of the plate. The initial stresses cannot exceed the strength of the material(s), which does not exceed a few percent of Young's modulus of the material(s), see Sect. 4.5, in particular Table 4.1. Thus, the initial stresses are much smaller than the elastic constants in absolute value.

The case where the initial stresses are of the same order as the elastic constants of the material is not something rare. On the contrary, it is a frequently encountered case. Such is, for example, the case of thermal stresses (arising both from thermal heating of the composite as a whole Annin et al. 2022 and from heating of the components of the composite material, for example, by electric current Andrianov et al. 2024). Additional information on the thermal properties of the composite, including thermal stresses in the composite, can be found in Kolpakov and Rakin (1986), Kalamkarov and Kolpakov (1997). An interesting prediction of the homogenization theory is the possibility of producing a composite possessing negative coefficient of thermal expansion from materials with positive coefficients of thermal expansion (Kolpakov and Rakin 1986; Kalamkarov and Kolpakov 1997; Miller et al. 2009; Kelly et al. 2005).

7.1 Statement of the Problem

We consider a thin elastic body G_ε of periodic structure, described in Sect. 6.2.3. As $\varepsilon \to 0$, the three-dimensional domain G_ε tends to two-dimensional region D on Ox_1x_2-plane (Fig. 6.5)—a two-dimensional plate.

We start from the model of a body with initial stresses, described in Sect. 4.1: the equilibrium equations of the plate as a three-dimensional elastic body written in the

A. Kolpakov, *Thin-Walled Structures with Initial Stresses*, Mathematical Engineering,
https://doi.org/10.1007/978-3-032-07247-4_7

form

$$\frac{\partial \sigma_{ij}^{\varepsilon}}{\partial x_j} + f_i^{\varepsilon}(x_1, x_2, \mathbf{x}/\varepsilon) = 0 \text{ in } G_{\varepsilon}; \tag{7.1}$$

and the boundary conditions taken in the form

$$\begin{aligned} \sigma_{ij}^{\varepsilon} n_j^{\varepsilon} &= h_i^{\varepsilon}(x_1, x_2, \mathbf{x}/\varepsilon) \text{ on } S_{\varepsilon}; \\ \mathbf{u}^{\varepsilon}(\mathbf{x}) &= 0 \text{ on } S_u. \end{aligned} \tag{7.2}$$

The notations are the same as in Sect. 6.2.3, see Fig. 6.5. In this chapter, we consider initial stresses which has the same order in ε as the elastic constants. It means that in the case under consideration, the quantities h_{ijmn} have the form

$$h_{ijmn}(x_1, x_2, \mathbf{x}/\varepsilon) = \varepsilon^{-3}(c_{ijmn}(\mathbf{x}/\varepsilon) + b_{ijmn}(x_1, x_2, \mathbf{x}/\varepsilon)); \tag{7.3}$$

where

$$b_{ijmn}(x_1, x_2, \mathbf{x}/\varepsilon) = \sigma_{jn}^{*(-3)}(x_1, x_2, \mathbf{x}/\varepsilon)\delta_{im}. \tag{7.4}$$

For $h_{ijmn}(x_1, x_2, \mathbf{y})$ defined by (7.3), the relationship between the additional stresses σ_{ij} and the additional displacements $\mathbf{u}^{\varepsilon}$ takes the form

$$\sigma_{ij}^{\varepsilon} = \varepsilon^{-3} h_{ijmn}(x_1, x_2, \mathbf{x}/\varepsilon) \frac{\partial u_m^{\varepsilon}}{\partial x_n} \tag{7.5}$$

and can be treated as local governing equations for a body with initial stresses.

In the case under consideration, it is necessary to require that the resultant initial in-plane stresses and the resultant moments of the initial stresses be equal to zero, which can be written as

$$\langle \sigma_{\alpha\beta}^{*(-3)} \rangle = 0; \; \langle \sigma_{\alpha\beta}^{*(-3)} y_3 \rangle = 0 \; (\alpha, \beta = 1, 2). \tag{7.6}$$

7.2 Two-Scale Expansions and Homogenized Equilibrium Equations

We will use the general asymptotic expansions (6.16) and (6.17) for the displacements of the plate and the local stresses in the plate. We write these expansions here for convenience:

asymptotic expansion for displacement

$$\mathbf{u}^{\varepsilon} = \mathbf{u}^{(0)}(x_1, x_2) + \sum_{k=1}^{\infty} \varepsilon^k \mathbf{u}^{(k)}(x_1, x_2, \mathbf{y}), \tag{7.7}$$

asymptotic expansion for stresses

$$\sigma_{ij}^{\varepsilon} = \sum_{p=-3}^{\infty} \varepsilon^p \sigma_{ij}^{(p)}(x_1, x_2, \mathbf{y}). \tag{7.8}$$

The functions on the right-hand sides (7.7) and (7.8) are assumed to be periodic in y_1, y_2 with periodicity cell T.

For the functions from (7.7) and (7.8), the differentiation rule preserves the form (6.18).

As above, the analysis of the problem (7.1)–(7.4) is carried out in two stages. The first is to obtain equilibrium equations for the plate considered as a two-dimensional structure. This stage does not use local governing equations (Eq. (7.5) in the case under consideration) and is the same for any governing equations.

For the case under consideration, the homogenized equilibrium equations may be obtained as in Sect. 6.2.2 and have the following form: equations with respect to the resultant in-plane forces (i, j=1,2,3 and $\alpha, \beta = 1, 2$)

$$N_{\alpha\beta,\beta x}^{(-3)} = 0 \ (p = -3); \tag{7.9}$$

$$\begin{aligned} N_{\alpha\beta,\beta x}^{(-2)} + \langle f_{\alpha}^{\varepsilon} \rangle + \langle h_{\alpha}^{\varepsilon} \rangle &= 0 \ (p = -2); \\ N_{3\beta,\beta x}^{(-1)} + \langle f_{3}^{\varepsilon} \rangle + \langle h_{3}^{\varepsilon} \rangle_S &= 0; \end{aligned} \tag{7.10}$$

and equations with respect to the resultant moments

$$M_{\beta\alpha,\alpha x}^{(-3)} - N_{\beta 3}^{(-2)} = 0 \ (p = -3); \tag{7.11}$$

$$M_{\beta\alpha,\alpha x}^{(-2)} - N_{\beta 3}^{(-1)} + \langle f_{\alpha}^{\varepsilon} y_3 \rangle + \langle h_{\alpha}^{\varepsilon} y_3 \rangle_S = 0 \ (p = -2). \tag{7.12}$$

The equations corresponding to $p > -2$ are not used further and therefore are not presented.

Substituting (7.8) into the equilibrium equations (7.2), we obtain taking into account the differentiation rule (6.18)

$$\sum_{p=-3}^{\infty} \varepsilon^p (\sigma_{i\alpha,\alpha x}^{(p)} + \varepsilon^{-1} \sigma_{ij,jy}^{(p)}) + f_i^{\varepsilon} = 0. \tag{7.13}$$

By equating the terms with the identical power of ε in (7.13), we obtain the following local equations:

$$\begin{aligned}
&\sigma_{ij,jy}^{(-3)} = 0 \text{ in } Y \ \ (p=-3);\\
&\sigma_{ij,jy}^{(-2)} + \sigma_{i\beta,\beta x}^{(-3)} = 0 \text{ in } Y \ \ (p=-2);\\
&\sigma_{ij}^{(p)} n_j = 0 \text{ on } S(p=-3,-2,...).
\end{aligned} \tag{7.14}$$

The equations corresponding to $p = -1, 0, \ldots$ are not used further, and therefore they are not given.

7.2.1 *Homogenized Constitutive Equations*

The second stage of the problem analysis is obtaining the homogenized constitutive equations for the plate. It is seen that the constitutive equations (7.5) differ from the constitutive equations (6.37). For this reason, we have to investigate the problem independently in order to obtain the homogenized constitutive equations.

Substituting (7.7), (7.8) into the local constitutive equation (7.5) and taking into account the differentiation rule (9.8), we obtain,

$$\sum_{p=-3}^{\infty} \varepsilon^p \sigma_{ij}^{(p)} = \sum_{k=0}^{\infty} \varepsilon^{k-3} h_{ijmn}(x_1, x_2, \mathbf{y})(u_{m,\beta x}^{(k)} \delta_{m\beta} + \varepsilon^{-1} u_{m,ny}^{(k)}). \tag{7.15}$$

Equating the terms with the identical power of ε in (7.15), we obtain

$$\sigma_{ij}^{(p)} = h_{ijm\beta}(x_1, x_2, \mathbf{y}) u_{m,\beta x}^{(p+3)} + h_{ijmn}(x_1, x_2, \mathbf{y}) u_{m,ny}^{(p+4)} \tag{7.16}$$

$(p = -3, -2, -1, \ldots)$.

Consider problem (7.14) and (7.16) with $p = -3$. It can be written as

$$\begin{cases}
(h_{ijmn}(x_1, x_2, \mathbf{y}) u_{m,ny}^{(1)} + h_{ijm\alpha}(x_1, x_2, \mathbf{y}) u_{m,\alpha x}^{(0)}(x_1, x_2))_{,jy} = 0 \text{ in } G_\varepsilon;\\
(h_{ijmn}(x_1, x_2, \mathbf{y}) u_{m,ny}^{(1)} + h_{ijm\alpha}(x_1, x_2, \mathbf{y}) u_{m,\alpha x}^{(0)}(x_1, x_2)) n_j = 0 \text{ on } S_\varepsilon.
\end{cases} \tag{7.17}$$

Due to the fact that the function of variables x_1, x_2 plays the role of a parameter in problems written in variables $\mathbf{y}$, and $\mathbf{u}^{(0)}$ depends only on x_1, x_2, the solution to problem (7.17) has the form

$$\mathbf{u}^{(1)} = \mathbf{K}^{0m\alpha}(\mathbf{y}) u_{m,\alpha x}^{(0)}(x_1, x_2) + \mathbf{V}(x_1, x_2). \tag{7.18}$$

In (7.18) $\mathbf{V}(x_1, x_2)$ is a function of the variables x_1, x_2, which will be defined below; $\mathbf{K}^{0m\alpha}(\mathbf{y})$ is solution to the periodicity cell problem

$$\begin{cases} (h_{ijmn}(x_1, x_2, \mathbf{y})K^{0m\alpha}_{m,ny} + h_{ijm\alpha}(x_1, x_2, \mathbf{y})_{,jy} = 0 \text{ in } Y; \\ (h_{ijmn}(x_1, x_2, \mathbf{y})K^{0m\alpha}_{m,ny} + h_{ijm\alpha}(x_1, x_2, \mathbf{y}))n_j = 0 \text{ on } S; \\ \mathbf{K}^{0m\alpha}(\mathbf{y}) \text{ periodic in } y_1, y_2 \text{ with periodicity cell } T. \end{cases} \quad (7.19)$$

$\mathbf{n}$ means the external normal to S, see Fig. 6.5 ($\alpha, \beta = 1, 2; m = 1, 2, 3$).

The following proposition is similar but not equivalent to Proposition 6.2. The difference is related to the dependence of the coefficients h_{ijmn} in the problem (7.19) on the initial tresses.

Proposition 7.1 *Let*

$$\sigma^{*(-3)}_{ij,jy} = 0 \text{ in } Y,\ \sigma^{*(-3)}_{ij} n_j = 0 \text{ on } S.$$

Then, for $m = 3$, the solution to the periodicity cell problem (7.21) is given by the following formula

$$\mathbf{K}^{03\alpha}(\mathbf{y}) = -y_3\mathbf{e}_\alpha, \quad (7.20)$$

$\{\mathbf{e}_i\}$ *are the basis vectors of the coordinate system.*

We can verify the validity of (7.20) by substituting (7.20) into the expression $h_{ijmn}(x_1, x_2, \mathbf{y})K^{03\alpha}_{m,ny} + h_{ijm\alpha}(x_1, x_2, \mathbf{y})$. After the substitution, we obtain, taking into account the definition of h_{ijmn} (7.3):

$$h_{ij\alpha 3} - h_{ij3\alpha} = c_{ij\alpha 3} + b_{ij\alpha 3} - (c_{ij3\alpha} + b_{ij3\alpha}) = \sigma^{*(-3)}_{j3}\delta_{i\alpha} - \sigma^{*(-3)}_{j\alpha}\delta_{i3}.$$

Here, the symmetry of the elastic constants $c_{ij3\alpha} = c_{ij\alpha 3}$ is taken into account (see condition C1, Sect. 3.1).

Let's denote $\sigma^0_{ij} = \sigma^{*(-3)}_{j3}\delta_{i\alpha} - \sigma^{*(-3)}_{j\alpha}\delta_{i3}$ and analyze it. In this notation, the periodicity cell problem (7.19) for $\mathbf{K}^{03\alpha}(\mathbf{y})$ may be written as

$$\sigma^0_{ij,jy} = 0 \text{ in } Y,\ \sigma^0_{ij}n_j = 0 \text{ on } S.$$

Let us check that these equations are indeed satisfied. We have

$$\sigma^0_{ij,jy} = \sigma^{*(-3)}_{j3,jy}\delta_{i\alpha} - \sigma^{*(-3)}_{j\alpha,jy}\delta_{i3} = \sigma^{*(-3)}_{3j,jy}\delta_{i\alpha} - \sigma^{*(-3)}_{\alpha j,jy}\delta_{i3} = 0 \text{ in } Y$$

due to the condition $\sigma^{*(-3)}_{ij,jy} = 0$ in Y. Here the symmetry of the initial stresses $\sigma^{*(-3)}_{ij}$ is essential.

Next, we have

$$\sigma^0_{ij}n_j = \sigma^{*(-3)}_{j3}n_j\delta_{i\alpha} - \sigma^{*(-3)}_{j\alpha}n_j\delta_{i3} = \sigma^{*(-3)}_{3j}n_j\delta_{i\alpha} - \sigma^{*(-3)}_{\alpha j}n_j\delta_{i3} = 0 \text{ on } S$$

due to the condition $\sigma_{ij}^{*(-3)} n_j = 0$ on S. Here again the symmetry of the initial stresses $\sigma_{ij}^{*(-3)}$ is used. Then, (7.20) is solution to the periodicity cell problem (7.19).

By virtue of Proposition 7.1, the following non-trivial problems remain from the problems (7.19) ($\alpha, \beta = 1, 2$):

$$\begin{cases} (h_{ijmn}(x_1, x_2, \mathbf{y}) K^{0\beta\alpha}_{m,ny} + h_{ijm\alpha}(x_1, x_2, \mathbf{y})_{,jy} = 0 \text{ in } Y; \\ (h_{ijmn}(x_1, x_2, \mathbf{y}) K^{0\beta\alpha}_{m,ny} + h_{ijm\alpha}(x_1, x_2, \mathbf{y}))n_j = 0 \text{ on } S; \\ \mathbf{K}^{0\beta\alpha}(\mathbf{y}) \text{ periodic in } y_1, y_2 \text{ with periodicity cell } T. \end{cases} \tag{7.21}$$

Note. The periodicity cell problems (7.19) and (7.21) are similar but not equivalent to the periodicity cell problems (6.47) and (6.57). The periodicity cell problem (7.19) and (7.21) have the coefficients h_{ijmn} differ from the coefficients c_{ijmn} of the periodicity cell problems (6.47) and (6.57).

Taking into account (7.20), we can write (7.18) as

$$\mathbf{u}^{(1)} = \mathbf{K}^{0\beta\alpha}(\mathbf{y}) u^{(0)}_{\beta,\alpha x}(x_1, x_2) - y_3 \mathbf{e}_\alpha u^{(0)}_{3,\alpha x}(x_1, x_2) + \mathbf{V}(x_1, x_2). \tag{7.22}$$

Substituting (7.22) into (7.16, $p = -3$) yields the following equations:

$$\begin{aligned} \sigma^{(-3)}_{ij} &= (h_{ij\alpha\beta}(x_1, x_2, \mathbf{y}) + h_{ijkl}(x_1, x_2, \mathbf{y}) K^{0\alpha\beta}_{k,ly}(\mathbf{y})) u^{(0)}_{\alpha,\beta x} \\ &\quad + (h_{ij3\alpha}(x_1, x_2, \mathbf{y}) - h_{ij\alpha 3}(x_1, x_2, \mathbf{y})) u^{(0)}_{3,\alpha x}. \end{aligned} \tag{7.23}$$

Averaging (7.23) over the periodicity cell Y with $ij = \gamma\delta$, we obtain, taking into account the definition (7.3) and Proposition 6.3.

$$\langle \sigma^{(-3)}_{\gamma\delta} \rangle = A_{\gamma\delta\alpha\beta}(\sigma) u^{(0)}_{\alpha,\beta x}, \tag{7.24}$$

where

$$A_{\gamma\delta\alpha\beta}(\sigma) = \langle h_{ij\alpha\beta}(x_1, x_2, \mathbf{y}) + h_{ijkl}(x_1, x_2, \mathbf{y}) K^{0\alpha\beta}_{k,ly}(\mathbf{y}) \rangle. \tag{7.25}$$

By virtue of the definition of the quantities h_{ijkl} (7.3), the following equalities hold:

$$\langle h_{ij\alpha\beta} \rangle = \langle c_{ij\alpha\beta}(\mathbf{y}) \rangle + \langle b_{ij\alpha\beta}(x_1, x_2, \mathbf{y}) \rangle,$$

$$\langle h_{ij\alpha\beta}(x_1, x_2, \mathbf{y}) K^{0\alpha\beta}_{k,ly}(\mathbf{y}) \rangle = \langle c_{ij\alpha\beta}(\mathbf{y}) K^{0\alpha\beta}_{k,ly}(\mathbf{y}) \rangle + \langle b_{ij\alpha\beta}(x_1, x_2, \mathbf{y}) K^{0\alpha\beta}_{k,ly}(\mathbf{y}) \rangle.$$

We assume that the initial stresses satisfy the condition of Proposition 6.3. Under conditions of Proposition 6.3 we have: $\langle b_{ij\alpha\beta}(x_1, x_2, \mathbf{y})\rangle = 0$ and $\langle b_{ij\alpha\beta}(x_1, x_2, K^{0\alpha\beta}_{k,ly}(\mathbf{y})\rangle = 0$ and formula (7.25) takes the form

$$A_{\gamma\delta\alpha\beta}(\sigma) = \langle c_{ij\alpha\beta}(\mathbf{y}) + c_{ijkl}(\mathbf{y})K^{0\alpha\beta}_{k,ly}(\mathbf{y})\rangle. \tag{7.26}$$

According to the homogenized equilibrium equations (7.9, $p = -3$)

$$(A_{\gamma\delta\alpha\beta}(\sigma)u^{(0)}_{\alpha,\beta x})_{,\delta x} = 0 \text{ in } D. \tag{7.27}$$

The boundary conditions for $u^{(0)}_\alpha$ $(\alpha = 1, 2)$ follow from (7.2) and (7.7):

$$u^{(0)}_\alpha(x_1, x_2) = 0 \text{ on } \partial D. \tag{7.28}$$

In general, it cannot be claimed that the coefficients $h_{\gamma\delta\alpha\beta}(\sigma)$ form a positive definite tensor. However, in practice, the initial stresses $\sigma^{*(-3)}_{ij}$ are small compared to the elastic constants c_{ijkl}, see Sect. 4.5. As a result, in practice, we are dealing with a small perturbation of the elastic constants which preserves the positive definiteness of $h_{\gamma\delta\alpha\beta}(\sigma)$ (it is known that for $\sigma^{*(-3)}_{ij} = 0$ the tensor of the homogenized elastic constants $h_{\gamma\delta\alpha\beta}(0) = C_{\gamma\delta\alpha\beta}$ of body with no initial stresses is positive defined Caillerie 1984; Kalamkarov and Kolpakov 1997). Then the solution to the problem (7.27) and (7.28) is $u^{(0)}_\alpha = 0$ $(\alpha = 1, 2)$, and the equalities (7.16, $p = -3, -2$) take the form

$$\begin{aligned} \sigma^{(-3)}_{ij} &= -h_{ij\alpha 3}(x_1, x_2, \mathbf{y})u^{(0)}_{3,\alpha x} + h^{(0)}_{ij3\alpha}(x_1, x_2, \mathbf{y})u^{(0)}_{3,\alpha x} \\ &= (\sigma^{*(-3)}_{j3}(x_1, x_2, \mathbf{y})\delta_{i\alpha} - \sigma^{*(-3)}_{j\alpha}(x_1, x_2, \mathbf{y})\delta_{i3})u^{(0)}_{3,\alpha x}, \end{aligned} \tag{7.29}$$

$$\begin{aligned} \sigma^{(-2)}_{ij} = {} & h_{ijmn}(x_1, x_2, \mathbf{y})u^{(2)}_{m,ny} - h_{ij\alpha\beta}(x_1, x_2, \mathbf{y})y_3 u^{(0)}_{3,\alpha x\beta x}(x_1, x_2) \\ & + h_{ijm\beta}(x_1, x_2, \mathbf{y})V_{m,\beta x}(x_1, x_2). \end{aligned} \tag{7.30}$$

Let us consider the problem (7.14, $p = -2$) with $\sigma^{(-2)}_{ij}$ (7.30). Taking into account (7.29), it can be written as

$$\begin{cases} h_{ijmn}(x_1, x_2, \mathbf{y})u^{(2)}_{m,ny} - h_{ij\alpha\beta}(x_1, x_2, \mathbf{y})y_3 u^{(0)}_{3,\alpha x\beta x}(x_1, x_2) \\ + h_{ijm\beta}(x_1, x_2, \mathbf{y})V_{m,\beta x}(x_1, x_2))_{,jy} \\ + \sigma^{*(-3)}_{\beta\alpha}(x_1, x_2, \mathbf{y})\delta_{i3}u^{(0)}_{3,\alpha x\beta x} = 0 \text{ in } G_\varepsilon; \\ (h_{ijmn}(x_1, x_2, \mathbf{y})u^{(2)}_{m,ny} - h_{ij\alpha\beta}(x_1, x_2, \mathbf{y})y_3 u^{(0)}_{3,\alpha x\beta x}(x_1, x_2) \\ + h_{ijm\beta}(x_1, x_2, \mathbf{y})V_{m,\beta x}(x_1, x_2))n_j \text{ on } S_\varepsilon. \end{cases} \tag{7.31}$$

To solve problems of this type, we introduce a periodicity cell problem corresponding to the bending of a plate,

$$\begin{cases} (h_{ijmn}(x_1, x_2, \mathbf{y})K^{1\alpha\beta}_{m,ny} - y_3 h_{ij\alpha\beta}(x_1, x_2, \mathbf{y}))_{,jy} = 0 \text{ in } Y; \\ (h_{ijmn}(x_1, x_2, \mathbf{y})K^{1\alpha\beta}_{m,ny} - y_3 h_{ij\alpha\beta}(x_1, x_2, \mathbf{y}))n_j = 0 \text{ on } S; \\ \mathbf{K}^{1\alpha\beta}(\mathbf{y}) \text{ periodic in } y_1,\ y_2 \text{ with periodicity cell } T. \end{cases} \quad (7.32)$$

and the periodicity cell problem corresponding to the initial stresses

$$\begin{cases} (h_{ijmn}(x_1, x_2, \mathbf{y})Z^{\alpha\beta}_{m,ny})_{,jy} + \sigma^{*(-3)}_{\beta\alpha}(x_1, x_2, \mathbf{y})\delta_{i\alpha} = 0 \text{ in } Y; \\ h_{ijmn}(x_1, x_2, \mathbf{y})Z^{\alpha\beta}_{m,ny} n_j = 0 \text{ on } S; \\ \mathbf{Z}^{\alpha\beta}(\mathbf{y}) \text{ periodic in } y_1,\ y_2 \text{ with periodicity cell } T. \end{cases} \quad (7.33)$$

The solution to the problem (7.31) can be expressed through the functions $\mathbf{K}^{\nu\alpha\beta}(\mathbf{y})$ $(\nu = 0, 1)$ – solutions to the periodicity cell problems (7.21) and (7.33), and function $\mathbf{Z}^{\alpha\beta}(\mathbf{y})$ - solution to the periodicity cell problem (7.33), in the form:

$$\begin{aligned} \mathbf{u}^{(2)} = {} & \mathbf{K}^{1\alpha\beta}(\mathbf{y})u^{(0)}_{3,\alpha x\beta x}(x_1, x_2) + \mathbf{Z}^{1\alpha\beta}(\mathbf{y})u^{(0)}_{3,\alpha x\beta x}(x_1, x_2) \\ & - y_3 V_{3,\beta x}(x_1, x_2)\mathbf{e}_\beta + \mathbf{K}^{0\alpha\beta}(\mathbf{y})V_{\alpha,\beta x}(x_1, x_2). \end{aligned} \quad (7.34)$$

Writing (7.34), we used Proposition 7.1.

Substituting (7.34) into (7.30), we have

$$\begin{aligned} \sigma^{(-2)}_{ij} = {} & (h_{ijmn}(x_1, x_2, \mathbf{y})K^{0\alpha\beta}_{m,ny} + h_{ij\alpha\beta}(x_1, x_2, \mathbf{y}))V_{\alpha,\beta x}(x_1, x_2) \\ & + (h_{ijmn}(x_1, x_2, \mathbf{y})K^{1\alpha\beta}_{m,ny} - y_3 h_{ij\alpha\beta}(x_1, x_2, \mathbf{y}))u^{(0)}_{3,\alpha x\beta x}(x_1, x_2) \\ & + h_{ijmn}(x_1, x_2, \mathbf{y})Z^{1\alpha\beta}_{m,ny}(\mathbf{y})u^{(0)}_{3,\alpha x\beta x}(x_1, x_2) \\ & + (h_{ij3\beta}(x_1, x_2, \mathbf{y}) - h_{ij\beta 3}(x_1, x_2, \mathbf{y}))V_{3,\beta x} \end{aligned} \quad . \quad (7.35)$$

7.2.2 *Homogenized Constitutive Equations for in-Plane Resultant Forces and Resultant Moments*

Averaging (7.35) with $ij = \gamma\delta$ $(\gamma, \delta = 1, 2)$ over the periodicity cell Y, we obtain

$$N^{(-2)}_{\gamma\delta} = A^0_{\gamma\delta\alpha\beta}(\sigma)V_{\alpha,\beta x} + A^1_{\gamma\delta\alpha\beta}(\sigma)u^{(0)}_{3,\alpha x\beta x} + R^0_{\gamma\delta\alpha\beta}(\sigma)u^{(0)}_{3,\alpha x\beta x}. \quad (7.36)$$

Here we used the fact that $-h_{\gamma\delta\alpha 3} + h_{\gamma\delta 3\alpha} = -\sigma^{*(-3)}_{j3}\delta_{i\alpha} + \sigma^{*(-3)}_{j\alpha}\delta_{i3}$. Then for $ij = \gamma\delta$ by virtue of Proposition 6.3

$$\langle -h_{ij\alpha 3}(x_1, x_2, \mathbf{y}) + h_{ij3\alpha}(x_1, x_2, \mathbf{y})\rangle = -\langle \sigma^{*(-3)}_{\delta 3}(x_1, x_2, \mathbf{y})\rangle \delta_{\gamma\alpha} = 0.$$

Multiplying (7.35) with $ij = \gamma\delta$ ($\gamma,\delta = 1, 2$) by $-y_3$ and averaging the result over the periodicity cell Y, we obtain the following equation

$$M^{(-2)}_{\gamma\delta} = A^1_{\gamma\delta\alpha\beta}(\sigma)V_{\alpha,\beta x}(\sigma) + A^2_{\gamma\delta\alpha\beta}(\sigma)u^{(0)}_{3,\alpha x\beta x} + R^1_{\gamma\delta\alpha\beta}(\sigma)u^{(0)}_{3,\alpha x\beta x}. \tag{7.37}$$

The coefficients in Eqs. (7.36) and (7.37) are

$$A^{\mu+\nu}_{\gamma\delta\alpha\beta}(\sigma) = \langle((-1)^\nu h_{\gamma\delta\alpha\beta}(x_1, x_2, \mathbf{y})y_3 + h_{\gamma\delta mn}(x_1, x_2, \mathbf{y})K^{\nu\alpha\beta}_{m,ny})(-1)^\mu y_3^\mu\rangle \tag{7.38}$$

and

$$R^\mu_{\gamma\delta\alpha\beta}(\sigma) = \langle h_{\gamma\delta mn}(x_1, x_2, \mathbf{y})Z^{\alpha\beta}_{m,ny}(-1)^\mu y_3^\mu\rangle \tag{7.39}$$

($\nu, \mu = 0, 1$).

The obtained Eqs. (7.36) and (7.37) are the homogenized governing equations of the plate considered as a two-dimensional object. The coefficients $A^{\mu+\nu}_{\gamma\delta\alpha\beta}(\sigma)$ (7.38) are the stiffnesses of the plate. They are expressed through the solutions to the periodicity cell problems (7.21) and (7.32). We also see in (7.36) and (7.37) the coefficients $R^\mu_{\gamma\delta\alpha\beta}(\sigma)$ (7.39) which are expressed through solution to the periodicity cell problems (7.33). The coefficients $R^\mu_{\gamma\delta\alpha\beta}(\sigma)$ vanish if initial stresses $\sigma^{*(-3)}_{ij} = 0$. We introduce the coefficients $A^{\mu+\nu}_{\gamma\delta\alpha\beta}(\sigma) + R^\mu_{\gamma\delta\alpha\beta}(\sigma)$ to shorten the notation of the homogenized governing equations.

The homogenized governing equations (7.36) and (7.37) determine the resultant in-plane forces and the resultant moments, but not the shear forces.

7.2.3 Governing Equations for Shear Forces

In the case under consideration, $N^{(-1)}_{ij}$ are again asymmetric in i and j, and the method proposed in Sect. 5.3.3 is used to eliminate them. According to (7.16, $p = -1$) and definition of h_{ijmn} (7.3), we have

$$\begin{aligned}\sigma_{ij}^{(-1)} &= h_{ijm\gamma}(x_1, x_2, \mathbf{y})u_{m,\gamma x}^{(2)} + h_{ijmn}(x_1, x_2, \mathbf{y})u_{m,ny}^{(3)} \\ &= c_{ijm\gamma}(\mathbf{y})u_{m,\gamma x}^{(2)} + c_{ijmn}(\mathbf{y})u_{m,ny}^{(3)} \\ &+b_{ijm\gamma}(x_1, x_2, \mathbf{y})u_{m,\gamma x}^{(2)} + b_{ijmn}(x_1, x_2, \mathbf{y})u_{m,ny}^{(3)}\end{aligned} \quad . \tag{7.40}$$

The first and second terms on the right-hand side of the equation (7.40) are symmetrical in i and j due to the symmetry of the elastic constants ($c_{ijmn} = c_{jimn}$, see condition C1 Sect. 3.1). Then

$$\begin{aligned}\sigma_{ij}^{(-1)} - \sigma_{ji}^{(-1)} &= b_{ijm\gamma}(x_1, x_2, \mathbf{y})u_{m,\gamma x}^{(2)} + b_{ijmn}(x_1, x_2, \mathbf{y})u_{m,ny}^{(3)} \\ &-b_{jim\gamma}(x_1, x_2, \mathbf{y})u_{m,\gamma x}^{(2)} - b_{jimn}(x_1, x_2, \mathbf{y})u_{m,ny}^{(3)}.\end{aligned} \tag{7.41}$$

Substituting $u^{(2)}$ according to (7.34) into (7.41), we obtain

$$\begin{aligned}\sigma_{ij}^{(-1)} - \sigma_{ji}^{(-1)} &= (b_{ijmn}(x_1, x_2, \mathbf{y}) - b_{jimn}(x_1, x_2, \mathbf{y}))u_{m,ny}^{(3)} \\ &+(b_{ijm\gamma}(x_1, x_2, \mathbf{y})K_m^{1\alpha\beta} - b_{jim\gamma}(x_1, x_2, \mathbf{y})K_m^{1\alpha\beta})u_{3,\gamma x\alpha x\beta x}^{(0)} \\ &+(b_{ijm\gamma}(x_1, x_2, \mathbf{y})Z_m^{1\alpha\beta} - b_{jim\gamma}(x_1, x_2, \mathbf{y}))Z_m^{1\alpha\beta})u_{3,\gamma x\alpha x\beta x}^{(0)} \\ &-(b_{ij\beta\gamma}(x_1, x_2, \mathbf{y}) - b_{ji\beta\gamma}(x_1, x_2, \mathbf{y}))y_3 V_{3,\gamma x\beta x} \\ &+(b_{ijm\gamma}(x_1, x_2, \mathbf{y}) - b_{jim\gamma}(x_1, x_2, \mathbf{y}))K_m^{0\alpha\beta} V_{\alpha,\beta x\gamma x}(x_1, x_2).\end{aligned} \tag{7.42}$$

Averaging equality (7.42) over the periodicity cell Y, we obtain

$$\begin{aligned}N_{ij}^{(-1)} - N_{ji}^{(-1)} &= \langle(b_{ijmn}(x_1, x_2, \mathbf{y}) - b_{jimn}(x_1, x_2, \mathbf{y}))u_{m,ny}^{(3)}\rangle \\ &+\langle(b_{ijm\gamma}(x_1, x_2, \mathbf{y}) - b_{jim\beta}(x_1, x_2, \mathbf{y}))K_m^{1\alpha\beta}\rangle u_{3,\gamma x\alpha x\beta x}^{(0)} \\ &+\langle(b_{ijm\gamma}(x_1, x_2, \mathbf{y}) - b_{jim\gamma}(x_1, x_2, \mathbf{y}))Z_m^{1\alpha\beta})\rangle u_{3,\gamma x\alpha x\beta x}^{(0)} \\ &-\langle(b_{ij\beta\gamma}(x_1, x_2, \mathbf{y}) - b_{ji\beta\gamma}(x_1, x_2, \mathbf{y}))y_3\rangle V_{3,\gamma x\beta x} \\ &+\langle(b_{ijm\gamma}(x_1, x_2, \mathbf{y}) - b_{jim\gamma}(x_1, x_2, \mathbf{y}))K_m^{0\alpha\beta}\rangle V_{\alpha,\beta x\gamma x}(x_1, x_2).\end{aligned}$$

Consider the terms on the right-hand side (7.43). We have

$$\langle b_{ijmn}(x_1, x_2, \mathbf{y})u_{m,ny}^{(3)}\rangle = 0, \langle b_{jimn}(x_1, x_2, \mathbf{y})u_{m,ny}^{(3)}\rangle = 0$$

by virtue of Proposition 6.3.

Then

$$
\begin{aligned}
&N_{ij}^{(-1)} - N_{ji}^{(-1)} = \\
&= \langle (b_{ijm\gamma}(x_1, x_2, \mathbf{y}) - b_{jim\gamma}(x_1, x_2, \mathbf{y})) K_m^{1\alpha\beta} \rangle u^{(0)}_{3,\gamma x\alpha x\beta x}(x_1, x_2) \\
&-\langle (b_{ij\beta\gamma}(x_1, x_2, \mathbf{y}) - b_{ji\beta\gamma}(x_1, x_2, \mathbf{y})) y_3 \rangle V_{3,\gamma\beta x}(x_1, x_2) \\
&+\langle (b_{ijm\gamma}(x_1, x_2, \mathbf{y}) - b_{jim\gamma}(x_1, x_2, \mathbf{y})) K_m^{0\alpha\beta} \rangle V_{\alpha,\beta x\gamma x}(x_1, x_2).
\end{aligned}
\tag{7.43}
$$

Note that the coefficient of $V_{3,\delta x\beta x}$ in (7.43) vanishes. Really,

$$
\langle b_{\mu\delta\beta}(x_1, x_2, \mathbf{y}) y_3 \rangle = \langle \sigma^*_{\mu\beta}(x_1, x_2, \mathbf{y}) y_3 \rangle \delta_{3\delta} = 0
$$

since $\delta_{3\delta} = 0$ ($\delta = 1, 2$), and

$$
\langle b_{\mu\delta\beta}(x_1, x_2, \mathbf{y}) y_3 \rangle = \langle \sigma^*_{3\beta}(x_1, x_2, \mathbf{y}) y_3 \rangle \delta_{\mu\delta} = 0
$$

by virtue of the Proposition 6.3. Then (7.43) takes the form

$$
\begin{aligned}
&N_{ij}^{(-1)} - N_{ji}^{(-1)} \\
&= \langle (b_{ijm\gamma}(x_1, x_2, \mathbf{y}) - b_{jim\gamma}(x_1, x_2, \mathbf{y})) K_m^{1\alpha\beta} \rangle u^{(0)}_{3,\gamma x\alpha x\beta x}(x_1, x_2) \\
&+\langle (b_{ijm\gamma}(x_1, x_2, \mathbf{y}) - b_{jim\gamma}(x_1, x_2, \mathbf{y})) K_m^{0\alpha\beta} \rangle V_{\alpha,\beta x\gamma x}(x_1, x_2).
\end{aligned}
\tag{7.44}
$$

For $i = 3$ and $j = \mu$, we obtain from (7.44):

$$
N_{3\mu}^{(-1)} - N_{\mu 3}^{(-1)} = K_\mu, \tag{7.45}
$$

where K_μ is defined by the formula

$$
K_\mu = a^0_{\mu\gamma\alpha\beta} u^{(0)}_{3,\gamma x\alpha x\beta x} + a^1_{\mu\gamma\alpha\beta} V_{\alpha,\gamma x\beta x} \tag{7.46}
$$

with the coefficients ($\alpha, \beta, \mu, \delta = 2, 3$; $\nu = 0, 1$)

$$
\begin{aligned}
&a^\nu_{\mu\gamma\alpha\beta} = \langle (b_{3\mu m\gamma}(x_1, x_2, \mathbf{y}) - b_{\mu 3m\gamma}(x_1, x_2, \mathbf{y})) K_m^{\nu\alpha\beta} \rangle \\
&= \langle (\sigma^{*(-3)}_{\mu\gamma}(x_1, x_2, \mathbf{y}) \delta_{3m} - \sigma^{*(-3)}_{3\gamma}(x_1, x_2, \mathbf{y}) \delta_{\mu m}) K_m^{\nu\alpha\beta} \rangle \\
&= \langle \sigma^{*(-3)}_{\mu\gamma}(x_1, x_2, \mathbf{y}) K_3^{\nu\alpha\beta} \rangle - \langle \sigma^{*(-3)}_{3\gamma}(x_1, x_2, \mathbf{y}) K_\mu^{\nu\alpha\beta} \rangle.
\end{aligned}
\tag{7.47}
$$

Here we use that the coefficients of $u^{(0)}_{3,\gamma x\alpha x\beta x}$ and $V_{\alpha,\delta x\beta x}$ in (7.46) and (7.47) are similar, as a result, these coefficients can be written using the same formula if to introduce the index $\nu = 0, 1$.

Note. The functions $\mathbf{K}^{\nu\alpha\beta}$ $(\nu = 0, 1)$ are defined up to an arbitrary constant. By Proposition 6.3 and condition (7.6), this constant does not affect the numerical value of $a^{\nu}_{\mu\delta\alpha\beta}$.

Note that all the quantities containing combinations of the form $\langle b_{ijmn} y_3 \rangle$ in the above calculations have vanished for one reason or another. As a result, the moments of the initial stresses were not included in the governing equations. This is consistent with the results of Kolpakov (2004).

7.2.4 Two-Dimensional Homogenized Problem

Now we can eliminate the quantities $N^{(-1)}_{3\gamma}$, $N^{(-1)}_{\gamma 3}$ from the equilibrium equations (7.10), (7.11) and obtain a closed two-dimensional model of the plate. Differentiating (7.11, $p = 2$) and using (7.10), (7.45), (7.46), we obtain

$$\begin{gathered}(A^1_{\gamma\delta\alpha\beta}(\sigma) V_{\alpha,\beta x} + A^2_{\gamma\delta\alpha\beta}(\sigma) u^{(0)}_{3,\alpha x\beta x})_{,\gamma x\delta x} \\ +(\langle f^{\varepsilon}_{\alpha} y_3 \rangle + \langle h^{\varepsilon}_{\alpha} y_3 \rangle_S)_{,\delta x} + \langle f_3 \rangle + \langle h_3 \rangle_S - K_{\gamma,\gamma x} = 0.\end{gathered} \tag{7.48}$$

Substituting into (7.48) K_γ in accordance with (7.46), we obtain

$$\begin{gathered}(A^1_{\gamma\delta\alpha\beta}(\sigma) V_{\alpha,\beta x} + A^2_{\gamma\delta\alpha\beta}(\sigma) u^{(0)}_{3,\alpha x\beta x})_{,\gamma x\delta x} \\ -(a^0_{\delta\gamma\alpha\beta} u^{(0)}_{3,\gamma x\delta x\beta x} + a^1_{\delta\gamma\alpha\beta} V_{\alpha,\gamma x\beta x})_{,\delta x} + \langle f_3 \rangle + \langle h_3 \rangle_S \\ +(\langle f^{\varepsilon}_{\alpha} y_3 \rangle + \langle h^{\varepsilon}_{\alpha} y_3 \rangle_S)_{,\delta x} = 0.\end{gathered} \tag{7.49}$$

Equalities (7.9, $p = -2$) and (7.36) yield

$$(A^0_{\gamma\delta\alpha\beta}(\sigma) V_{\alpha,\beta x} + A^1_{\gamma\delta\alpha\beta}(\sigma) u^{(0)}_{3,\alpha x\beta x})_{,\gamma x} = 0. \tag{7.50}$$

Boundary conditions

$$V_{\alpha}(x_1, x_2) = u^{(0)}_3(x_1, x_2) = \frac{\partial u^{(0)}_3}{\partial \mathbf{n}}(x_1, x_2) = 0 \text{ on } \partial D \tag{7.51}$$

are obtained by substituting (7.7) into the original boundary conditions (7.2) taking into account (7.22) ($\mathbf{n}$ is the outward normal to ∂D).

Problem (7.49)–(7.51) describes the deformation of a plate with initial stresses. In Eq. (7.49), all coefficients depend on the initial stresses.

If the coefficients in Eq. (7.49) do not depend on x_1, x_2, it takes the form

$$\begin{gathered}(A^2_{\gamma\delta\alpha\beta}(\sigma) - a^0_{\delta\gamma\alpha\beta}(\sigma)) u^{(0)}_{3,\alpha x\beta x\gamma x\delta x} + (A^1_{\gamma\delta\alpha\beta}(\sigma) - a^1_{\delta\gamma\alpha\beta}(\sigma)) V_{\delta,\gamma x\beta x\delta x} \\ +\langle f_3 \rangle + \langle h_3 \rangle_S + (\langle f^{\varepsilon}_{\alpha} y_3 \rangle + \langle h^{\varepsilon}_{\alpha} y_3 \rangle_S)_{,\delta x} = 0,\end{gathered}$$

and the influence of shear forces corresponding to bending can be reduced to the introduction of bending rigidity $A^2_{\gamma\delta\alpha\beta} - a^0_{\delta\gamma\alpha\beta}$. For the term corresponding to displacements, such a reduction cannot be done.

7.3 Case of Small Initial Stresses

In this section, we consider the case when the initial stresses are of the order of elastic constants. It means that both the initial stresses and elastic constants are of the order of ε^{-3}), but the initial stresses are significantly smaller than the elastic constants in absolute value. A discussion of such kind "small initial stresses" was presented in Sect. 4.5.

Let us denote by μ the ratio of the initial stresses in a structural element to the Young's modulus of the material from which the structural elements is made of. As noted in Sect. 4.5, see, in particular, Table 4.1, this for most materials is less than 0.02. This fact can be used to introduce once more small parameter and use classical small parameter method to obtain approximate formulas for calculating the plate stillnesses. The value of the parameter, as small as 0.02, makes it possible to save into computation only the first-order (linear) corrector.

Stiffnesses $A^{\mu+\nu}_{\gamma\delta\alpha\beta}$ in the homogenized governing equations (7.36) and (7.38) depend only on the functions $\mathbf{K}^{\nu\alpha\beta}$ and may be considered independently on other coefficients and $R^{\mu}_{\gamma\delta\alpha\beta}(\sigma)$ (7.39).

The general form of the periodicity cell problems (7.17) and (7.32) is

$$\begin{cases} (h_{ijmn}(x_1, x_2, \mathbf{y})H^{\alpha\beta}_{m,ny} + f^{\alpha\beta}_{ij}(x_1, x_2, \mathbf{y}))_{,jy} = 0 \text{ in } Y; \\ (h_{ijmn}(x_1, x_2, \mathbf{y})H^{\alpha\beta}_{m,ny} + f^{\alpha\beta}_{ij}(x_1, x_2, \mathbf{y}))n_j = 0 \text{ on } S; \\ \mathbf{H}^{\alpha\beta}(\mathbf{y}) \text{ is periodic in } y_1, y_2 \text{ with periodicity cell } T, \end{cases} \tag{7.52}$$

where

$$\begin{aligned} f^{\alpha\beta}_{ij}(x_1, x_2, \mathbf{y}) &= h_{ij\alpha\beta}(x_1, x_2, \mathbf{y}) \text{ for (7.17)}, \\ f^{\alpha\beta}_{ij}(x_1, x_2, \mathbf{y}) &= -h_{ij\alpha\beta}(x_1, x_2, \mathbf{y})y_3 \text{ for (7.32)}. \end{aligned} \tag{7.53}$$

We introduce the following linear combination of the functions (7.53):

$$f^{\alpha\beta}_{ij}(x_1, x_2, \mathbf{y}) = \eta h_{ij\alpha\beta}(x_1, x_2, \mathbf{y}) - h_{ij\alpha\beta}(x_1, x_2, \mathbf{y})y_3 \tag{7.54}$$

where η is a parameter (an arbitrary real number). Do not confuse this parameter with the small parameter μ. If μ has a physical meaning and characterizes the values of the initial stresses, the parameter η has no physical meaning and will be used only to shorten mathematical calculations.

Solution to the problem (7.52) with function $f_{ij}^{\alpha\beta}(x_1, x_2, \mathbf{y})$ (7.54) is

$$\mathbf{H}^{\alpha\beta} = \eta \mathbf{K}^{0\alpha\beta} + \mathbf{K}^{1\alpha\beta},$$

where $\mathbf{K}^{0\alpha\beta}$ and $\mathbf{K}^{1\alpha\beta}$ are solutions to (7.21) and (7.32), respectively. This equality is used significantly below.

The functions included in (7.52), (7.54) have the form

$$\begin{aligned} h_{ijmn}(x_1, x_2, \mathbf{y}) &= c_{ijmn}(\mathbf{y}) + \mu b_{ijmn}(x_1, x_2, \mathbf{y}); \\ f_{ij}^{\alpha\beta}(x_1, x_2, \mathbf{y}) &= f_{ij}^{\alpha\beta 1}(x_1, x_2, \mathbf{y}) + \mu f_{ij^{\alpha\beta 1}}(\mathbf{y}). \end{aligned} \tag{7.55}$$

7.3.1 Expansion of Solutions to the Periodicity Cell Problems in Series

We will seek a solution to the periodicity cell problem (7.52) in the form

$$\mathbf{H}^{\alpha\beta}(x_1, x_2, \mathbf{y}) = \sum_{s=0}^{\infty} \mu^s \mathbf{H}^{\alpha\beta s}(x_1, x_2, \mathbf{y}), \tag{7.56}$$

where $\mathbf{H}^{\alpha\beta s}(\mathbf{y})$ are periodic in y_1, y_2 with the periodicity cell T.

Substituting (7.56), (7.55) into (7.52) yields

$$\begin{cases} \Big((c_{ijmn}(\mathbf{y}) + \mu b_{ijmn}(x_1, x_2, \mathbf{y})) \sum_{s=0}^{\infty} (\mu^s H_{m,ny}^{\alpha\beta s} + f_{ij}^{\alpha\beta 0}(\mathbf{y}) \\ \quad + \mu f_{ij}^{\alpha\beta 1}(x_1, x_2, \mathbf{y})\Big)_{,jy} = 0 \text{ in } Y; \\ \Big((c_{ijmn}(\mathbf{y}) + \mu b_{ijmn}(x_1, x_2, \mathbf{y}) \sum_{s=0}^{\infty} (\mu^s H_{m,ny}^{\alpha\beta s} + f_{ij}^{\alpha\beta 0}(x_1, x_2, \mathbf{y}) \\ \quad + \mu f_{ij}^{\alpha\beta 1}(x_1, x_2, \mathbf{y})\Big) n_j = 0 \text{ on } S; \\ \mathbf{H}^{\alpha\beta \mathbf{s}}(\mathbf{y}) \text{ periodic in } y_1, y_2 \text{ with periodicity cell } T. \end{cases} \tag{7.57}$$

By equating the expressions for the same powers of μ in (7.57), we obtain an infinite sequence of equations, the first two of which have the following form: zero-order problem

$$
\begin{cases}
(c_{ijmn}(\mathbf{y})H^{\alpha\beta 0}_{m,ny} + f^{\alpha\beta 0}_{ij}(x_1, x_2, \mathbf{y}))_{,jy} = 0 \text{ in } Y; \\
(c_{ijmn}(\mathbf{y})H^{\alpha\beta 0}_{m,ny} + f^{\alpha\beta 0}_{ij}(x_1, x_2, \mathbf{y}))n_j = 0 \text{ on } S; \\
\mathbf{H}^{\alpha\beta 0}(\mathbf{y}) \text{ is periodic in } y_1,\, y_2 \text{ with periodicity cell } T;
\end{cases}
\tag{7.58}
$$

the first-order problem

$$
\begin{cases}
(c_{ijmn}(\mathbf{y})H^{\alpha\beta 1}_{m,ny} + b_{ijmn}(x_1, x_2, \mathbf{y})H^{\alpha\beta 0}_{m,ny} \\
+f^{\alpha\beta 1}_{ij}(x_1, x_2, \mathbf{y}))_{,jy} = 0 \text{ in } Y; \\
(c_{ijmn}(\mathbf{y})H^{\alpha\beta 1}_{m,ny} + b_{ijmn}(x_1, x_2, \mathbf{y})H^{\alpha\beta 0}_{m,ny} \\
+f^{\alpha\beta 1}_{ij}(x_1, x_2, \mathbf{y}))n_j = 0 \text{ on } S; \\
\mathbf{H}^{\alpha\beta 1}(\mathbf{y}) \text{ is periodic in } y_1,\, y_2 \text{ with periodicity cell } T.
\end{cases}
\tag{7.59}
$$

Formula (7.56) accounting the first-order corrector, only, takes the form

$$
\mathbf{H}^{\alpha\beta}(x_1, x_2, \mathbf{y}) = \mathbf{H}^{\alpha\beta 0}(x_1, x_2, \mathbf{y}) + \mu \mathbf{H}^{\alpha\beta 1}(x_1, x_2, \mathbf{y}).
\tag{7.60}
$$

7.3.2 *Representation of the Homogenized Stiffnesses as a Quadratic Functional*

Let us write the representation of the homogenized stiffnesses in the form of quadratic functional. This will simplify subsequent calculations. For this purpose, we consider the problem (7.52) with free term (7.53);

$$
\begin{cases}
(h_{ijmn}(x_1, x_2, \mathbf{y})H^{\eta\alpha\beta}_{m,ny} + \eta h_{ij\alpha\beta}(x_1, x_2, \mathbf{y}) \\
-h_{ij11}(x_1, \mathbf{y})y_\alpha)_{,jy} = 0 \text{ in } Y; \\
(h_{ijmn}(x_1, \mathbf{y})H^{\eta\alpha\beta}_{m,ny} + \eta h_{ij\alpha\beta}(x_1, x_2, \mathbf{y}) \\
-h_{ij\alpha\beta}(x_1, x_2, \mathbf{y})y_3)n_j = 0 \text{ on } S; \\
\mathbf{H}^{\eta\alpha\beta}(\mathbf{y}) \text{ is periodic in } y_1,\, y_2 \text{ with periodicity cell } T.
\end{cases}
\tag{7.61}
$$

Multiplying Eq. (7.61) by $H^{\eta\gamma\delta}_i$ and integrating the result by parts, we obtain

$$
\begin{aligned}
&\langle (h_{ijmn}(x_1, x_2, \mathbf{y}) H^{\eta\alpha\beta}_{m,ny} + \eta h_{ij\alpha\beta}(x_1, x_2, \mathbf{y}) - h_{ij\alpha\beta}(x_1, x_2, \mathbf{y}) y_3) H^{\eta\gamma\delta}_{i,jy} \rangle \\
&\quad = \int_S (h_{ijmn}(x_1, x_2, \mathbf{y}) H^{\eta\alpha\beta}_{m,ny} + \eta h_{ij\alpha\beta}(x_1, x_2, \mathbf{y}) \\
&\qquad\qquad - h_{ij\alpha\beta}(x_1, x_2, \mathbf{y}) y_3) H^{\eta\gamma\delta}_i n_j d\mathbf{y} \\
&\quad + \int_\omega (h_{ijmn}(x_1, x_2, \mathbf{y}) H^{\eta\alpha\beta}_{m,ny} + \eta h_{ij\alpha\beta}(x_1, x_2, \mathbf{y}) - \\
&\qquad\qquad - h_{ij\alpha\beta}(x_1, x_2, \mathbf{y}) y_3) H^{\eta\gamma\delta}_i n_j d\mathbf{y},
\end{aligned}
\tag{7.62}
$$

where ω are the opposite (parallel to the Oy_3-axis) parts of the periodicity cell boundary, see Fig. 6.5.

The first integral on the right-hand side (7.62) is equal to zero due to the boundary condition on S in (7.61). Due to the periodicity of $h_{ijmn}(\mathbf{y}) H^{\eta\alpha\beta}_{m,ny} + \eta h_{ij\alpha\beta}(\mathbf{y}) - h_{ij\alpha\beta}(\mathbf{y}) y_3$ and $\mathbf{H}^{\eta\alpha\beta}$, and the antiperiodicity of the vector-normal $\mathbf{n}$ with respect to y_1, y_2, the second integral on the right-hand side (7.62) also is equal to zero.

As a result,

$$
\langle (h_{ijmn}(x_1, x_2, \mathbf{y}) H^{\eta\alpha\beta}_{m,ny} + \eta h_{ij\alpha\beta}(x_1, x_2, \mathbf{y}) - h_{ij\alpha\beta}(x_1, x_2, \mathbf{y}) y_3) H^{\eta\gamma\delta}_{i,jy} \rangle = 0.
$$

By using that the average value of a sum is the sum of the average values, we rewrite this formula as

$$
\begin{aligned}
&\langle (h_{ijmn}(x_1, x_2, \mathbf{y}) H^{\eta\alpha\beta}_{m,ny}(\mathbf{y}) H^{\eta\gamma\delta}_{i,jy}(\mathbf{y}) \rangle \\
&+ \eta \langle h_{ij\alpha\beta}(x_1, x_2, \mathbf{y}) H^{\eta\gamma\delta}_{i,jy}(\mathbf{y}) \rangle - \langle h_{ij\alpha\beta}(x_1, x_2, \mathbf{y}) y_3 H^{\eta\gamma\delta}_{i,jy}(\mathbf{y}) \rangle = 0.
\end{aligned}
\tag{7.63}
$$

The first summand in (7.63) is a quadratic functional.

It is evident that

$$
\mathbf{H}^{\eta\gamma\delta}(\mathbf{y}) = \eta \mathbf{K}^{0\gamma\delta}(\mathbf{y}) + \mathbf{K}^{1\gamma\delta}(\mathbf{y}),
\tag{7.64}
$$

where $\mathbf{K}^{0\gamma\delta}$ is the solution to the periodicity cell problem (7.21), $\mathbf{K}^{1\gamma\delta}$ is the solution to the periodicity cell problem (7.32).

Let's consider the last two terms in (7.63). Taking into account (7.64), we have that the first of them is equal to

$$
\begin{aligned}
&\eta \langle h_{ij\alpha\beta}(x_1, x_2, \mathbf{y}) (\eta \mathbf{K}^{0\gamma\delta} + \mathbf{K}^{1\gamma\delta})_{i,jy}(\mathbf{y}) \rangle \\
&= \eta^2 \langle h_{ij\alpha\beta}(x_1, x_2, \mathbf{y}) K^{0\gamma\delta}_{i,jy}(\mathbf{y}) \rangle + \eta \langle h_{ij\alpha\beta}(x_1, x_2, \mathbf{y}) K^{1\gamma\delta}_{i,jy}(\mathbf{y}) \rangle.
\end{aligned}
\tag{7.65}
$$

The second of them in (7.63) is equal to

$$
\begin{aligned}
&\eta \langle h_{ij\alpha\beta}(x_1, x_2, \mathbf{y}) y_3 (\eta \mathbf{K}^{0\gamma\delta}(\mathbf{y}) + \mathbf{K}^{1\gamma\delta}(\mathbf{y}))_{i,jy} \rangle \\
&= \eta^2 \langle h_{ij\alpha\beta}(x_1, x_2, \mathbf{y}) y_3 K^{0\gamma\delta}_{i,jy}(\mathbf{y}) \rangle + \eta \langle h_{ij\alpha\beta}(x_1, x_2, \mathbf{y}) y_3 K^{1\gamma\delta}_{i,jy}(\mathbf{y}) \rangle.
\end{aligned}
\tag{7.66}
$$

Formulas (7.38) for calculating the homogenized stiffnesses of the plate can be written as

$$
\begin{aligned}
A^0_{\alpha\beta\gamma\delta}(\sigma) &= \langle h_{\alpha\beta\gamma\delta}(x_1, x_2, \mathbf{y})\rangle + \langle h_{\alpha\beta mn}(x_1, x_2, \mathbf{y}) K^{0\gamma\delta}_{m,ny}(\mathbf{y})\rangle;\\
A^1_{\alpha\beta\gamma\delta}(\sigma) &= \langle h_{\alpha\beta\gamma\delta}(x_1, x_2, \mathbf{y}) y_3\rangle + \langle h_{\alpha\beta mn}(x_1, x_2, \mathbf{y}) y_3 K^{0\gamma\delta}_{m,ny}(\mathbf{y})\rangle;\\
A^1_{\alpha\beta\gamma\delta}(\sigma) &= \langle h_{\alpha\beta\gamma\delta}(x_1, x_2, \mathbf{y}) y_3\rangle + \langle h_{\alpha\beta mn}(x_1, x_2, \mathbf{y}) K^{1\gamma\delta}_{m,ny}(\mathbf{y})\rangle;\\
A^2_{\alpha\beta\gamma\delta}(\sigma) &= \langle h_{\alpha\beta\gamma\delta}(x_1, x_2, \mathbf{y}) y_3^2\rangle + \langle h_{\alpha\beta mn}(x_1, x_2, \mathbf{y}) y_3 K^{0\gamma\delta}_{m,ny}(\mathbf{y})\rangle.
\end{aligned}
\tag{7.67}
$$

Taking into account (7.67) and equality $h_{ij\alpha\beta} = h_{\alpha\beta ij}$ (4.19), we obtain that the expression (7.65) is equal to

$$
\eta^2(A^0_{\alpha\beta\gamma\delta}(\sigma) - \langle h_{\alpha\beta\gamma\delta}(x_1, x_2, \mathbf{y})\rangle) + \eta(A^1_{\alpha\beta\gamma\delta}(\sigma) + \langle h_{\alpha\beta\gamma\delta}(x_1, x_2, \mathbf{y}) y_3\rangle)
$$

and expression (7.66) is equal to

$$
\eta(A^1_{\alpha\beta\gamma\delta}(\sigma) + \langle h_{\alpha\beta\gamma\delta}(x_1, x_2, \mathbf{y})\rangle) - (A^2_{\alpha\beta\gamma\delta}(\sigma) - \langle h_{\alpha\beta\gamma\delta}(x_1, x_2, \mathbf{y}) y_3^2\rangle).
$$

Then (7.63) takes the form

$$
\begin{aligned}
&\langle (h_{ijmn}(\mathbf{y}) H^{\eta\alpha\beta}_{m,ny}(\mathbf{y}) H^{\eta\gamma\delta}_{i,jy}(\mathbf{y})\rangle + \eta^2(A^0_{\alpha\beta\gamma\delta}(\sigma) - \langle h_{\alpha\beta\gamma\delta}(x_1, x_2, \mathbf{y})\rangle)\\
&\quad + 2\eta(A^1_{\alpha\beta\gamma\delta}(\sigma) + \langle h_{\alpha\beta\gamma\delta}(x_1, x_2, \mathbf{y}) y_3\rangle)\\
&\quad + (A^2_{\alpha\beta\gamma\delta}(\sigma) - \langle h_{\alpha\beta\gamma\delta} y_3^2(x_1, x_2, \mathbf{y})\rangle) = 0.
\end{aligned}
\tag{7.68}
$$

Thus, to obtain the expansions of the homogenized stiffnesses $A^0_{\alpha\beta\gamma\delta}(\sigma)$, $A^1_{\alpha\beta\gamma\delta}(\sigma)$ and $A^2_{\alpha\beta\gamma\delta}(\sigma)$ in the parameter μ, it is sufficient to obtain the expansion in μ of the first (quadratic) term on the left-hand side (7.68). Substituting the expansion (7.56) with $\mathbf{H}^{\alpha\beta}(\mathbf{y}) = \mathbf{H}^{\eta\alpha\beta}(\mathbf{y})$ into this term, we obtain that it is equal to

$$
\begin{aligned}
&\langle (c_{ijmn}(\mathbf{y}) + \mu b_{ijmn}(x_1, x_2, \mathbf{y}))(H^{\eta\alpha\beta 0}_{m,ny}(\mathbf{y}) + \mu H^{\eta\alpha\beta 1}_{m,ny}(\mathbf{y}) + \ldots)\\
&\quad \times (H^{\eta\gamma\delta 0}_{i,jy}(\mathbf{y}) + \mu H^{\eta\gamma\delta 1}_{i,jy}(\mathbf{y}) + \ldots)\rangle\\
&\quad = \langle c_{ijmn}(\mathbf{y}) H^{\eta\alpha\beta 0}_{m,ny}(\mathbf{y}) H^{\eta\gamma\delta 0}_{i,jy}(\mathbf{y})\rangle\\
&+\mu[\langle b_{ijmn}(\mathbf{y}) H^{\eta\alpha\beta 0}_{m,ny}(\mathbf{y}) H^{\eta\gamma\delta 0}_{i,jy}(\mathbf{y})\rangle + \langle c_{ijmn}(\mathbf{y}) H^{\eta\alpha\beta 0}_{m,ny}(\mathbf{y}) H^{\eta\gamma\delta 1}_{i,jy}(\mathbf{y})\rangle\\
&\quad + \langle c_{ijmn}(\mathbf{y}) H^{\eta\alpha\beta 1}_{m,ny}(\mathbf{y}) H^{\eta\gamma\delta 0}_{i,jy}(\mathbf{y})\rangle] + \ldots .
\end{aligned}
\tag{7.69}
$$

We can eliminate $\mathbf{H}^{\eta\gamma\delta 1}$, $\mathbf{H}^{\eta\alpha\beta 1}$ from (7.69). To this end, we multiply the equation from (7.59) by $H^{\eta\gamma\delta 0}_i$ and integrate the result by parts. Taking into account the periodicity of the functions $\mathbf{H}^{\eta\gamma\delta 1}$, $\mathbf{H}^{\eta\alpha\beta 1}$, and $\mathbf{H}^{\eta\gamma\delta 0}$ in y_1, y_2 (similar to the calculations of the integrals in (7.62)) we obtain

$$\langle c_{ijmn}(\mathbf{y})H^{\eta\alpha\beta 1}_{m,ny}H^{\eta\gamma\delta 0}_{i,jy}\rangle + \langle b_{ijmn}(x_1,x_2,\mathbf{y})H^{\eta\alpha\beta 0}_{m,ny}H^{\eta\gamma\delta 0}_{i,jy}\rangle + \\ +\langle f^{\alpha\beta 1}_{ij}(\mathbf{y})H^{\eta\gamma\delta 0}_{i,jy}\rangle = 0. \tag{7.70}$$

Then the last two summands in (7.70) are

$$\langle c_{ijmn}(\mathbf{y})H^{\eta\alpha\beta 0}_{m,ny}(\mathbf{y})H^{\eta\gamma\delta 1}_{i,jy}(\mathbf{y})\rangle \\ = \langle b_{ijmn}(x_1,x_2,\mathbf{y})H^{\eta\alpha\beta 0}_{m,ny}(\mathbf{y})H^{\eta\gamma\delta 0}_{i,jy}(\mathbf{y})\rangle - \langle f^{\gamma\delta 1}_{ij}(\mathbf{y})H^{\eta\alpha\beta 0}_{i,jy}(\mathbf{y})\rangle;$$

$$\langle c_{ijmn}(\mathbf{y})H^{\eta\alpha\beta 1}_{m,ny}(\mathbf{y})H^{\eta\gamma\delta 0}_{i,jy}(\mathbf{y})\rangle \\ = \langle b_{ijmn}(x_1,x_2,\mathbf{y})H^{\eta\alpha\beta 0}_{m,ny}(\mathbf{y})H^{\eta\gamma\delta 0}_{i,jy}(\mathbf{y})\rangle - \langle f^{\alpha\beta 1}_{ij}(\mathbf{y})H^{\eta\gamma\delta 0}_{i,jy}(\mathbf{y})\rangle.$$

Then the multiplier at μ in (7.69) can be written as

$$\langle b_{ijmn}(x_1,x_2,\mathbf{y})H^{\eta\alpha\beta 0}_{m,ny}(\mathbf{y})H^{\eta\gamma\delta 0}_{i,jy}(\mathbf{y})\rangle \\ -\langle b_{ijmn}(x_1,x_2,\mathbf{y})H^{\eta\alpha\beta 0}_{m,ny}(\mathbf{y})H^{\eta\gamma\delta 0}_{i,jy}(\mathbf{y})\rangle \\ -\langle f^{\gamma\delta 1}_{ij}(\mathbf{y})H^{\eta\alpha\beta 0}_{i,jy}(\mathbf{y})\rangle - \langle b_{ijmn}(x_1,x_2,\mathbf{y})H^{\eta\alpha\beta 0}_{m,ny}(\mathbf{y})H^{\eta\gamma\delta 0}_{i,jy}(\mathbf{y})\rangle \\ -\langle f^{\alpha\beta 1}_{ij}(\mathbf{y})H^{\eta\gamma\delta 0}_{i,jy}(\mathbf{y})\rangle \\ = \langle b_{ijmn}(x_1,x_2,\mathbf{y})H^{\eta\alpha\beta 0}_{m,ny}(\mathbf{y})H^{\eta\gamma\delta 0}_{i,jy}(\mathbf{y})\rangle \\ -\langle f^{\gamma\delta 1}_{ij}(\mathbf{y})H^{\eta\alpha\beta 0}_{i,jy}(\mathbf{y})\rangle - \langle f^{\alpha\beta 1}_{ij}(\mathbf{y})H^{\eta\gamma\delta 0}_{i,jy}(\mathbf{y})\rangle. \tag{7.71}$$

In our computations

$$f^{\alpha\beta}_{ij}(x_1,x_2,\mathbf{y}) = \eta b_{ij\alpha\beta}(x_1,x_2,\mathbf{y}) - b_{ij\alpha\beta}(x_1,x_2,\mathbf{y})y_3. \tag{7.72}$$

Substituting (7.72) into (7.71) and taking into account definition of b_{ijkl} (7.4), we obtain that the multiplier of μ in (7.69) is

$$\langle \sigma^{*(-3)}_{jn}(x_1,x_2,\mathbf{y})H^{\eta\alpha\beta 0}_{m,ny}(\mathbf{y})H^{\eta\gamma\delta 0}_{m,jy}(\mathbf{y})\rangle \\ -\eta[\langle \sigma^{*(-3)}_{j\delta}(x_1,x_2,\mathbf{y})H^{\eta\alpha\beta 0}_{\gamma,jy}(\mathbf{y})\rangle + \langle \sigma^{*(-3)}_{j\beta}(x_1,x_2,\mathbf{y})H^{\eta\gamma\delta 0}_{\alpha,jy}(\mathbf{y})\rangle] \\ +\langle \sigma^{*(-3)}_{j\delta}(x_1,x_2,\mathbf{y})y_3H^{\eta\alpha\beta 0}_{\gamma,jy}(\mathbf{y})\rangle - \langle \sigma^{*(-3)}_{j\beta}(x_1,x_2,\mathbf{y})y_3H^{\eta\gamma\delta 0}_{\alpha,jy}(\mathbf{y})\rangle.$$

By virtue of Proposition 6.3

$$\langle \sigma^{*(-3)}_{j\delta}(x_1,x_2,\mathbf{y})H^{\eta\alpha\beta 0}_{\gamma,jy}\rangle = 0,\ \langle \sigma^{*(-3)}_{j\beta}(x_1,x_2,\mathbf{y})H^{\eta\gamma\delta 0}_{\alpha,jy}\rangle = 0.$$

As a result, we obtain that the multiplier of μ in (7.69) is equal to

$$\begin{aligned}&\langle\sigma^*_{jn}(x_1,x_2,\mathbf{y})H^{\eta\alpha\beta0}_{m,ny}H^{\eta\gamma\delta0}_{m,jy}\rangle\\&+\langle\sigma^{*(-3)}_{j\delta}(x_1,x_2,\mathbf{y})y_3H^{\eta\alpha\beta0}_{\gamma,jy}\rangle-\langle\sigma^{*(-3)}_{j\beta}(x_1,x_2,\mathbf{y})y_3H^{\eta\gamma\delta0}_{\alpha,jy}\rangle.\end{aligned}\tag{7.73}$$

The right side of the formula (7.69) takes the form

$$\begin{aligned}&\langle c_{ijmn}(x_1,x_2,\mathbf{y})H^{\eta\alpha\beta0}_{m,ny}(\mathbf{y})H^{\eta\gamma\delta0}_{i,jy}(\mathbf{y})\rangle\\&+\mu[-\langle\sigma^{*(-3)}_{jn}(x_1,x_2,\mathbf{y})H^{\eta\alpha\beta0}_{m,ny}(\mathbf{y})H^{\eta\gamma\delta0}_{m,jy}(\mathbf{y})\rangle\\&+\langle\sigma^{*(-3)}_{j\delta}(x_1,x_2,\mathbf{y})y_3H^{\eta\alpha\beta0}_{\gamma,jy}(\mathbf{y})\rangle-\langle\sigma^{*(-3)}_{j\beta}(x_1,x_2,\mathbf{y})y_3H^{\eta\gamma\delta0}_{\alpha,jy}(\mathbf{y})\rangle].\end{aligned}\tag{7.74}$$

From (7.64) and (7.58) it follows that

$$\mathbf{H}^{\eta\alpha\beta0}(\mathbf{y})=\eta\mathbf{N}^{0\alpha\beta}(\mathbf{y})+\mathbf{N}^{1\alpha\beta}(\mathbf{y}),\tag{7.75}$$

where $\mathbf{N}^{0\alpha\beta}$, $\mathbf{N}^{1\alpha\beta}$ are solutions to the periodicity cell problems (7.21) and (7.30) for $\sigma^{*(-3)}_{ij}=0$ (that is, solutions to the periodicity cell problems (9.32) and (9.45)). Bearing in mind the equality (7.75), we can rewrite the formula (7.69) as

$$\begin{aligned}&\eta^2\langle c_{ijmn}(\mathbf{y})N^{0\alpha\beta}_{m,ny}(\mathbf{y})N^{0\gamma\delta}_{i,jy}(\mathbf{y})\rangle+\eta(\langle c_{ijmn}(\mathbf{y})N^{1\alpha\beta}_{m,ny}(\mathbf{y})N^{0\gamma\delta}_{i,jy}(\mathbf{y})\rangle\\&+\langle c_{ijmn}(\mathbf{y})N^{0\alpha\beta}_{m,ny}N^{1\gamma\delta}_{i,jy}(\mathbf{y})\rangle)+\langle c_{ijmn}(\mathbf{y})N^{1\alpha\beta}_{m,ny}N^{1\gamma\delta}_{i,jy}(\mathbf{y})\rangle\\&+\mu\{-\eta^2\langle\sigma^{*(-3)}_{jn}(x_1,x_2,\mathbf{y})N^{0\alpha\beta}_{m,ny}(\mathbf{y})N^{0\gamma\delta}_{m,jy}(\mathbf{y})\rangle\\&-\eta[\langle\sigma^{*(-3)}_{jn}(x_1,x_2,\mathbf{y})N^{1\alpha\beta}_{m,ny}(\mathbf{y})N^{0\gamma\delta}_{m,jy}(\mathbf{y})\rangle\\&+\langle\sigma^{*(-3)}_{jn}(x_1,x_2,\mathbf{y})N^{0\alpha\beta}_{m,ny}(\mathbf{y})N^{1\gamma\delta}_{m,jy}(\mathbf{y})\rangle]\\&-\langle\sigma^{*(-3)}_{jn}(x_1,x_2,\mathbf{y})N^{1\alpha\beta}_{m,ny}(\mathbf{y})N^{1\gamma\delta}_{m,jy}(\mathbf{y})\rangle\\&+\eta\langle\sigma^{*(-3)}_{j\delta}(x_1,x_2,\mathbf{y})y_3N^{0\alpha\beta}_{\gamma,jy}(\mathbf{y})\rangle-\eta\langle\sigma^{*(-3)}_{j\beta}(\mathbf{y})y_3N^{0\gamma\delta}_{\alpha,jy}(\mathbf{y})\rangle\\&+(\langle\sigma^{*(-3)}_{j\delta}(x_1,x_2,\mathbf{y})y_3N^{1\alpha\beta0}_{\gamma,jy}(\mathbf{y})\rangle-\langle\sigma^{*(-3)}_{j\beta}(x_1,x_2,\mathbf{y})y_3N^{1\gamma\delta}_{\alpha,jy}(\mathbf{y})\rangle)\}\\&+\eta^2(A^0_{\alpha\beta\gamma\delta}(\sigma)-\langle h_{\alpha\beta\gamma\delta}(x_1,x_2,\mathbf{y})\rangle)\\&+2\eta(A^1_{\alpha\beta\gamma\delta}(\sigma)-\langle h_{\alpha\beta\gamma\delta}(x_1,x_2,\mathbf{y})y_3\rangle)\\&+(A^2_{\alpha\beta\gamma\delta}(\sigma)-\langle h_{\alpha\beta\gamma\delta}(x_1,x_2,\mathbf{y})y_3^2\rangle).\end{aligned}\tag{7.76}$$

When equating in (7.76) the terms with the same powers of η^0, η^1, η^2 (recall that η is an arbitrary parameter), we obtain

$$
\begin{aligned}
&\langle c_{ijmn}(\mathbf{y})N^{1\alpha\beta}_{m,ny}N^{1\gamma\delta}_{i,jy}(\mathbf{y})\rangle+\mu\{\langle\sigma^{*(-3)}_{jn}(x_1,x_2,\mathbf{y})N^{1\alpha\beta}_{m,ny}(\mathbf{y})N^{1\gamma\delta}_{m,jy}(\mathbf{y})\rangle\\
&-\langle\sigma^{*(-3)}_{j\delta}(x_1,x_2,\mathbf{y})y_3N^{1\alpha\beta}_{\gamma,jy}(\mathbf{y})\rangle-\langle\sigma^{*(-3)}_{j\beta}(x_1,x_2,\mathbf{y})y_3N^{1\gamma\delta}_{\alpha,jy}\rangle\}\\
&+(A^2_{\alpha\beta\gamma\delta}-\langle h_{\alpha\beta\gamma\delta}(x_1,x_2,\mathbf{y})y_3^2\rangle)=0;\\
&\langle c_{ijmn}(\mathbf{y})N^{1\alpha\beta}_{m,ny}(\mathbf{y})N^{0\gamma\delta}_{i,jy}(\mathbf{y})\rangle+\langle c_{ijmn}(\mathbf{y})N^{0\alpha\beta}_{m,ny}(\mathbf{y})N^{1\gamma\delta}_{i,jy}(\mathbf{y})\rangle\\
&+\mu\{\langle\sigma^{*(-3)}_{jn}(x_1,x_2,\mathbf{y})N^{1\alpha\beta}_{m,ny}(\mathbf{y})N^{0\gamma\delta}_{m,jy}(\mathbf{y})\rangle\\
&+\langle\sigma^{*(-3)}_{jn}(x_1,x_2,\mathbf{y})N^{0\alpha\beta}_{m,ny}(\mathbf{y})N^{1\gamma\delta}_{m,jy}(\mathbf{y})\rangle+\\
&+\langle\sigma^{*(-3)}_{j\delta}(x_1,x_2,\mathbf{y})y_3N^{1\alpha\beta0}_{\gamma,jy}\rangle+\langle\sigma^{*(-3)}_{j\beta}(x_1,x_2,\mathbf{y})y_3N^{1\gamma\delta}_{\alpha,jy}(\mathbf{y})\rangle\}\\
&+(A^1_{\alpha\beta\gamma\delta}-\langle h_{\alpha\beta\gamma\delta}(x_1,x_2,\mathbf{y})y_3\rangle)=0;\\
&\langle c_{ijmn}(\mathbf{y})N^{0\alpha\beta}_{m,ny}(\mathbf{y})N^{0\gamma\delta}_{i,jy}(\mathbf{y})\rangle+\mu\{\langle\sigma^{*(-3)}_{jn}(x_1,x_2,\mathbf{y})N^{0\alpha\beta}_{m,ny}(\mathbf{y})N^{0\gamma\delta}_{m,jy}(\mathbf{y})\rangle\}\\
&+(A^0_{\alpha\beta\gamma\delta}-\langle h_{\alpha\beta\gamma\delta}(x_1,x_2,\mathbf{y})\rangle)=0.
\end{aligned}
\tag{7.77}
$$

From Eqs. (7.77), we can express $A^0_{\alpha\beta\gamma\delta}$, $A^1_{\alpha\beta\gamma\delta}$, and $A^2_{\alpha\beta\gamma\delta}$ as follows:

$$
\begin{aligned}
&A^2_{\alpha\beta\gamma\delta}(\sigma)=\langle h_{\alpha\beta\gamma\delta}(x_1,x_2,\mathbf{y})y_3^2\rangle+\langle c_{ijmn}(\mathbf{y})N^{1\alpha\beta}_{m,ny}(\mathbf{y})N^{1\gamma\delta}_{i,jy}(\mathbf{y})\rangle\\
&+\mu\{\langle\sigma^{*(-3)}_{jn}(x_1,x_2,\mathbf{y})N^{1\alpha\beta}_{m,ny}(\mathbf{y})N^{1\gamma\delta}_{m,jy}(\mathbf{y})\rangle\\
&-\langle\sigma^{*(-3)}_{j\delta}(x_1,x_2,\mathbf{y})y_3N^{1\alpha\beta}_{\gamma,jy}(\mathbf{y})\rangle-\langle\sigma^{*(-3)}_{j\beta}(x_1,x_2,\mathbf{y})y_3N^{1\gamma\delta}_{\alpha,jy}(\mathbf{y})\rangle\};\\
&A^1_{\alpha\beta\gamma\delta}(\sigma)=\langle h_{\alpha\beta\gamma\delta}(x_1,x_2,\mathbf{y})y_3\rangle\\
&+\langle c_{ijmn}(\mathbf{y})N^{1\alpha\beta}_{m,ny}(\mathbf{y})N^{0\gamma\delta}_{i,jy}(\mathbf{y})\rangle+\langle c_{ijmn}(\mathbf{y})N^{0\alpha\beta}_{m,ny}N^{1\gamma\delta}_{i,jy}(\mathbf{y})\rangle\\
&+\mu\{\langle\sigma^{*(-3)}_{jn}(x_1,x_2,\mathbf{y})N^{1\alpha\beta}_{m,ny}(\mathbf{y})N^{0\gamma\delta}_{m,jy}(\mathbf{y})\rangle\\
&+\langle\sigma^{*(-3)}_{jn}(x_1,x_2,\mathbf{y})N^{0\alpha\beta}_{m,ny}(\mathbf{y})N^{1\gamma\delta}_{m,jy}(\mathbf{y})\rangle\\
&+\langle\sigma^{*(-3)}_{j\delta}(x_1,x_2,\mathbf{y})y_3N^{1\alpha\beta0}_{\gamma,jy}(\mathbf{y})\rangle-\langle\sigma^{*(-3)}_{j\beta}(x_1,x_2,\mathbf{y})y_3N^{1\gamma\delta}_{\alpha,jy}(\mathbf{y})\rangle\};\\
&A^0_{\alpha\beta\gamma\delta}(\sigma)=\langle h_{\alpha\beta\gamma\delta}(x_1,x_2,\mathbf{y})\rangle+\langle c_{ijmn}(\mathbf{y})N^{0\alpha\beta}_{m,ny}(\mathbf{y})N^{0\gamma\delta}_{i,jy}(\mathbf{y})\rangle\\
&+\mu\langle\sigma^{*(-3)}_{jn}(x_1,x_2,\mathbf{y})N^{0\alpha\beta}_{m,ny}(\mathbf{y})N^{0\gamma\delta}_{m,jy}(\mathbf{y})\rangle.
\end{aligned}
\tag{7.78}
$$

The first two terms in each formula in (7.78) are the stiffnesses $A^0_{\alpha\beta\gamma\delta}$, $A^1_{\alpha\beta\gamma\delta}$, $A^2_{\alpha\beta\gamma\delta}$of the plate without initial stresses, given by formulas (9.54). Then we can rewrite formulas (7.78) as

$$A^2_{\alpha\beta\gamma\delta}(\sigma) = A^2_{\alpha\beta\gamma\delta} + \mu\{\langle\sigma^{*(-3)}_{jn}(x_1, x_2, \mathbf{y})N^{1\alpha\beta}_{m,ny}(\mathbf{y})N^{1\gamma\delta}_{m,jy}(\mathbf{y})\rangle$$

$$-\langle\sigma^{*(-3)}_{j\delta}(x_1, x_2, \mathbf{y})y_3N^{1\alpha\beta}_{\gamma,jy}(\mathbf{y})\rangle - \langle\sigma^{*(-3)}_{j\beta}(x_1, x_2, \mathbf{y})y_3N^{1\gamma\delta}_{\alpha,jy}(\mathbf{y})\rangle\};$$

$$A^1_{\alpha\beta\gamma\delta}(\sigma) = A^1_{\alpha\beta\gamma\delta} + \mu\{\langle\sigma^{*(-3)}_{jn}(x_1, x_2, \mathbf{y})N^{1\alpha\beta}_{m,ny}(\mathbf{y})N^{0\gamma\delta}_{m,jy}(\mathbf{y})\rangle$$

$$+\langle\sigma^{*(-3)}_{jn}(x_1, x_2, \mathbf{y})N^{0\alpha\beta}_{m,ny}(\mathbf{y})N^{1\gamma\delta}_{m,jy}(\mathbf{y})\rangle + \langle\sigma^{*(-3)}_{j\delta}(x_1, x_2, \mathbf{y})y_3N^{1\alpha\beta0}_{\gamma,jy}(\mathbf{y})\rangle \tag{7.79}$$

$$-\langle\sigma^{*}_{j\beta}(x_1, x_2, \mathbf{y})y_3N^{1\gamma\delta}_{\alpha,jy}(\mathbf{y})\rangle\};$$

$$A^0_{\alpha\beta\gamma\delta}(\sigma) = A^0_{\alpha\beta\gamma\delta} + \mu\langle\sigma^{*(-3)}_{jn}(x_1, x_2, \mathbf{y})N^{0\alpha\beta}_{m,ny}(\mathbf{y})N^{0\gamma\delta}_{m,jy}(\mathbf{y})\rangle.$$

7.4 Other Parameters of the Homogenized Model

The limit model parameters of the order of initial stresses include $R^{\mu}_{\gamma\delta\alpha\beta}(\sigma)$ (7.39) and a^{ν}_{α}. In the absence of initial stresses, they are equal to zero.

7.4.1 Coefficients $R^{\mu}_{\gamma\delta\alpha\beta}(\sigma)$

In accordance to (7.39), the coefficients

$$R^{\mu}_{\gamma\delta\alpha\beta}(\sigma) = \langle h_{\gamma\delta mn}(x_1, x_2, \mathbf{y})Z^{\alpha\beta}_{m,ny})(-1)^{\mu}y^{\mu}_3\rangle \ (\mu = 0, 1) \tag{7.80}$$

depend on solution $\mathbf{Z}^{\alpha\beta}(\mathbf{y})$ to the periodicity cell problem (7.33) and do not involve solutions to other periodicity cells. For small initial stresses, $\mathbf{Z}^{\alpha\beta 0}(\mathbf{y}) = 0$ and $\mathbf{Z}^{\alpha\beta 1}(\mathbf{y}) = 0$ are determined from solution to the problem

$$\begin{cases} (c_{ijmn}(x_1, x_2, \mathbf{y})Z^{\alpha\beta 1}_{m,ny})_{,jy} + \sigma^{*(-3)}_{\beta\alpha}(x_1, x_2, \mathbf{y})\delta_{i3} = 0 \text{ in } Y; \\ c_{ijmn}(x_1, x_2, \mathbf{y})Z^{\alpha\beta}_{m,ny}n_j = 0 \text{ on } S; \\ \mathbf{Z}^{\alpha\beta 1}(\mathbf{y}) \text{ periodic in } y_1, y_2 \text{ with periodicity cell } T. \end{cases} \tag{7.81}$$

Correspondingly, (7.80) becomes

$$R^{\mu}_{\gamma\delta\alpha\beta}(\sigma) = \langle c_{\gamma\delta mn}(x_1, x_2, \mathbf{y})Z^{\alpha\beta 1}_{m,ny})(-1)^{\mu}y^{\mu}_3\rangle \ (\mu = 0, 1). \tag{7.82}$$

7.4.2 Coefficients a_α^ν

In (7.46), coefficients

$$a^\nu_{\mu\delta\alpha\beta} = \langle \sigma^{*(-3)}_{\mu\delta}(x_1, x_2, \mathbf{y}) K_3^{\nu\alpha\beta} \rangle \\ - \langle \sigma^{*(-3)}_{3\gamma}(x_1, x_2, \mathbf{y}) K_\mu^{\nu\alpha\beta} \rangle. \tag{7.83}$$

We substitute (7.56) into (7.83). In this case, we need to replace $\sigma^{*(-3)}_{ij}$ in (7.46) for $\mu\sigma^{*(-3)}_{ij}$ and $\mathbf{K}^{\nu\alpha\beta}$ for $\mathbf{K}^{\nu\alpha\beta 0} + \mu\mathbf{K}^{\nu\alpha\beta 1} = \mathbf{N}^{\nu\alpha\beta} + \mu\mathbf{K}^{\nu\alpha\beta 1}$, see (7.60). Keeping the term linear in μ, we obtain

$$a^\nu_{\mu\delta\alpha\beta} = \mu(\langle \sigma^{*(-3)}_{\mu\delta}(x_1, x_2, \mathbf{y}) N_3^{\nu\alpha\beta}(\mathbf{y}) \rangle \\ - \langle \sigma^{*(-3)}_{3\gamma}(x_1, x_2, \mathbf{y}) N_\mu^{\nu\alpha\beta}(\mathbf{y}) \rangle), \tag{7.84}$$

where $\mathbf{N}^{0\alpha\beta}$, $\mathbf{N}^{1\alpha\beta}$ are solutions of the periodicity cell problems (7.21), (7.30) for $\sigma^{*(-3)}_{ij} = 0$ (i.e. for a plate with no initial stresses).

7.5 Uniform Plates

We call uniform plates of constant thickness made of homogeneous material.

Let the plate be made of homogeneous isotropic material and have constant thickness. The solutions of the periodicity cell problems for such a plate are known, see, e.g., Sect. 5.3.5, formula (5.75), and have the form ($\nu = 0, 1; \alpha, \beta = 1, 2$)

$$\mathbf{N}^{\nu\alpha\beta} = \mathbf{C}^{\nu\alpha\beta} y_3^n, \tag{7.85}$$

where n is an integer and $\mathbf{C}^{\nu\alpha\beta}$ are constants.

Substituting (7.85) into the terms of order of μ in (7.79) and (7.84) leads to linear combinations of expressions of the form $\langle \sigma^{*(-3)}_{i3} y_3^m \rangle$, m is an integer. By virtue of Proposition 6.3 such expressions are equal to zero (it is sufficient to set in Proposition 6.3 $Z = y_3^{m+1}/(m+1)$. Thus, in a classical uniform plate the above-mentioned effects cannot take place.

7.6 Local Stresses

The total stresses $\sigma^{total}_{ij}(\mathbf{x})$ in the plate are the sum of the initial stresses $\sigma^{*\varepsilon}_{ij}(\mathbf{x})$ and additional stresses $\sigma^{\varepsilon}_{ij}(\mathbf{x})$:

$$\begin{aligned}\sigma_{ij}^{total}(\mathbf{x})) &= \sigma_{ij}^{*\varepsilon}(\mathbf{x}) + \sigma_{ij}^{\varepsilon}(\mathbf{x}) \\ &= \sigma_{ij}^{*\varepsilon}(\mathbf{x}) + \varepsilon^{-3}\sigma_{ij}^{-3}(\mathbf{x}) + \varepsilon^{-2}\sigma_{ij}^{-2}(\mathbf{x}).\end{aligned} \tag{7.86}$$

In accordance with (7.29), the additional stresses

$$\sigma_{ij}^{(-3)} = (\sigma_{j3}^{*(-3)}(x_1, x_2, \mathbf{y})\delta_{i\alpha} - \sigma_{j\alpha}^{*(-3)}(x_1, x_2, \mathbf{y})\delta_{i3})u_{3,\alpha x}^{(0)}, \tag{7.87}$$

and in accordance with (7.35)

$$\begin{aligned}\sigma_{ij}^{(-2)} &= (h_{ijmn}(x_1, x_2, \mathbf{y})K_{m,ny}^{0\alpha\beta} + h_{ij\alpha\beta}(x_1, x_2, \mathbf{y}))V_{\alpha,\beta x}(x_1, x_2) \\ &+(h_{ijmn}(x_1, x_2, \mathbf{y})K_{m,ny}^{1\alpha\beta} - y_3 h_{ij\alpha\beta}(x_1, x_2, \mathbf{y}))u_{3,\alpha x\beta x}^{(0)}(x_1, x_2) \\ &+(h_{ijmn}(x_1, x_2, \mathbf{y})Z_{m,ny}^{1\alpha\beta}(\mathbf{y}))u_{3,\alpha x\beta x}^{(0)}(x_1, x_2) \\ &+(h_{ij3\beta}(x_1, x_2, \mathbf{y}) - h_{ij\beta 3}(x_1, x_2, \mathbf{y}))V_{3,\beta x}.\end{aligned} \tag{7.88}$$

Thus, the total local stresses in the plate are given by the formula

$$\begin{aligned}\sigma_{ij}^{total} &= \sigma_{ij}^{*\varepsilon}(\mathbf{x}) + \varepsilon^{-3}(\sigma_{j3}^{*(-3)}(x_1, x_2, \mathbf{y})\delta_{i\alpha} - \sigma_{j\alpha}^{*(-3)}(x_1, x_2, \mathbf{y})\delta_{i3})u_{3,\alpha x}^{(0)} \\ &+\varepsilon^{-2}[(c_{ijmn}N_{m,ny}^{01} + c_{ij11})V_{1,1x}(x_1) \\ &+(c_{ijmn}N_{m,ny}^{1\alpha} + c_{ij11}y_\alpha)u_{\alpha,1x1x}^{(0)}(x_1) \\ &+(c_{ijmn}X_{m,ny}^{3} + c_{ij\gamma 1}s_\gamma y_\Gamma)\phi_{,1x}(x_1)].\end{aligned} \tag{7.89}$$

References

Andrianov IV, Kolpakov AG, Rakin SI (2024) Justification of the averaged Joule-Lenz law for composite materials. Appl Mech Tech Phys 65(2):300–307

Annin BD, Kolpakov AG, Rakin SI (2022) Boundary layers on the upper/lower surfaces of reinforced plates. Vestnik St Petersburg Univ 55(3):257–266

Caillerie D (1984) Thin elastic, periodic plates. Math Meth Appl Sci 6:159–191

Kalamkarov AL, Kolpakov AG (1997) Analysis, design and optimization of composite structures. Wiley, Chichester

Kelly A, McCartney LN, Clegg WJ, Stearn RJ (2005) Controlling thermal expansion to obtain negative expansivity using laminated composites. Comput Sci Technol 65(1):47–59

Kolpakov AG (2004) Stressed composite structures: homogenized models for thin-walled nonhomogeneous structures with initial stresses. Springer, Berlin

Kolpakov AG, Rakin SI (1986) On problem of sintsesys of composite material of unidirectional structure with prescribed characteristics. J Appl Mech Tech Phys 6:143–150

Miller W, Smith CW, Mackenzie DS, Evans KE (2009) Negative thermal expansion: a review. J Mater Sci 44:5441–5451

Chapter 8
Uniform Beams with Initial Stresses

The following chapters of this book are devoted to the asymptotic analysis of beams—structural elements occupying small diameter domains.

In this chapter, we consider beams of constant cross-section made of homogeneous material (hereinafter referred to as uniform beams). Such kind beams are traditional and widely used structural elements. The generally accepted opinion is that Euler and Bernoulli were the first to formulate a complete theory of beams. Their predecessors in attempts to construct a theory of beams were Galilei and da Vinci. It is known that the Euler-Bernoulli beam theory does not take into account the transverse deformation of the beam as a result of the Poisson effect, thus, does not correspond to the physics reality. Nevertheless, the Euler-Bernoulli beam theory does give correct formulas for calculating the bending stiffness of the beam. A rigorous analysis of the beam bending and torsion problem on the basis of the elasticity theory was carried out by de Saint-Venant. The historical details may be found in Timoshenko (1983), see also Euler (1759), Bernoulli (1751), de Saint-Venant (1856).

Asymptotic analysis of the uniform beam was given in Tutek and Aganovich (1987). Analysis of non-uniform and/or inhomogeneous beam of arbitrary structure was started in Kolpakov (1991), Trabucho and Viano (1996), Rodriguez and Viano (1997). Analysis of non-uniform and/or inhomogeneous beam with initial stresses was started in Kolpakov (1992).

8.1 Statement of the Problem

Consider a cylindrical body G_ε of small diameter εR located along the Ox_1-axis, see Fig. 8.1. The smallness of ε is formalized as $\varepsilon \to 0$. Denote cross-section of the cylinder G_ε by εC. In the uniform beam, the cross-section does not depend on x_1. When $\varepsilon \to 0$, the three-dimensional beam tends to a segment $[-a, a]$ on the Ox_1-

A. Kolpakov, *Thin-Walled Structures with Initial Stresses*, Mathematical Engineering,
https://doi.org/10.1007/978-3-032-07247-4_8

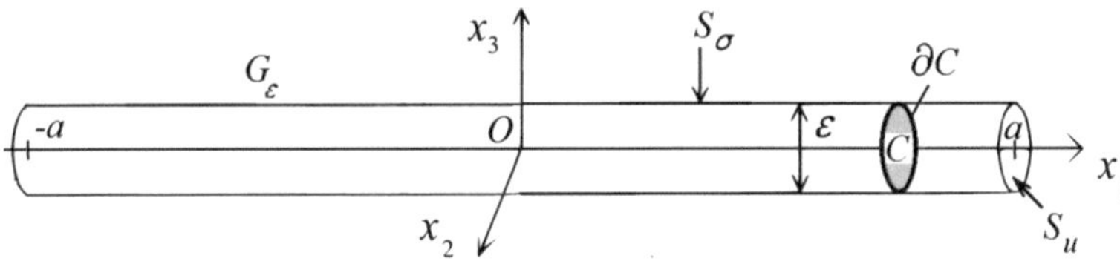

Fig. 8.1 Three-dimensional small-diameter body—uniform beam RISUNOK DO -a

axis (see Fig. 8.1)—a one-dimensional beam. Our goal is to derive a model describing this one-dimensional beam.

The starting point of our investigation is the three-dimensional elasticity theory problem for a body with initial stresses presented in Sect. 4.1. In accordance with this model, the equilibrium equations of a three-dimensional elastic body with initial stresses are written in the form

$$\frac{\partial}{\partial x_j}\left(h_{ijmn}(x_1, \mathbf{x}/\varepsilon)\frac{\partial u_m^\varepsilon}{\partial x_n}\right) + f^\varepsilon(\mathbf{x}) = 0 \text{ in } G_\varepsilon. \tag{8.1}$$

The boundary conditions can be written in the form

$$h_{ijmn}(x_1, \mathbf{x}/\varepsilon)\frac{\partial u_m^\varepsilon}{\partial x_n} n_j = h_i^\varepsilon(\mathbf{x}) \text{ on } S_\sigma; \tag{8.2}$$

$$\mathbf{u}^\varepsilon(\mathbf{x}) = 0 \text{ on } S_u.$$

The mass force is taken in the form

$$\mathbf{f}^\varepsilon(\mathbf{x}) = (\varepsilon^{-2} f_1(\mathbf{x}/\varepsilon), \quad \varepsilon^{-2} f_2(\mathbf{x}/\varepsilon), \quad \varepsilon^{-3} f_3(\mathbf{x}/\varepsilon))).$$

The surface force is taken in the form

$$\mathbf{h}^\varepsilon(\mathbf{x}) = (\varepsilon^{-1} h_1(\mathbf{x}/\varepsilon), \quad \varepsilon^{-1} h_2(\mathbf{x}/\varepsilon), \quad \varepsilon^{-2} h_3(\mathbf{x}/\varepsilon))).$$

Such choice of the mass and surface forces guaranties that the limit model will be not trivial.

Here, S_σ is the lateral surface of the beam, $\mathbf{n}$ is a normal to S_σ; the beam is fastened along the surface S_u (see Fig. 8.1), and $\mathbf{u}^\varepsilon(\mathbf{x})$ are the displacements. The coefficients h_{ijmn} are combination of the tensor of elasticity constants $\varepsilon^{-4} c_{ijmn}$ and initial stresses σ_{ij}^* described in Sect. 4.2, see (4.14). For beams, is taken in the form,

$$h_{ijmn}(x_1, \mathbf{x}/\varepsilon) = \varepsilon^{-4} c_{ijmn} + \varepsilon^{-4} b_{ijmn}^{(-4)}(x_1, x_2/\varepsilon, x_3/\varepsilon) \tag{8.3}$$

$$+ \varepsilon^{-3} b_{ijmn}^{(-3)}(x_1, x_2/\varepsilon, x_3/\varepsilon) + \varepsilon^{-2} b_{ijmn}^{(-2)}(x_1, x_2/\varepsilon, x_3/\varepsilon),$$

where

$$b^{(p)}_{ijmn}(x_1, x_2/\varepsilon, x_2/\varepsilon) = \sigma^{*(p)}_{jn}(x_1, x_2/\varepsilon, x_2/\varepsilon)\delta_{im} \ (p = -3, -2). \tag{8.4}$$

8.2 Beam Subjected to Axial Resultant Initial Stresses

We start our consideration with the comment on the choice of the powers of the parameter ε in (8.3) and (8.4).

Note on the coupling of material and geometrical characteristics in the model of beam. In (8.4), one sees the product of the powers of ε (geometrical characteristic ε has the meaning of the diameter of the beam) and material constants c_{ijmn} and combinations of initial stresses $\sigma^{*(-4)}_{jn}$, $\sigma^{*(-3)}_{jn}$ and $\sigma^{*(-2)}_{jn}$. In (8.3) the multiplier ε^{-4} at c_{ijmn} guarantees that the bending and torsional stiffnesses of the beam will be nonzero, as $\varepsilon \to 0$. The term $\varepsilon^{-2}b^{(-2)}_{ijmn} = \varepsilon^{-1}\sigma^{*(-2)}_{jn}\delta_{im}$ corresponds to the resultant axial forces in the beam independent on ε. If the axial resultant force is zero,

$$\langle\sigma^{*(-2)}_{11}\rangle = 0, \tag{8.5}$$

the local initial stresses in the three-dimensional beam can be nonzero. In order to take into account this fact, we introduce the term $\varepsilon^{-3}b^{(-3)}_{ijmn} = \varepsilon^{-3}\sigma^{*(-3)}_{jn}\delta_{im}$, corresponding to the resultant moments of initial stresses. The term $\varepsilon^{-4}b^{(-)}_{ijmn}$ corresponds to the initial stresses of order of elastic constants.

We meet here, like in the case of thin plates with initial stresses, see Sect. 5.1, the phenomenon of coupling of material and geometrical characteristics in the limit model of beam. This coupling is known from the classical theory, see, e.g., Timoshemko and Goodier (1951). For example formula $E\pi D^4/64$ for computation of the bending stiffnesses of uniform circular beam (Timoshemko and Goodier 1951) involves Young's modulus E and diameter D of the beam, which are characteristics of different nature.

8.2.1 Two-Scale Expansions and Homogenized Equations of Equilibrium

In the problem under consideration, we naturally distinguish two scales. Due to the small diameter of the body shown in Fig. 8.1, the stress-strain state in this body in changes significantly at a distance of the order of $\varepsilon \ll 1$ in the directions Ox_2 and Ox_3. In the direction Ox_1, the stress-strain state changes significantly at a distance of the order of unit.

Thus, the problem under consideration is described by the functions of arguments $(x_1, x_2/\varepsilon, x_2/\varepsilon)$. We introduce the "slow" variable x_1, and the "fast" variables $y_2 = x_2/\varepsilon$ and $y_3 = x_3/\varepsilon$. In accordance with this note, the elastic constants and initial stresses will be written in the form c_{ijmn} and $\sigma_{jn}^{*(p)}(x_1, y_2, y_3)$, correspondingly.

Additional stresses $\sigma_{ij}^{\varepsilon}$ are related with additional displacements $\mathbf{u}^{\varepsilon}$ by the formula,

$$\sigma_{ij}^{\varepsilon} = h_{ijmn}(x_1, x_2/\varepsilon, x_3/\varepsilon)\frac{\partial u_m^{\varepsilon}}{\partial x_n}. \tag{8.6}$$

Formula (8.6) can be considered as a local governing equation for the body under consideration.

We will study global deformations of the beams, as $\varepsilon \to 0$. To do this, we use the following asymptotic expansion:

asymptotic expansion for displacements

$$\mathbf{u}^{\varepsilon} = \mathbf{u}^{(0)}(x_1) + \sum_{k=1}^{\infty} \varepsilon^k \mathbf{u}^{(k)}(x_1, y_2, y_3); \tag{8.7}$$

asymptotic expansion for stresses

$$\sigma_{ij}^{\varepsilon} = \sum_{p=-4}^{\infty} \varepsilon^p \sigma_{ij}^{(p)}(x_1, y_2, y_3). \tag{8.8}$$

Similarly to the plates, see Sect. 5.3.1, the leading terms in the series (8.7) depends only on "slow" variable, thus, the deflection of beam have limit value, as $\varepsilon \to 0$. All the terms in the series (8.8), including the leading term, depend on the "fast" variables. This is in accordance with the classical concepts of deformation of uniform beams, see. e.g., Timoshenko and Woinowsky-Krieger (1959).

The analysis of the problem (8.1)–(8.4) breaks down into two stages. The first stage entails the deriving the equilibrium equations for the beam, considered as a one-dimensional structure. As in the case of plates (see Chap. 6), this stage is not related to local governing equations (Eq. (8.6) in the case under consideration), and it is carried out in the same way for any governing equations.

Using the two-scale expansion, the differential operators are represented in the form of the sum of operators in the "slow" and "fast" variables. For a function $Z(x_1, y_2, y_3)$ of the arguments x_1 and y_2, y_3, like on the right-hand side of equations (8.7), (8.8), the differential operators are represented in the form

$$\frac{\partial Z}{\partial x_1} \to Z_{,1x}\,, \quad \frac{\partial Z}{\partial x_\alpha} \to \varepsilon^{-1} Z_{,\alpha y} \quad (\alpha = 2, 3). \tag{8.9}$$

Here and below in this chapter, the Latin indexes take the values 1, 2, 3 and Greek indexes take the values 2, 3; subscript, αx means $\partial/\partial x_\alpha$ and , $1y$ means $\partial/\partial y_1$.

We introduce

$$\langle\rangle = \int_C dy_2 dy_3, \quad \langle\rangle_S = \int_{\partial C} dy_2 dy_3 \tag{8.10}$$

– the average values over the cross-section C of the of the beam and over the lateral surface $S = \partial C$ of the cross-section C in the "fast" variables y_2, y_3, see Fig. 8.1.

Proposition 8.1 *For a function* $Z(x_1, y_2, y_3)$

$$\varepsilon^{-2} \int_{G_\varepsilon} Z(x_1, x_2/\varepsilon, x_2/\varepsilon) d\mathbf{x} = \int_{-a}^{a} \langle Z\rangle(x_1) dx_1;$$

$$\int_{S_\varepsilon} Z(x_1, x_2/\varepsilon, x_2/\varepsilon) d\mathbf{x} = \int_{-a}^{a} \langle Z\rangle_S(x_1) dx_1, \; as\ \varepsilon \to 0.$$

Proposition 8.1 follows from Fubini theorem (Rudin 1967).

To obtain the homogenized equilibrium equations, we write a weak form of problem (8.1)

$$-\int_{G_\varepsilon} \sigma_{ij}^{\varepsilon} \frac{\partial v_i}{\partial x_j} d\mathbf{x} + \int_{S_\varepsilon} \mathbf{h}^\varepsilon \mathbf{v} d\mathbf{x} + \int_{G_\varepsilon} \mathbf{f}^\varepsilon \mathbf{v} d\mathbf{x} = 0 \text{ for any } \mathbf{v} \in V_\varepsilon, \tag{8.11}$$

where

$$V_\varepsilon = \{\mathbf{v} \in H^1(G_\varepsilon)^3 : \mathbf{v}(\mathbf{x}) = 0 \text{ on } S_u\}$$

is the set of admissible displacements of the beam considered as three-dimensional elastic body. The additional stresses $\sigma_{ij}^{\varepsilon}$ are determined by (8.81).

Substituting (8.8) in (8.11), we obtain, with allowance for the differentiation rule (8.9)

$$\sum_{p=-4}^{\infty} \int_{G_\varepsilon} \varepsilon^p \sigma_{ij}^{(p)} (v_{i,1x}\delta_{j1} + \varepsilon^{-1} v_{i,jy}\delta_{j\alpha}) d\mathbf{x} \tag{8.12}$$

$$= \int_{S_\varepsilon} \mathbf{h}^\varepsilon \mathbf{v} dx + \int_{G_\varepsilon} \mathbf{f}^\varepsilon \mathbf{v} d\mathbf{x} \quad \text{for any } \mathbf{v} \in U_\varepsilon.$$

We use the functional space

$$H = H^{01}([-a, a]) = \{\mathbf{w}(x_1) \in H^1([-a, a])^3 : \mathbf{w}(\pm a) = 0\}, \tag{8.13}$$

which is the set of admissible displacements of the beam considered as one-dimensional elastic object. It is obvious that $H \subset U_\varepsilon$.

We take $\mathbf{v} = \mathbf{w}(x_1) \in H$ in (8.14). For this choice, $w_{i,\alpha y} = 0$, and (8.12) takes the form

$$\sum_{p=-4}^{\infty} \int_{G_\varepsilon} \varepsilon^p \sigma_{i1}^{(p)} w_{i,1x} d\mathbf{x} = \int_{S_\varepsilon} \mathbf{h}^\varepsilon \mathbf{w} d\mathbf{x} + \int_{G_\varepsilon} \mathbf{f}^\varepsilon \mathbf{w} d\mathbf{x} \tag{8.14}$$

for any $\mathbf{w} \in H$.

Applying Proposition 8.1 to the equality (8.14), we obtain

$$\sum_{p=-4}^{\infty} \int_{-a}^{a} \varepsilon^p \langle \sigma_{i1}^{(p)} \rangle w_{i,1x} dx_1 = \int_{-a}^{a} \langle \mathbf{h}^\varepsilon \rangle_S \mathbf{w} dx_1 + \int_{-a}^{a} \langle \mathbf{f}^\varepsilon \rangle \mathbf{w} dx_1 \tag{8.15}$$

for any $\mathbf{w} \in H$.

Formula (8.15) indicates that in the case under consideration one has to use the following classical definition of the resultant stresses ($i, j = 1, 2, 3$; $p = -4, -3, ...$):

$$N_{i1}^{(p)}(x_1) = \langle \sigma_{i1}^{(p)}(x_1, y_2, y_3) \rangle = \int_C \sigma_{ij}^{(p)}(x_1, y_2, y_3) dy_2 dy_3. \tag{8.16}$$

Using the notation (8.16), we rewrite (8.15) as follows:

$$\sum_{p=-4}^{\infty} \int_{-a}^{a} \varepsilon^p N_{i1}^{(p)} w_{i,1x} dx_1 \tag{8.17}$$

$$= \int_{-a}^{a} \langle \mathbf{h}^\varepsilon \rangle_S \mathbf{w} dx_1 + \int_{-a}^{a} \langle \mathbf{f}^\varepsilon \rangle \mathbf{w} dx_1 \quad \text{for any } \mathbf{w} \in H.$$

Integrating (8.17) by parts and taking into account that $\mathbf{w}(x_1)$ is an arbitrary function, we obtain the following set of the limit equilibrium equations with respect to the resultant stresses:

$$N_{11,1x}^{(-4)} = 0; \tag{8.18}$$

$$N_{11,1x}^{(-3)} + \langle f_1^\varepsilon \rangle + \langle h_1^\varepsilon \rangle_S = 0;$$

$$N_{\beta 1,1x}^{(-2)} + \langle f_\beta^\varepsilon \rangle + \langle h_\beta^\varepsilon \rangle_S = 0. \tag{8.19}$$

The equations with $p > -2$ are not used in the following analysis.

The Eqs. (8.18) and (8.19) are limit because they are derived under condition $\varepsilon \to 0$ and do not depend on ε.

Now take trial function in the form $\mathbf{v} = \mathbf{w}(x_1) y_\alpha$, where $\mathbf{w}(x_1) \in H$. Substituting (8.8) in (8.11), we obtain, with allowance for the differentiation rule (8.9)

$$\sum_{p=-4}^{\infty} \int_{G_\varepsilon} (\varepsilon^p \sigma_{i1}^{(p)} w_{i,1x} y_\alpha + \varepsilon^{p-1} \sigma_{i\alpha}^{(p)} w_i) d\mathbf{x} \tag{8.20}$$

$$= \int_{S_\varepsilon} \mathbf{h}^\varepsilon \mathbf{w} y_\alpha d\mathbf{x} + \int_{G_\varepsilon} \mathbf{f}^\varepsilon \mathbf{w} y_\alpha d\mathbf{x} \quad \text{for any } \mathbf{w} \in H.$$

With regard to Proposition 8.1, we obtain from (8.20)

$$\sum_{p=-4}^{\infty} \int_{-a}^{a} (\varepsilon^p \langle \sigma_{i1}^{(p)} y_\alpha \rangle w_{i,1x} + \varepsilon^{p-1} \langle \sigma_{i\alpha}^{(p)} \rangle w_i) d\mathbf{x} \tag{8.21}$$

$$= \int_{-a}^{a} \langle \mathbf{h}^\varepsilon y_\alpha \rangle \mathbf{w} d\mathbf{x} + \int_{-a}^{a} \langle \mathbf{f}^\varepsilon y_\alpha \rangle \mathbf{w} d\mathbf{x} \quad \text{for any } \mathbf{w} \in H.$$

Formula (8.21) indicates that in the case under consideration one has to use the following classical definition of the resultant moments ($i = 1, 2, 3$, $\beta = 2, 3$, and $p = -4, -3, \ldots$):

$$M_{i\beta}^{(p)} = -\langle \sigma_{i1}^{(p)} y_\beta \rangle = -\int_C \sigma_{ij}^{(p)}(x_1, y_2, y_3) y_\beta dy_2 dy_3.$$

Changing in (8.21) $\langle \sigma_{i1}^{(p)} y_\alpha \rangle$ for $M_{i\alpha}$ and $\langle \sigma_{i\alpha}^{(p)} \rangle$ for $N_{i\alpha}^{(p)}$, correspondingly, we obtain

$$\sum_{p=-4}^{\infty} \int_{-a}^{a} (\varepsilon^p M_{i\alpha}^{(p)} w_{i,1x} + \varepsilon^{p-1} N_{i\alpha}^{(p)} w_i) dx_1 \tag{8.22}$$

$$= \int_{-a}^{a} \langle \mathbf{h}^\varepsilon y_\alpha \rangle_S \mathbf{w} d\mathbf{x} + \int_{-a}^{a} \langle \mathbf{f}^\varepsilon y_\alpha \rangle \mathbf{w} d\mathbf{x} \quad \text{for any } \mathbf{w} \in H.$$

Integrating in (8.22) by parts, we obtain the following set of limit equilibrium equations:

$$M_{i\beta,1x}^{(-4)} - N_{i\beta}^{(-3)} = 0; \tag{8.23}$$

$$M_{i\beta,1x}^{(-3)} - N_{i\beta}^{(-2)} + \delta_{i1}(\langle f_1^\varepsilon y_\beta \rangle + \langle h_1^\varepsilon y_\beta \rangle_S) = 0; \tag{8.24}$$

$$N_{i\alpha}^{(-3)} = \langle \sigma_{i\alpha}^{(-3)} \rangle = 0. \tag{8.25}$$

Equations (8.24) can be broken down into the following two groups: equilibrium equations for the bending moments $M_{1\beta}^{-3}$ ($\beta = 2, 3$), which follows from (8.24) with $i = 1$:

$$M_{1\beta,1x}^{(-3)} - N_{1\beta}^{(-2)} + \langle f_1^\varepsilon y_\beta \rangle + \langle h_1^\varepsilon y_\beta \rangle_S = 0, \tag{8.26}$$

and equilibrium equation for the torsion moment $M = M_{32}^{-3} - M_{23}^{-3}$:

$$M_{,1x} + (N_{32}^{(-2)} - N_{23}^{(-2)}) = 0, \tag{8.27}$$

which are derived from (8.24) with $i = 2, 3$:

$$M_{2\beta,1x}^{(-3)} - N_{2\beta}^{(-2)} = 0, \tag{8.28}$$

$$M_{3\beta,1x}^{(-3)} - N_{3\beta}^{(-2)} = 0.$$

Equations (8.18), (8.19), (8.23), and (8.25) look like the classical ones. But in contrast to the classical equations, in (8.19), (8.26)

$$N_{\beta 1}^{(-2)} \neq N_{1\beta}^{(-2)}.$$

and in (8.27)

$$N_{32}^{(-2)} \neq N_{23}^{(-2)}$$

due to asymmetry of h_{ijmn} in (8.3).

Substituting expansion (8.8) in the equilibrium equations (8.1), we obtain, with allowance for (8.9),

$$\sum_{p=-4}^{\infty} (\varepsilon^p \sigma_{i1,1x}^{(p)} + \varepsilon^{p-1} \sigma_{i\alpha,\alpha y}^{(p)}) + f^{\varepsilon}(\mathbf{x}) = 0 \text{ in } G_{\varepsilon}; \tag{8.29}$$

$$\sum_{p=-4}^{\infty} \varepsilon^p \sigma_{i\alpha}^{(p)} n_{\alpha} = h^{\varepsilon}(\mathbf{x}) \text{ on } S_{\varepsilon}.$$

We use here that the normal to cylinder displayed in Fig. 8.1 has the form $\mathbf{n} = (0, n_2, n_3)$.

Equating in (8.26) the terms corresponding to the identical negative power of ε, we obtain in particular,

$$\sigma_{i\alpha,\alpha y}^{(-4)} = 0 \text{ in } C \ (p = -4), \tag{8.30}$$

$$\sigma_{i\alpha,\alpha y}^{(-3)} + \sigma_{i1,1x}^{(-4)} = 0 \text{ in } C \ (p = -3),$$

$$\sigma_{i\alpha}^{(p)} n_{\alpha} = 0 \text{ on } S = \partial C \ (p = -4, -3, \dots).$$

Here ∂C means the lateral surface of the cross-section C in the "fast" variables; $\mathbf{n}$ is the normal to $S = \partial C$ (see Fig. 8.1).

8.2.2 Limit Governing Equations

The second stage of the analysis of the problem consists of obtaining the governing equations for the beam considered as a one-dimensional structure and elimination of unknown quantities from the limit equilibrium equations. In contrast to the first stage, this stage does involve local governing equations.

In this section, we consider only the case,

$$\sigma_{ij}^{*(-3)} = 0 \text{ and } \sigma_{11}^{*(-2)} \neq 0$$

corresponding to nonzero initial resultant axial force applied to the beam, see Note after (8.4).

Substituting (8.7) and (8.8) in local governing equation (8.6), with allowance for the differentiation rule (8.9), and equating the terms with identical power of ε, we obtain in particular,

$$\sigma_{ij}^{(p)} = c_{ijm1} u_{m,1x}^{(p+4)} + c_{ijm\alpha} u_{m,\alpha y}^{(p+5)} \ (p = -4, -3); \tag{8.31}$$

$$\sigma_{ij}^{(-2)} = c_{ijm1} u_{m,1x}^{(2)} + b_{ijm1}^{(-3)}(x_1, \mathbf{y}) u_{m,1x}^{(3)} \tag{8.32}$$
$$+ c_{ijm\alpha} u_{m,\alpha y}^{(3)} + b_{ijmn}^{(-3)}(x_1, \mathbf{y}) u_{m,nx}^{(1)}.$$

Let us consider the problem (8.30, $p = -4$), (8.31, $p = -4$). It can be written in the form

$$\begin{cases} (c_{i\beta m\alpha} u_{m,\alpha y}^{(1)} + c_{i\beta m1} u_{m,1x}^{(0)}(x_1))_{,\beta y} = 0 \text{ in } C; \\ (c_{i\beta m\alpha} u_{m,\alpha y}^{(1)} + c_{i\beta m1} u_{m,1x}^{(0)}(x_1)) n_\beta = 0 \text{ on } \partial C. \end{cases} \tag{8.33}$$

In order to solve problems of this kind, we introduce the functions $\mathbf{N}^{0p}$ $(p = 1, 2, 3)$ as solutions to the following problems:

$$\begin{cases} (c_{i\beta m\alpha} N_{m,\alpha y}^{0p} + c_{i\beta p1})_{,\beta y} = 0 \text{ in } C; \\ (c_{i\beta m\alpha} N_{m,ny}^{0p} + c_{i\beta m1}) n_\beta = 0 \text{ on } \partial C. \end{cases} \tag{8.34}$$

Proposition 8.2 *For $p = \alpha = 2, 3$, solutions to the problems (8.34) are given by the formula*

$$\mathbf{N}^{0\alpha} = -y_\alpha \mathbf{e}_1, \tag{8.35}$$

where $\{\mathbf{e}_i\}$ are basis vectors of the coordinate system.

To verify equality (8.35), let's substitute $\mathbf{N}^{0\alpha}$ given by (8.35) into (8.36). As a result, we have

$$c_{ijm\gamma} N_{m,\gamma y}^{0\alpha} + c_{ij\alpha 1} = -c_{ij1\alpha} + c_{ij\alpha 1} = 0.$$

The last equality follows from the symmetry of the elastic constants (see condition C1, Sect. 3.1).

As a result, from (8.34) the following non-trivial problem remains:

$$\begin{cases} (c_{i\beta m\alpha} N^{01}_{m,\alpha y} + c_{i\beta 11})_{,\beta y} = 0 \text{ in } C; \\ \\ (c_{i\beta m\alpha} N^{01}_{m,ny} + c_{i\beta 11}) n_\beta = 0 \text{ on } \partial C. \end{cases} \tag{8.36}$$

Proposition 8.3 *The problem (8.36) is degenerate. The kernel of problem (8.36) is generated by the function* **const** $= (C_1, C_2, C_3)$ *and the function*

$$\mathbf{U} = y_\Gamma s_\gamma \mathbf{e}_\gamma, \tag{8.37}$$

where $\{\mathbf{e}_i\}$ *are basis vectors of the coordinate system.*

In (8.37) and below, we use the notations

$$s_1 = 0, s_2 = -1, s_3 = 1; \tag{8.38}$$

$$\Gamma = 2 \ \text{ if } \gamma = 3 \text{ and } \Gamma = 3 \ \text{ if } \gamma = 2.$$

The summation in the pair of the indexes γ, Γ is assumed (if otherwise not indicated directly) in accordance with the rule $x_\Gamma y_\gamma = x_3 y_2 + x_2 y_3$.

To verify (8.37), we compute $c_{ijmn} U_{m,ny}$, where $\mathbf{U}$ is given by (8.37). It is equal to $c_{ij\gamma\Gamma} s_\gamma = c_{ij23} - c_{ij32} = 0$.

The solution $\mathbf{U}(y_2, y_3)$ (8.37) corresponds to the macroscopic torsion about the axis of the beam (Ox_1-axis in the case under consideration). From the mathematical point of view, torsion is a result of the degeneration of the problem (8.36).

Allowing for the fact that the function $\mathbf{u}^{(0)}(x_1)$ of the argument x_1 plays the role of a parameter in the problems in the "fast" variable $\mathbf{y}$, solution to the problem (8.33) can be written in the form,

$$\mathbf{u}^{(1)} = -y_\alpha \mathbf{e}_1 u^{(0)}_{\alpha,1x}(x_1) + \mathbf{N}^{01}(y_2, y_3) u^{(0)}_{1,1x}(x_1) \tag{8.39}$$

$$+ y_\Gamma s_\gamma \mathbf{e}_\gamma \phi(x_1) + \mathbf{V}(x_1).$$

Here, $\phi(x_1)$ and $\mathbf{V}(x_1)$ are arbitrary functions of the argument x_1, which will be determined later. The mechanical meanings of these functions will be discussed below.

From (8.31, $p = -4$) and (8.39), we have

$$\sigma^{(-4)}_{ij} = (c_{ijm\gamma} N^{01}_{m,\gamma y} + c_{ij11}) u^{(0)}_{1,1x}(x_1). \tag{8.40}$$

Averaging (8.40) with $ij = 11$, we obtain

$$N_{11}^{(-4)} = Au_{1,1x}^{(0)}(x_1), \tag{8.41}$$

where

$$A = \langle c_{1111} + c_{11m\gamma} N_{m,\gamma y}^{01}(y_2, y_3) \rangle. \tag{8.42}$$

In accordance with the homogenized equilibrium equation (8.18) from (8.35), we obtain

$$(Au_{1,1x}^{(0)})_{,1x} = 0. \tag{8.43}$$

The boundary condition for $u_1^{(0)}$ follows from (8.2) and expansion (8.7), and has the form

$$u_1^{(0)}(-a) = u_1^{(0)}(a). \tag{8.44}$$

Solution to problem (8.43), (8.44) is $u_1^{(0)} = 0$. Then (8.39) takes a form

$$\mathbf{u}^{(1)} = -y_\alpha \mathbf{e}_1 u_{\alpha,1x}^{(0)}(x_1) + y_\Gamma s_\gamma \mathbf{e}_\gamma \phi(x_1) + \mathbf{V}(x_1). \tag{8.45}$$

Substitution of equality (8.45) into (8.39) gives the following equations:

$$\sigma_{ij}^{(-4)} = 0, \tag{8.46}$$

$$\sigma_{ij}^{(-3)} = c_{ijmn} u_{m,ny}^{(2)} - c_{ij11} y_\alpha u_{\alpha,1x1x}^{(0)}(x_1) \tag{8.47}$$

$$+ c_{ij11} V_{1,1x}(x_1).$$

Let us examine problem (8.30, $p = -3$), (8.47). It can be written as

$$\begin{cases} (c_{ijmn} u_{m,ny}^{(2)} - c_{ij11} y_\alpha u_{\alpha,1x1x}^{(0)}(x_1) + c_{ij11} V_{1,1x}(x_1))_{,\beta y} = 0 \text{ in } G_\varepsilon; \\ (c_{ijmn} u_{m,ny}^{(2)} - c_{ij11} y_\alpha u_{\alpha,1x1x}^{(0)}(x_1) + c_{ij11} V_{1,1x}(x_1)) n_{\beta y} = 0 \text{ on } S_\varepsilon. \end{cases} \tag{8.48}$$

In order to solve the boundary-value problem (8.48), we introduce functions $\mathbf{N}^{1\alpha}(y_2, y_3)$ ($\alpha = 2, 3$) and $\mathbf{X}^3(y_2, y_3)$ which are determined by solutions of the following problems (corresponding to bending and torsion of beam):

$$\begin{cases} (c_{i\beta m\gamma} N_{m,\gamma y}^{1\alpha} - c_{i\beta 11} y_\alpha)_{,\beta y} = 0 \text{ in } C; \\ (c_{i\beta m\gamma} N_{m,\gamma y}^{1\alpha} - c_{i\beta 11} y_\alpha) n_\beta = 0 \text{ on } \partial C. \end{cases} \tag{8.49}$$

$$\begin{cases} (c_{i\beta m\alpha} X^3_{m,\alpha y} + c_{i\beta\gamma 1} s_\gamma y_\Gamma)_{,\beta y} = 0 \text{ in } C; \\ (c_{i\beta m\alpha} X^3_{m,\alpha y} + c_{i\beta\gamma 1} s_\gamma y_\Gamma) n_\beta = 0 \text{ on } \partial C. \end{cases} \tag{8.50}$$

Solution to problem (8.48) can be expressed through the functions $\mathbf{N}^{01}$, $\mathbf{N}^{1\alpha}(y_2, y_3)$, and $\mathbf{X}^3(y_2, y_3)$, and the functions introduced in Propositions 8.2 and 8.3, as follows:

$$\mathbf{u}^{(2)} = \mathbf{N}^{1\alpha}(y_2, y_3) u^{(0)}_{\alpha,1x1x}(x_1) - y_\alpha V_{\alpha,1x}(x_1)\mathbf{e}_1 + \mathbf{N}^{01}(y_2, y_3) V_{1,1x}(x_1)$$

$$+ \mathbf{X}^3(y_2, y_3)\phi_{,1x}(x_1) + \mathbf{U}(x_1) + s_\gamma y_\Gamma \mathbf{e}_\gamma \psi(x_1). \tag{8.51}$$

Writing (8.51), we use the equality $c_{ij\gamma\Gamma s_\gamma} = 0$ following from the symmetry of the elastic constants (see condition C1, Sect. 3.1). In (8.51), $\psi(x_1)$ is a new function (that will not influence the problem).

Substituting (8.51) in (8.46), we obtain

$$\sigma^{(-3)}_{ij} = (c_{ijm\gamma} N^{01}_{m,\gamma y} - c_{ij11}) V_{1,1x}(x_1) \tag{8.52}$$

$$+ (c_{ijm\gamma} N^{1\alpha}_{m,\gamma y} + c_{ij11} y_\alpha) u^{(0)}_{\alpha,1x1x}(x_1)$$

$$+ (c_{ijm\gamma} X^3_{m,\gamma y} + c_{ij\gamma 1} s_\gamma y_\Gamma)\phi_{,1x}(x_1).$$

Note that the functions V_α are present in (6.58) but absent in (6.59). This is because the functions V_α appear in (6.58) multiplied by the function $y_\alpha \mathbf{e}_1$, which plays a special role in the asymptotic theory of beams.

Averaging (8.52) with $ij = 11$ over the cross-section C of the beam, we obtain the the following equations for the resultant forces:

$$N^{(-3)}_{11} = A V_{1,1x} + A^1_{\alpha\beta} u^{(0)}_{\alpha,1x1x} + b\phi(x_1). \tag{8.53}$$

Multiplying (8.52) with $j = 1$ by y_β and averaging over the cross-section C, we obtain equations, which can be broken down into two groups: the limit governing equations for the bending moments $M^{(-3)}_{1\beta}$

$$M^{(-3)}_{1\beta} = \langle \sigma^{-3}_{11} y_\beta \rangle,$$

and the limit governing equations equations for torsion moments,

$$M = M^{(-3)}_{\Gamma\gamma} s_\gamma = M^{(-3)}_{32} - M^{(-3)}_{23} = \langle \sigma^{-3}_{\Gamma 1} y_\gamma s_\gamma \rangle = \langle \sigma^{-3}_{31} y_2 - \sigma^{-3}_{21} y_3 \rangle.$$

They are the following:

$$M^{(-3)}_{1\beta} = A_\beta V_{1,1x} + A^2_{\beta\alpha} u^{(0)}_{\alpha,1x1x} + B_\beta \phi(x_1), \tag{8.54}$$

$$M = A^1 V_{1,1x} + A^2_\alpha u^{(0)}_{\alpha,1x1x} + B\phi(x_1), \tag{8.55}$$

The coefficients in (8.53)–(8.55) are

$$A = \langle c_{1111} + c_{11\delta m\gamma} N^{01}_{m,\gamma y}(y_2, y_3)\rangle; \tag{8.56}$$

$$A^1_\alpha = \langle c_{1111} y_\alpha + c_{11m\gamma} N^{1\alpha\beta}_{m,\gamma y}(y_2, y_3)\rangle;$$

$$b = \langle c_{11\gamma 1} s_\gamma y_\Gamma + c_{11\,m\gamma} X^3_{m,\gamma y}(y_2, y_3)\rangle,$$

$${}^1A_{\alpha\beta} = \langle (c_{1111} + c_{11m\gamma} N^{01}_{m,\gamma y}(y_2, y_3)) y_\beta\rangle;$$

$$A^2_{\alpha\beta} = \langle (c_{1111} y_\alpha + c_{11\,m\gamma} N^{1\alpha}_{m,\gamma y}(y_2, y_3)) y_\beta\rangle;$$

$$B_\beta = \langle (c_{11\gamma 1} s_\gamma y_\Gamma + c_{11\,m\gamma} X^3_{m,\gamma y}(y_2, y_3)) y_\beta\rangle s_\beta;$$

$$A^1 = \langle (c_{3111} + c_{31m\gamma} N^{01}_{m,\gamma y}(y_2, y_3)) y_2 - (c_{2111} + c_{21m\gamma} N^{01}_{m,\gamma y}(y_2, y_3)) y_3\rangle;$$

$$A^2_\alpha = \langle (c_{3111} y_\alpha + c_{31m\gamma} N^{1\alpha}_{m,\gamma y}(y_2, y_3)) y_2$$

$$- c_{2111} y_\alpha + c_{21m\gamma} N^{1\alpha}_{m,\gamma y}(y_2, y_3)) y_3\rangle$$

$(\alpha, \beta, \gamma, \delta = 2, 3)$.

The Eqs. (8.53)–(8.55) are asymptotically exact governing equations of the beam considered as an one-dimensional object. The coefficients in the right-hand parts of (8.53)–(8.55) are the homogenized stiffnesses of the beam (also called the homogenized rigidities). They are expressed through solutions to the problems (8.36), (8.49), and (8.50). Note that the derived limit governing equations (8.53)–(8.55) coincide with the classical ones in form.

Let us denote the right-hand parts of the governing equations (8.53)–(8.55) by

$$N(V_{1,1x}, u^{(0)}_{\alpha,1x1x}, \phi_{,1x}) = AV_{1,1x} + A^1_{\alpha\beta} u^{(0)}_{\alpha,1x1x} + b\phi(x_1)_{,1x};$$

$$M^1_\beta(V_{1,1x}, u^{(0)}_{\alpha,1x1x}, \phi_{,1x}) =^1 A_\beta V_{1,1x} + A^2_{\beta\alpha} u^{(0)}_{\alpha,1x1x} + B_\beta \phi(x_1);$$

$$M(V_{1,1x}, u^{(0)}_{\alpha,1x1x}, \phi_{,1x}) = A^1 V_{1,1x} + A^2_\alpha u^{(0)}_{\alpha,1x1x} + B\phi(x_1)_{,1x}.$$

The meaning of V_1, $u^{(0)}_\alpha$ ($\alpha = 2, 3$), and $\phi(x_1)$ are clear seen from the governing equations. They are the axial displacement, the normal deflections, and torsion angle, correspondingly.

8.2.3 Elimination of Shear Forces

Now we come to the point of fundamental difference between the asymptotic theory of beams with initial stresses and classical theories and the asymptotic theory of beam with no initial stresses. The difference is related to the problem of elimination of the resultant shear forces in the limit equilibrium equations.

We collect the non-trivial equilibrium equations (8.19), (8.26), and (8.27). They are the following: the equilibrium equations for the resultant axial forces

$$N^{(-3)}_{11,1x} + \langle f^{\varepsilon}_1 \rangle + \langle h^{\varepsilon}_1 \rangle_S = 0; \tag{8.57}$$

$$N^{(-2)}_{\beta 1,1x} + \langle f^{\varepsilon}_\beta \rangle + \langle h^{\varepsilon}_\beta \rangle_S = 0; \tag{8.58}$$

and the equilibrium equations for the resultant moments

$$M^{(-3)}_{1\beta,1x} - N^{(-2)}_{1\beta} + \langle f^{\varepsilon}_1 y_3 \rangle + \langle h^{\varepsilon}_1 y_3 \rangle_S = 0; \tag{8.59}$$

$$M_{,1x} + (N^{(-2)}_{32} - N^{(-2)}_{23}) = 0. \tag{8.60}$$

In the absence of initial stresses, the additional stresses are symmetric: $\sigma^{(-2)}_{ij} = \sigma^{(-2)}_{ji}$. Due to this symmetry, the resultant forces are also symmetric: $N^{(-2)}_{ij} = N^{(-2)}_{ji}$, and we can eliminate $N^{(-2)}_{1\beta}$ and $N^{(-2)}_{\beta 1}$ from Eqs. (8.58) and (8.59), as well as $N^{(-2)}_{32} - N^{(-2)}_{23}$ from (8.59), using this symmetry, only, as done in both the classical theory of beams (Timoshenko and Woinowsky-Krieger 1959) and in the asymptotic theory of beams with no initial stresses (Kolpakov 1991; Trabucho and Viano 1996).

In our case $N^{(-2)}_{ij}$ *is not symmetrical* in i and j due to (8.29) because $b^{(-3)}_{ijmn}$ (8.6) are not symmetric in i and j. As a result, the classical method of elimination of shear forces does not work. To overcome the difficulty, it is necessary to study $N^{(-2)}_{ij}$ in detail.

Let us insert $\mathbf{u}^{(1)}$ into (8.32) in accordance with (8.45). Then, we obtain

$$\sigma^{(-2)}_{ij} = c_{ijm1} u^{(2)}_{m,1x} + c_{ijmn} u^{(3)}_{m,ny} \tag{8.61}$$

$$+ b^{(-3)}_{ij\alpha 1}(x_1, \mathbf{y}) u^{(0)}_{\alpha,1x} + b^{(-3)}_{ij\gamma\Gamma m}(x_1 \mathbf{y}) s_\gamma \phi(x_1).$$

The first and second terms on the right-hand side of equation (8.51) are symmetric with respect to i and j by virtue of the symmetry of the elastic constants c_{ijkl} in i and j. Then the antisymmetric part of the tensor $\sigma^{(-2)}_{ij}$ (8.62) is

$$\sigma^{(-2)}_{ij} - \sigma^{(-2)}_{ji} = b^{(-3)}_{ij\alpha 1}(x_1, \mathbf{y}) u^{(0)}_{3,\beta x}(x_1) - b^{(-3)}_{ji\alpha 1}(x_1, \mathbf{y}) u^{(0)}_{\alpha,1x}(x_1) \tag{8.62}$$

$$+ b^{(-3)}_{ij\gamma\Gamma}(x_1, \mathbf{y}) s_\gamma \phi(x_1) - b^{(-3)}_{ji\gamma\Gamma}(x_1, \mathbf{y}) s_\gamma \phi(x_1).$$

Averaging (8.62) over the cross-section C of the beam, we obtain

$$N_{ij}^{(-2)} - N_{ji}^{(-2)} = \langle b_{ij\alpha 1}^{(-3)}(x_1, \mathbf{y})\rangle u_{3,\beta x}^{(0)}(x_1) - \langle b_{ji\alpha 1}^{(-3)}(x_1, \mathbf{y})\rangle u_{\alpha,1x}^{(0)}(x_1)$$

$$+ \langle b_{ij\gamma\Gamma}^{(-3)}(x_1, \mathbf{y})\rangle s_\gamma \phi(x_1) - \langle b_{ji\gamma\Gamma}^{(-3)}(x_1, \mathbf{y})\rangle s_\gamma \phi(x_1). \tag{8.63}$$

To continue our investigation, we need some facts about the average values of the initial stresses.

Proposition 8.4 *Let initial stresses* σ_{ij}^*, *which can be represented as* $\sigma_{ij}^* = \varepsilon^{-3}\sigma_{ij}^{*(-3)} + \varepsilon^{-2}\sigma_{ij}^{*(-2)} + \cdots$, *satisfy the equilibrium equations,*

$$\sigma_{ij,j}^* + \varepsilon^A f_i(x_1, \mathbf{y}) = 0 \ \textit{in}\ G_\varepsilon,$$

$$\sigma_{ij}^* n_j = \varepsilon^B h_i(x_1, \mathbf{y}) = 0 \ \textit{on}\ S_\varepsilon,$$

Then,

(1) $\langle b_{ijm\alpha}^{*(-3)}(x_1, \mathbf{y}) Z_{,\alpha y}\rangle = 0$ *if* $A < -4,\ B < -3$;
(2) $\langle b_{ijm\alpha}^{*(-2)}(x_1, \mathbf{y}) Z_{,\alpha y}\rangle = 0$ *if* $\sigma_{ij}^{*(-3)} = 0$ *and* $A < -3,\ B < -2$,

for every differentiable function $Z(y_2, y_3)$.

Substituting the expansion $\sigma_{ij}^* = \varepsilon^{-3}\sigma_{ij}^{*(-3)} + \varepsilon^{-2}\sigma_{ij}^{*(-2)} + \cdots$ into (8.64), we obtain

$$\begin{cases} \sigma_{i\alpha,\alpha y}^{*(p)} = 0 \text{ in } Y, \\ \sigma_{ij\alpha}^{*(p)} n_\alpha = 0 \text{ on } \partial C \ (p = -4, -3, -2). \end{cases} \tag{8.64}$$

Taking into account the definition of $b_{ijmn}^{(-3)}$, it is sufficient to verify that $\langle \sigma_{j\alpha}^{*(-3)} Z_{,\alpha y}\rangle = 0$. Let us consider $\langle \sigma_{j\alpha}^* Z_{,\alpha y}\rangle$. Multiplying equation in (8.64) by $Z(y_2, y_3)$, integrating by parts, and accounting the definition of the average value over the cross-section C, we find that this value is equal to

$$\int_C \sigma_{j\alpha,\alpha y}^{*(-3)} Z(\mathbf{y}) dy_2 y_3 + \int_{\partial C} \sigma_{j\alpha}^{*(-3)} n_\alpha Z(\mathbf{y}) dy_2 y_3.$$

Taking into account that the integrals over C and ∂C are equal to zero due to (8.64), we obtain the required.

Proposition 8.5 *Under the conditions of Proposition 8.4,* $\langle \sigma_{i\alpha}^{*(-3)}\rangle = 0$ *and* $\langle \sigma_{1\alpha}^{*(-3)} y_\alpha\rangle = 0$. *In addition,* $\langle \sigma_{i\alpha}^{*(-2)}\rangle = 0$ *when* $\langle \sigma_{ij}^{*(-3)}\rangle = 0$.

Proposition 8.5 is a consequence of Proposition 8.4. To prove the first and the third equations, one set $Z = y_\alpha$, and $Z = y_\alpha^2$ to prove the second one.

8.2.4 Bending

Now, we can eliminate shear forces $N_{\beta 1}^{(-2)}$ and $N_{1\beta}^{(-2)}$ from the equilibrium equations.

Consider the following equilibrium equations (8.58) and (8.59). To exclude $N_{\beta 1}^{(-2)}$ and $N_{1\beta}^{(-2)}$, apply the relations, which follow from (8.63) and the definition of $b_{ijmn}^{(-2)}$ (8.4):

$$N_{1\beta}^{(-2)} - N_{\beta 1}^{(-2)} = K_\beta, \tag{8.65}$$

where K_β is defined by the formula

$$\begin{aligned} K_\beta &= \langle b_{\beta 1\alpha 1}^{(-3)}(x_1, y_2, y_3)\rangle u_{\alpha,1x}^{(0)}(x_1) - \langle b_{1\beta\alpha 1}^{(-3)}(x_1, y_2, y_3)\rangle u_{\alpha,1x}^{(0)}(x_1) \\ &= \langle b_{\beta 1\gamma\Gamma}^{(-3)}(x_1, y_2, y_3)\rangle s_\gamma \phi(x_1) - \langle b_{1\beta\gamma\Gamma}^{(-3)}(x_1, y_2, y_3)\rangle s_\gamma \phi(x_1) \\ &= \langle \sigma_{11}^{*(-2)}(x_1, y_2, y_3)\rangle \delta_{\beta\alpha} u_{\alpha,1x}^{(0)}(x_1) + \langle \sigma_{1\Gamma}^{*(-2)}(x_1, y_2, y_3)\rangle \delta_{\beta\gamma} s_\gamma \phi(x_1). \end{aligned} \tag{8.66}$$

Differentiating (8.59), we obtain

$$M_{1\beta,1x1x}^{(-3)} - N_{1\beta,1x}^{(-2)} + (\langle f_1^\varepsilon y_3\rangle + \langle h_1^\varepsilon y_3\rangle_S)_{,1x} = 0. \tag{8.67}$$

and (8.58) is

$$N_{\beta 1,1x}^{(-2)} + \langle f_\beta^\varepsilon\rangle + \langle h_\beta^\varepsilon\rangle_S = 0. \tag{8.68}$$

From (8.67) and (8.68), we obtain

$$\begin{aligned} &M_{1\beta,1x1x}^{(-3)} - N_{1\beta,1x}^{(-2)} + N_{\beta 1,1x}^{(-2)} \\ &+ (\langle f_1^\varepsilon y_3\rangle + \langle h_1^\varepsilon y_3\rangle_S)_{,1x} + \langle f_\beta^\varepsilon\rangle + \langle h_\beta^\varepsilon\rangle_S = 0. \end{aligned} \tag{8.69}$$

By virtue of (8.65), the difference $-N_{1\beta,1x}^{(-2)} + N_{\beta 1,1x}^{(-2)} = -K_\beta$. As a result, (8.69) takes the form

$$(M_{1\beta,1x}^{(-3)} + \langle f_1^\varepsilon y_\beta\rangle + \langle h_1^\varepsilon y_\beta\rangle)_{,1x} + \langle f_\beta^\varepsilon\rangle + \langle h_\beta^\varepsilon\rangle_S = K_{\beta,1x}. \tag{8.70}$$

Under the conditions of Proposition 8.4 ($A < -3$, $B < -2$)

$$K_\beta = \langle \sigma_{11}^{*(-2)}(x_1, y_2, y_3)\rangle u_{\beta,1x}^{(0)}(x_1) = N_{11}^*(x_1) u_{\beta,1x}^{(0)}(x_1),$$

where $N_{11}^* = \langle \sigma_{11}^{*(-2)}(x_1, \mathbf{y})\rangle$ is the initial resultant axial force. Then Eq. (8.70) takes the classical form,

$$M^{(-3)}_{1\beta,1x1x} + \langle f_\beta \rangle + \langle h_\beta \rangle_S + (\langle f_1 y_\beta \rangle + \langle h_1 y_\beta \rangle_S)_{,1x} \tag{8.71}$$

$$= (N^*_{11}(x_1) u^{(0)}_{\beta,1x}(x_1))_{,1x}.$$

8.2.5 Torsion

Here, as above, we are faced with a situation of asymmetry of the resulting forces $N^{(-2)}_{ij}$. Due to the asymmetry, we have an unknown (and non-zero) term $N^{(-2)}_{32} - N^{(-2)}_{23}$ in (8.60). In accordance with (8.63) and the definition of $b^{(-2)}_{ijmn}$ (8.4), we can write

$$N^{(-2)}_{23} = N^{(-2)}_{32} + K, \tag{8.72}$$

where

$$N^{(-2)}_{32} - N^{(-2)}_{23} = \langle b^{(-3)}_{32\alpha 1}(x_1, y_2, y_3) \rangle u^{(0)}_{3,\beta x}(x_1) - \langle b^{(-3)}_{23\alpha 1}(x_1, y_2, y_3) \rangle u^{(0)}_{\alpha,1x}(x_1)$$

$$+ \langle b^{(-3)}_{32\gamma\Gamma}(x_1, y_2, y_3) \rangle s_\gamma \phi(x_1) - \langle b^{(-3)}_{23\gamma\Gamma}(x_1, y_2, y_3) \rangle s_\gamma \phi(x_1) \tag{8.73}$$

$$= \langle \sigma^{*(-2)}_{31}(x_1, y_2, y_3) \rangle \delta_{2,\alpha} u^{(0)}_{\alpha,1x}(x_1) - \langle \sigma^{*(-2)}_{3\Gamma}(x_1, y_2, y_3) \rangle \delta_{2\gamma} \phi(x_1).$$

Then, (8.60) can be rewritten in the form,

$$M_{1,1x} - K = 0. \tag{8.74}$$

Under the conditions of Proposition 8.4 ($a \le -3$, $b \le -2$), $K = 0$. This means, that the initial stresses of the order ε^{-2} do not influence the torsion of the beam.

8.2.6 The Limit Problem

Substituting the limit governing equations (8.53)–(8.55) to the limit equations of equilibrium (8.57), (8.59), (8.74), we arrive at the following equations:

$$(AV_{1,1x} + A^1_{\alpha\beta} u^{(0)}_{\alpha,1x1x} + b\phi_{,1x})_{,1x} + \langle f^\varepsilon_1 \rangle + \langle h^\varepsilon_1 \rangle_S = 0; \tag{8.75}$$

$$({}^1A_\beta V_{1,1x} + A^2_{\beta\alpha} u^{(0)}_{\alpha,1x1x} + B_\beta \phi_{,1x})_{,1x1x} \tag{8.76}$$

$$+ \langle f_\beta \rangle + \langle h_\beta \rangle_S + (\langle f^\varepsilon_1 y_\beta \rangle + \langle h^\varepsilon_1 y_\beta \rangle_S)_{,1x} = (N^*_{11} u^{(0)}_{\alpha,1x})_{,1x};$$

$$(A^1 V_{1,1x} + A^2_\alpha u^{(0)}_{\alpha,1x1x} + B\phi_{,1x})_{,1x} = 0. \tag{8.77}$$

Here, $N_{11}^* = \langle \sigma_{11}^{*(-2)} \rangle$.

The boundary conditions are the following:

$$V_1(-a) = V_1(a) = 0; \tag{8.78}$$

$$u_\alpha^{(0)}(-a) = u_\alpha^{(0)}(a) = u_{\alpha,1x}^{(0)}(-a) = u_{\alpha,1x}^{(0)}(a) = 0;$$

$$\phi(-a) = \phi(a) = 0.$$

Problem (8.75)–(8.77) with boundary conditions (8.78) is a one-dimensional model of a beam with respect to the axial displacement $u_\alpha^{(0)}(x_1)$ (α), normal deflections $V_1(x_1)$, and torsion angle $\phi(x_1)$.

8.3 Beams Subjected to Moments of Initial Stresses

In this section, we consider the case when $\sigma_{ij}^{*(-2)} = 0$, i.e. there are no initial stresses that create a resultant axial force, and $\sigma_{ij}^{*(-3)} \neq 0$. The initial stresses $\sigma_{ij}^{*(-3)}$ create resultant moments. It is assumed that the stresses $\sigma_{ij}^{*(-3)}$ do not create a resultant axial force, i.e., they satisfy the condition

$$\langle \sigma_{11}^{*(-3)} \rangle = 0, \tag{8.79}$$

but, generally $\langle \sigma_{i1}^{*(-3)} y_\alpha \rangle \neq 0$.

This means that we consider beams subjected exclusively to the action of the resultant moments of the initial stresses. As demonstrated in Sect. 8.2.1, the homogenized equations of equilibrium do not depend on the local governing equation (8.6). Then, in the case under consideration, we must concentrate on the problem of the limit governing equations.

8.3.1 Limit Governing Equations

We use asymptotic expansions (8.7) and (8.8). Substituting (8.7) and (8.8) in local governing equation (8.6), with allowance for the differentiation rule (8.9), and equating the terms with identical powers of ε, we obtain, in particular,

$$\sigma_{ij}^{(-4)} = c_{ijm1}u_{m,1x}^{(0)} + c_{ijm\gamma}u_{m,\gamma y}^{(1)}; \tag{8.80}$$

$$\sigma_{ij}^{(-3)} = c_{ijm1}u_{m,1x}^{(1)} + b_{ijm1}^{(-3)}(x_1, y_2, y_3)u_{m,1x}^{(0)} \tag{8.81}$$
$$+ c_{ijm\gamma}(y_2, y_3)u_{m,\gamma y}^{(2)} + b_{ijmn}^{(-3)}(x_1, y_2, y_3)u_{m,ny}^{(1)};$$

$$\sigma_{ij}^{(-?)} = c_{ijm1}(y_2, y_3)u_{m,1x}^{(2)} + b_{ijm1}^{(-3)}(x_1, y_2, y_3)u_{m,1x}^{(1)} \tag{8.82}$$

$$+ c_{ijm\gamma}(y_2, y_3)u_{m,\gamma y}^{(3)} + b_{ijm\gamma}^{(-3)}(x_1, y_2, y_3)u_{m,\gamma y}^{(2)}.$$

One clearly see the similarities and the differences between the Eqs. (8.80), (8.81) and (8.102), (8.103)

Let us consider problem (8.30, $p = -4$) with $\sigma_{ij}^{(-4)}$ given by (8.80). This problem coincides with problem (8.30, $p = -4$), (8.31, $p = -4$) studied in Sect. 8.2.2. This happens because equation (8.80) coincides with equation (8.31, $p = -4$). As a result, we conclude that solution $\mathbf{u}^{(1)}$ to the problem (8.30, $p = -4$) with $\sigma_{ij}^{(-4)}$ is given by (8.45).

Substituting $\mathbf{u}^{(1)}$ (8.45) in (8.80) and (8.4), we obtain the following equations:

$$\sigma_{ij}^{(-4)} = 0; \tag{8.83}$$

$$\sigma_{ij}^{(-3)} = (c_{ijm\gamma}u_{m,\gamma y}^{2} - c_{ij11}y_\alpha u_{\alpha,1x1x}^{(0)}(x_1) \tag{8.84}$$
$$+ (c_{ij11}V_{1,1x}(x_1) + c_{ij\gamma 1}s_\gamma y_\Gamma)\phi_{,1x}(x_1)$$

$$+ (b_{ij\alpha 1}^{(-3)}(x_1, y_2, y_3) - b_{ij1\alpha}^{(-3)}(x_1, y_2, y_3))u_{\alpha,1x}^{(0)} + b_{ij\gamma\Gamma}^{(-3)}(x_1, y_2, y_3)s_\gamma\phi(x_1).$$

Consider the problem (8.30, $p = -3$) with $\sigma_{ij}^{(-3)}$ given by (8.84). If

$$b_{i\gamma mn,\gamma y}^{(-3)} = \sigma_{n\gamma,\gamma y}^{*(-3)}\delta_{im} = 0 \text{ in } Y, \text{ and } b_{i\gamma mn}^{(-3)}n_\gamma = \sigma_{n\gamma}^{*(-3)}\delta_{im}n_\gamma = 0 \text{ on } S,$$

then $b_{ijmn}^{(-3)}$ in (8.84) have no influence on the solution to the problem (8.30, $p = -3$) with $\sigma_{ij}^{(-3)}$ (8.84). Solution to the mentioned problem has the form

$$\sigma_{ij}^{(-3)} = c_{ijm\gamma}(y_2, y_3)u_{m,\gamma y}^{(2)} - c_{ij11}y_\alpha u_{\alpha,1x1x}^{(0)}(x_1) \tag{8.85}$$

$$+ c_{ij11}V_{1,1x}(x_1) + c_{ij\gamma 1}s_\gamma y_\Gamma\phi_{,1x}(x_1)$$

$$+ b_{ij\alpha 11}^{(-3)}(x_1, y_2, y_3)V_{1,1x}(x_1) + c_{ij\gamma 1}s_\gamma y_\Gamma\phi_{,1x}(x_1).$$

This expression on the right-hand side of (8.85) coincides with (8.47). Then the solution to the problem under consideration is given by (8.51). Substituting $\mathbf{u}^{(2)}$ (8.51) in (8.80), we obtain

$$\sigma_{ij}^{(-3)} = (c_{ijm\gamma} N^{01}_{m,\gamma y} + c_{ij11}) V_{1,1x}(x_1) \tag{8.86}$$

$$+ (c_{ijm\gamma} N^{1\alpha}_{m,\gamma y} + c_{ij11} y_\alpha) u^{(0)}_{\alpha,1x1x}(x_1)$$

$$+ (c_{ijm\gamma} X^{3}_{m,\gamma y} + c_{ij\gamma 1} s_\gamma y_\Gamma) \phi_{,1x}(x_1)$$

$$+ (b^{(-3)}_{ij\alpha 1}(x_1, \mathbf{y}) - b^{(-3)}_{ij1\alpha}(x_1, y_2, y_3)) u^{(0)}_{\alpha,1x}(x_1)$$

$$+ b^{(-3)}_{ij\gamma\Gamma}(x_1, y_2, y_3) s_\gamma \psi(x_1).$$

Average (8.86) with $ij = 11$ over the cross-section C of the beam taking into account the definition of $b^{(-3)}_{ijmn}$ (8.4) and Proposition 8.5. As a result, we obtain

$$N_{11}^{(-3)} = N(V_{1,1x}, u^{(0)}_{\alpha,1x1x}, \phi_{,1x}), \tag{8.87}$$

where $N(V_{1,1x}, u^{(0)}_{\alpha,1x1x}, \phi_{,1x})$ is determined in Sect. 8.2.2, see (8.53).

Multiply (8.86) with $j = 1$ by y_β and average over the cross-section C of the beam. Taking into account the definition of $b^{(-3)}_{ijmn}$ (8.4), the definition of the bending and torsion moments $M^{(-3)}_{1\beta}$ and $M^{(-3)}$, condition (8.79), and Propositions 8.5, we obtain

$$M^{(-3)}_{1\beta} = M_\beta(V_{1,1x}, u^{(0)}_{\alpha,1x1x}, \phi_{,1x}) + \mathbf{C}_{\alpha\beta} u^{(0)}_{\alpha,1x}, \tag{8.88}$$

$$M^{(-3)} = M(V_{1,1x}, u^{(0)}_{\alpha,1x1x}, \phi_{,1x}) + \mathbf{c}_{\alpha\beta} u^{(0)}_{\alpha,1x} + J\psi. \tag{8.89}$$

Here $M_{1\beta}(V_{1,1x}, u^{(0)}_{\alpha,1x1x}, \phi_{,1x})$ and $M(V_{1,1x}, u^{(0)}_{\alpha,1x1x}, \phi_{,1x})$ are determined in Sect. 8.2.2 by (8.54) and (8.55).

The coefficients on the right-hand side of (8.88) and (8.89) are the following:

$$\mathbf{C}_{\alpha\beta} = \langle (b^{(-3)}_{11\alpha 1}(x_1, \mathbf{x}) - b^{(-3)}_{111\alpha}(x_1, \mathbf{x})) y_\beta \rangle = -\langle \sigma^{*(-3)}_{1\alpha}(x_1, \mathbf{x}) y_\beta \rangle;$$

$$\mathbf{c}_\alpha = \langle (b^{(-3)}_{31\alpha 1}(x_1, \mathbf{x}) - b^{(-3)}_{311\alpha}(x_1, \mathbf{x})) y_2 - (b^{(-3)}_{21\alpha 1}(x_1, \mathbf{x}) - b^{(-3)}_{211\alpha}(x_1, \mathbf{x})) y_3 \rangle$$

$$= \langle \sigma^{*(-3)}_{11}(x_1, \mathbf{x}) y_2 \rangle \delta_{3\alpha} - \langle \sigma^{*(-3)}_{11}(x_1, \mathbf{x}) y_3 \rangle \delta_{2\alpha}; \tag{8.90}$$

$$\mathbf{J} = \langle b^{(-3)}_{31\gamma\Gamma}(x_1, \mathbf{x}) s_\gamma y_2 \rangle - \langle b^{(-3)}_{21\gamma\Gamma}(x_1, \mathbf{x}) s_\gamma y_3 \rangle$$

$$= -\langle \sigma^{*(-3)}_{12}(x_1, \mathbf{x}) y_2 \rangle - \langle \sigma^{*(-3)}_{13}(x_1, \mathbf{x}) y_3 \rangle.$$

Deriving (8.88), (8.89), and (8.90), we take into account the definition of $b^{(-3)}_{ijkl}$ (8.4) and use the equality

$$\langle b^{(-3)}_{31\gamma\Gamma}(x_1,\mathbf{x})s_\gamma y_2\rangle - \langle b^{(-3)}_{21\gamma\Gamma}(x_1,\mathbf{x})s_\gamma y_3\rangle$$

$$= -\langle \sigma^{*(-3)}_{12}(x_1,\mathbf{x})y_2\rangle - \langle \sigma^{*(-3)}_{13}(x_1,\mathbf{x})y_3\rangle = 0$$

following from Proposition 8.5.

8.3.2 Elimination of Shear Forces

Let us exclude the quantities $N^{(-2)}_{ij}$ from equilibrium equations (8.57)–(8.60). To do this, we examine $N^{(-2)}_{ij}$. In the case under consideration, see (8.81), $\sigma^{(-2)}_{ij}$ is the sum of the terms symmetrical with respect to i and j—the first and third terms in (8.82), and the terms asymmetrical with respect to i and j—the second and fourth terms in (8.82). One can find that the following relation holds:

$$N^{(-2)}_{ij} = N^{(-2)}_{ji} + \langle (b^{(-3)}_{ij11}(x_1,\mathbf{x}) - b^{(-3)}_{ji11}(x_1,\mathbf{x}))y_\alpha\rangle u^{(1)}_{\alpha,1x1x} \tag{8.91}$$

$$+ \langle (b^{(-3)}_{ij\gamma 1}(x_1,\mathbf{x}) - b^{(-3)}_{ji\gamma 1}(x_1,\mathbf{x}))y_\Gamma\rangle s_\gamma \phi_{,1x}(x_1)$$

$$+ \langle b^{(-3)}_{ijm1}(x_1,\mathbf{x}) - b^{(-3)}_{jim1}(x_1,\mathbf{x})\rangle V_{m,1x}(x_1)$$

$$+ \langle (b^{(-3)}_{ijmn}(x_1,\mathbf{x}) - b^{(-3)}_{jimn}(x_1,\mathbf{x}))u^{(2)}_{m,ny}\rangle.$$

Proposition 8.6 *Under the conditions of Proposition 8.4,*

$$\langle (b^{(-3)}_{ijmn} - b^{(-3)}_{jimn})u^{(2)}_{m,ny}\rangle = 0.$$

To prove the proposition, it is enough to put $Z = u^{(2)}_m$ into Proposition 8.4.

Substituting $\mathbf{u}^{(1)}$ (8.45) into (8.91) and taking into account Proposition 8.6, we can rewrite (8.91) as

$$N^{(-2)}_{ij} = N^{(-2)}_{ji} + \langle (b^{(-3)}_{ij11}(x_1,\mathbf{x}) - b^{(-3)}_{ji11}(x_1,\mathbf{x}))y_\alpha\rangle u^{(1)}_{\alpha,1x1x} \tag{8.92}$$

$$+ \langle (b^{(-3)}_{ij\gamma 1}(x_1,\mathbf{x}) - b^{(-3)}_{ji\gamma 1}(x_1,\mathbf{x}))y_\Gamma\rangle s_\gamma \phi_{,1x}(x_1)$$

$$+ \langle (b^{(-3)}_{ijm1}(x_1,\mathbf{x}) - b^{(-3)}_{jim1}(x_1,\mathbf{x}))\rangle V_{m,1x}(x_1).$$

In accordance with condition (8.79), the last term on the right-hand side of (8.92) is equal to zero. Applying (8.92), we can exclude $N^{(-2)}_{ij}$ from the equilibrium equations (8.57)–(8.59). The elimination procedure has some differences for the equilibrium equations corresponding to the bending and the torsion of beam. For this reason, we consider the bending and torsion of the beam separately.

8.3.3 Bending

Taking into account (8.88), (8.92), we can exclude $N^{(-2)}_{1\beta}$ and $N^{(-2)}_{\beta 1}$ from (8.58), (8.59) and obtain the following equation for the bending moments:

$$(M_{1\beta}(V_{1,1x}, u^{(0)}_{\alpha,1x1x}, \phi_{,1x}) + \mathbf{C}_{\alpha\beta} u^{(0)}_{\alpha,1x})_{,1x1x} = K_{\beta,1x}. \tag{8.93}$$

Here, we denote

$$K_\beta = N^{(-2)}_{1\beta} - N^{(-2)}_{\beta 1} = k_{\beta\alpha} u^{(0)}_{\alpha,1x1x} + k_\beta \phi_{,1x}, \tag{8.94}$$

$M_\beta(V_{1,1x}, u^{(0)}_{\alpha,1x1x}, \phi_{,1x})$ is determined in Sect. 8.1, see formula (8.54).
In (8.94), the following notations are used:

$$k_{\beta\alpha} = \langle (b^{(-3)}_{1\beta 11}(x_1, \mathbf{x}) - b^{(-3)}_{\beta 111}(x_1, \mathbf{x}) y_\alpha \rangle = \langle \sigma^{*(-3)}_{\beta 1}(x_1, \mathbf{x}) y_\alpha \rangle; \tag{8.95}$$

$$k_\beta = \langle (b^{(-3)}_{1\beta\gamma 1}(x_1, \mathbf{x}) - b^{(-3)}_{\beta 1\gamma 1}(x_1, \mathbf{x}) y_\Gamma \rangle s_B = \langle \sigma^{*(-3)}_{11}(x_1, \mathbf{x}) y_\beta \rangle s_B$$

($B = 3$ if $\beta = 2$, and $B = 2$ if $\beta = 3$; no sum in β, B here).

8.3.4 Torsion

For the torsion moment $M = M^{(-3)}_{23} - M^{(-3)}_{32}$, we can derive from (8.60), (8.89), and (8.92) the following equation:

$$(M(V_{1,1x}, u^{(0)}_{\alpha,1x1x}, \phi_{,1x}) + \mathbf{C}_\alpha u^{(0)}_{\alpha,1x} + \mathbf{J}\phi_{,1x})_{,1x} = K. \tag{8.96}$$

Here,

$$K = N^{(-2)}_{32} - N^{(-2)}_{23} = k\phi_{,1x}, \tag{8.97}$$

where the following notations are used:

$$k = \langle (b^{(-3)}_{32\gamma 1}(x_1, \mathbf{x}) - b^{(-3)}_{23\gamma 1}(x_1, \mathbf{x})) y_\Gamma \rangle s_\gamma \tag{8.98}$$

$$= \langle \sigma^{*(-3)}_{21}(x_1, \mathbf{x}) y_2 + \sigma^{*(-3)}_{31}(x_1, \mathbf{x}) y_3 \rangle,$$

$M(V_{1,1x}, u^{(0)}_{\alpha,1x1x}, \phi_{,1x})$ is determined in Sect. 8.1, see formula (8.55).
Note that all the coefficients determined by (8.90), (8.94), and (8.97) could be written as the combinations of the moments of initial stresses $M^{*(-3)}_{i\alpha} = \langle \sigma^{*(-3)}_{1\alpha} y_\alpha \rangle$ in the following form:

$$\mathbf{C}_{\beta\alpha} = M^{*(-3)}_{\beta\alpha}; \tag{8.99}$$

$$\mathbf{C}_{\alpha} = M^{*(-3)}_{12}\delta_{3\alpha} - M^{*(-3)}_{13}\delta_{2\alpha};$$

$$\mathbf{J}_{\alpha} = M^{*(-3)}_{22} - M^{*(-3)}_{33};$$

$$k_{\beta\alpha} = M^{*(-3)}_{\beta\alpha} = \mathbf{C}_{\beta\alpha};$$

$$k_{\beta} = -M^{*(-3)}_{1\beta} s_B \text{ (no sum in } \beta, B \text{ here)};$$

$$k = M^{*(-3)}_{22} + M^{*(-3)}_{33}.$$

By virtue of the symmetry of initial stresses σ^*_{ij} with respect to i and j and Proposition 8.5, $M^{*(-3)}_{\beta\beta} = 0$. Then, $\mathbf{C}_{\beta\beta} = 0$, $J = 0$, $k = 0$ and the nonzero coefficients in (8.99) are the following:

$$\mathbf{C}_{\beta B} = M^{*(-3)}_{\beta B}, \tag{8.100}$$

$$\mathbf{c}_{\alpha} = M^{*(-3)}_{12}\delta_{3\alpha} - M^{*(-3)}_{13}\delta_{2\alpha};$$

$$k_{\beta B} = M^{*(-3)}_{\beta B} = \mathbf{C}_{\beta B},$$

$$k_{\beta} = -M^{*(-3)}_{1\beta} s_B$$

no sum in β, B in (8.100) (index B is related with β like Γ and γ in (8.38)).

8.3.5 The Limit One-Dimensional Problem

Let us write the one-dimensional equations of equilibrium and boundary conditions obtained. The equations of the limit model can be written in the following form:

$$(AV_{1,1x} + A^1_{\alpha\beta}u^{(0)}_{\alpha,1x1x} + b\phi_{,1x})_{,1x} + \langle f^{\varepsilon}_1\rangle + \langle h^{\varepsilon}_1\rangle_S = 0; \tag{8.101}$$

$$\begin{aligned}&({}^1A_{\beta}V_{1,1x} + A^2_{\beta\alpha}u^{(0)}_{\alpha,1x1x} + B_{\beta}\phi_{,1x})_{,1x1x}\\&+ \langle f_{\beta}\rangle + \langle h_{\beta}\rangle_S + (\langle f^{\varepsilon}_1 y_{\beta}\rangle + \langle h^{\varepsilon}_1 y_{\beta}\rangle_S)_{,1x}\\&= (-\mathbf{C}_{\beta B}u^{(0)}_{B,1x})_{,1x1x} + (k_{\beta B}u^{(0)}_{B,1x1x} + k_{\beta}\phi_{,1x})_{,1x};\end{aligned} \tag{8.102}$$

$$(A^1V_{1,1x} + A^2_{\alpha}u^{(0)}_{\alpha,1x1x} + B\phi_{,1x})_{,1x} = (\mathbf{c}_{\alpha}u^{(0)}_{\alpha,1x})_{,1x}. \tag{8.103}$$

$\mathbf{C}_{\beta B}$, $\mathbf{c}_{\alpha}$, $k_{\beta B}$, k_{β} are determined by (8.100).

The boundary conditions are

$$V_1(-a) = V_1(a) = 0; \tag{8.104}$$

$$u_\alpha^{(0)}(-a) = u_\alpha^{(0)}(a) = u_{\alpha,1x}^{(0)}(-a) = u_{\alpha,1x}^{(0)}(a) = 0;$$

$$\phi(-a) = \phi(a) = 0.$$

8.4 Local Stresses

The total stresses $\sigma_{ij}^{total}(\mathbf{x})$ in the body with initial stresses are the sum of the initial stresses $\sigma_{ij}^{*\varepsilon}(\mathbf{x})$ and additional stresses $\sigma_{ij}^{\varepsilon}(\mathbf{x})$:

$$\sigma_{ij}^{total}(\mathbf{x})) = \sigma_{ij}^{*\varepsilon}(\mathbf{x}) + \sigma_{ij}^{\varepsilon}(\mathbf{x}). \tag{8.105}$$

In accordance with (8.8),

$$\sigma_{ij}^{\varepsilon} = \varepsilon^{-4}\sigma_{ij}^{(-4)}(x_1, y_2, y_3) + \varepsilon^{-3}\sigma_{ij}^{(-3)}(x_1, y_2, y_3) + ..., \tag{8.106}$$

where "..." means the terms smaller than ε^{-3}.

8.4.1 Case of Resultant Axial Initial Force

For a beam subjected to the resultant axial initial force, in accordance with (8.46), $\sigma_{ij}^{(-4)} = 0$, thus the local stresses in the beam are determined by the term $\sigma_{ij}^{(-3)}$ given by (8.52), and we may accept the formula

$$\begin{aligned}\sigma_{ij}^{total} &= \sigma_{ij}^{*\varepsilon} + \sigma_{ij}^{(-3)} \\ &= \sigma_{ij}^{*\varepsilon} + (c_{ijm\gamma}N_{m,\gamma y}^{01} + c_{ij11})V_{1,1x}(x_1) \\ &+ (c_{ijm\gamma}N_{m,\gamma y}^{1\alpha} + c_{ij11}y_\alpha)u_{\alpha,1x1x}^{(0)}(x_1) \\ &+ (c_{ijm\gamma}X_{m,\gamma y}^{3} + c_{ij\gamma 1}s_\gamma y_\Gamma)\phi_{,1x}(x_1).\end{aligned} \tag{8.107}$$

8.4.2 Case of Moments of Initial Stresses

For a beam subjected to the moments of the initial stresses, in accordance with (8.83), $\sigma_{ij}^{(-4)} = 0$, thus the local stresses in the beam are determined by the term $\sigma_{ij}^{(-3)}$. For the beam subjected to the moments of the initial stresses $\sigma_{ij}^{(-3)}$ is given by (8.86), and we may accept the formula

$$\begin{aligned}\sigma_{ij}^{total} &= \sigma_{ij}^{*\varepsilon} + \sigma_{ij}^{(-3)} \\ &= \sigma_{ij}^{*\varepsilon} + (c_{ijm\gamma} N_{m,\gamma y}^{01} + c_{ij11}) V_{1,1x}(x_1) \\ &+ (c_{ijm\gamma} N_{m,\gamma y}^{1\alpha} + c_{ij11} y_\alpha) u_{\alpha,1x1x}^{(0)}(x_1) \\ &+ (c_{ijm\gamma} X_{m,\gamma y}^{3} + c_{ij\gamma 1} s_\gamma y_\Gamma) \phi_{,1x}(x_1) \\ &+ (b_{ij\alpha 1}^{(-3)}(x_1, \mathbf{y}) - b_{ij1\alpha}^{(-3)}(x_1, \mathbf{y})) u_{\alpha,1x}^{(0)}(x_1) + b_{ij\gamma\Gamma}^{(-3)}(x_1, \mathbf{y}) s_\gamma \psi(x_1).\end{aligned} \tag{8.108}$$

Substituting $b_{ijmn}^{(-3)}$ in accordance with the definition (8.4), we have

$$\begin{aligned}\sigma_{ij}^{total} &= \sigma_{ij}^{*\varepsilon} + \sigma_{ij}^{(-3)} \\ &= \sigma_{ij}^{*\varepsilon} + (c_{ijm\gamma} N_{m,\gamma y}^{01} + c_{ij11}) V_{1,1x}(x_1) \\ &+ (c_{ijm\gamma} N_{m,\gamma y}^{1\alpha} + c_{ij11} y_\alpha) u_{\alpha,1x1x}^{(0)}(x_1) \\ &+ (c_{ijm\gamma} X_{m,\gamma y}^{3} + c_{ij\gamma 1} s_\gamma y_\Gamma) \phi_{,1x}(x_1) \\ &+ (\delta_{i\alpha} \sigma_{j1}^{*(-3)}(x_1, \mathbf{y}) - \delta_{i1} \sigma_{j\alpha}^{*(-3)}(x_1, \mathbf{y})) u_{\alpha,1x}^{(0)}(x_1) + \delta_{i\gamma} \sigma_{j\Gamma}^{*(-3)}(x_1, \mathbf{y}) s_\gamma \psi(x_1).\end{aligned} \tag{8.109}$$

References

Bernoulli D (1751) De Vibrationibus et Sono Laminarum Elasticarum. Commentarii Academiae Scientiarum Imperialis Petropolitanae 13:105–120

de Saint-Venant B (1856) Memoire sur la flexion des prismes, sur les glissements transversaux et longitudinaux qui l'accompagnent lorsqu'elle ne s'opere pas uniformement ou en arc de cercle, et sur la forme courbe affectee alors par leurs sections transversales primitivement planes. J Math Pure Appl (J Math Liouville) 1:89–189

de Saint-Venant B (1856) Memoire sur la torsion des prismes: avec des considerations sur leur flexion ainsi que sur l'equilibre interieur des solides elastiques en general, et des formules pratiques pour le calcul de leur resistance a divers efforts s'exercant simultanement. Mem Savants Etrang 14:233–560

Euler L (1759) Sur la force des colonnes. Histoire de l'Academie Royal des sciences et belle-lettres avec les Memoires. Tires des Registres de cette Academie 13:252–282
Kolpakov AG (1991) Calculation of the characteristics of thin elastic rods of periodic structure. J Appl Math Mech 57:440–448
Kolpakov AG (1992) On problem of beam with initial stress. Appl Mech Tech Phys 6:139–144
Rodriguez JM, Viano JM (1997) Asymptotic derivation of a general linear model for thin-walled elastic rods. Comput Methods Appl Mech Eng 147(3–4):287–321
Rudin W (1967) Principles of mathematical analysis, 3rd edn. McGraw-Hill, New York
Timoshemko S, Goodier JN (1951) Theory of elasticity. McGrow-Hill, New York
Timoshenko S (1983) History of strength of materials (with a brief account of the history of elasticity and theory of structures). Dover, New York
Timoshenko S, Woinowsky-Krieger K (1959) Theory of plates and shells. McGraw-Hill, New York
Trabucho L, Viano JM (1996) Mathematical modeling of rods. In: Ciarlet PG, Lions J-L (eds) Handbook of numerical analysis, vol IV. Elsevier, Amsterdam, pp 487–974
Tutek Z, Aganovich I (1987) A justification of the one-dimensional linear model of elastic beam. Math Methods Appl Sci 8:502–515

Chapter 9
Non-homogeneous Beams with Initial Stresses

In this chapter, we will consider non-uniform beams of periodic structure. The non-uniformity of a beam can be a result of non-homogeneity of the material from which it is made (the beams made of composite materials Johnson 2004; Agarwal et al. 2006; Kollar and Springer 2009; Kapuria et al. 2004; Zhen and Wanji 2008) or made of different constitutive elements (Pshenichnov 1993; Panasenko 1993), and non-uniformity of the beam geometry (complex geometry beam) (Qiao et al. 2017; Cartraud and Messager 2006; Vinot et al. 2001; Treyssede and Cartraud 2022; Gesualdo 2018), or both of them (Buannic and Cartraud 2001).

There were developed homogenization techniques for beams made of several materials (in particular, laminated beams) (Kapuria et al. 2004; Zhen and Wanji 2008), beams of varying cross-section (when the period of the thickness oscillation is compared with the beam diameter) (Gesualdo 2018; Levinski 1995), porous and frame-like beams (Groen et al. 2020; Araujo et al. 2023; Hebiba et al. 2022), variational principles and estimates for the homogenized stiffnesses of beams (Kolpakov 1998), etc., and corresponding numerical techniques (Cartraud and Messager 2006; Vinot et al. 2001). In the mentioned papers, plates with no initial stresses were investigated.

In this book only elastic beams are considered. The combinations of elastic effects with other effects (thermal, piezoelectric, etc., and control devices) in thin beams were studied in Kuczma (2010), Andrianov et al. (2024), He et al. (2023), Yadav (2023), Jiang and Li (2007), Kondaiah et al. (2012), Kalamkarov and Kolpakov (1997), Kolpakov and Kolpakov (2006).

9.1 Statement of the Problem

We consider an elastic body G_ε of periodic structure obtained by repeating a certain small periodicity cell εY along the Ox_1-axis, see Fig. 9.1. Here, ε is the periodicity

A. Kolpakov, *Thin-Walled Structures with Initial Stresses*, Mathematical Engineering,
https://doi.org/10.1007/978-3-032-07247-4_9

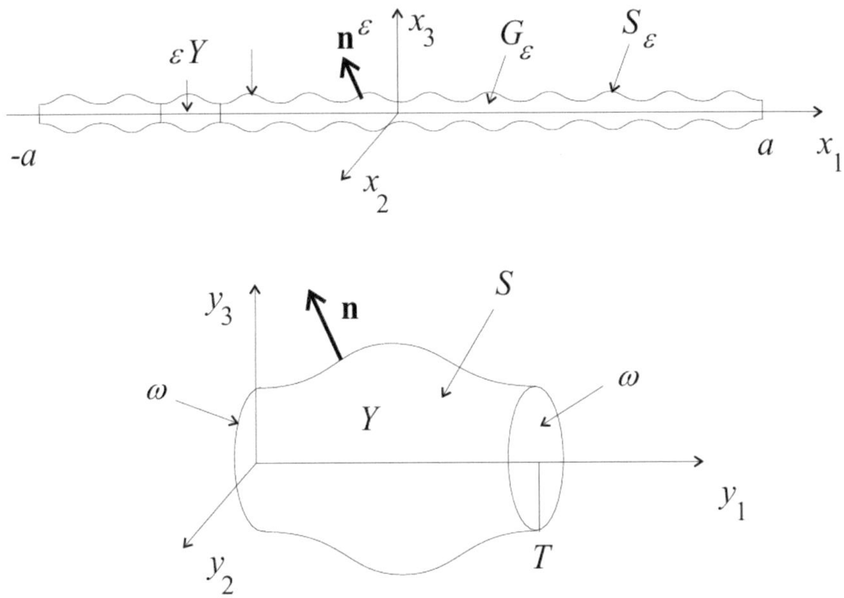

Fig. 9.1 A beam of periodic structure (top) and its periodicity cell in "fast" variables (bottom)

cell's characteristic dimension, which is assumed to be small (that is formalized in the form $\varepsilon \to 0$). As a result, we have a body of periodic structure with a small diameter—a three-dimensional beam. For $\varepsilon \to 0$, the three-dimensional beam tends to the segment $[-a, a]$ on the Ox_1-axis (Fig. 9.1)—a one-dimensional beam. Our aim is to derive a model describing a one-dimensional beam.

The starting point of our investigation is the three-dimensional elasticity problem for a body with initial presented in Sect. 4.1. In accordance with this model, the equilibrium equations of the beam as a three-dimensional elastic body with initial stresses can be written in the form,

$$\frac{\partial \sigma_{ij}^{\varepsilon}}{\partial x_j} + f^{\varepsilon}(\mathbf{x}) = 0 \text{ in } G_{\varepsilon}, \tag{9.1}$$

where $\sigma_{ij}^{\varepsilon}$ are additional stresses determined by the formula,

$$\sigma_{ij}^{\varepsilon} = h_{ijmn}(x_1, \mathbf{x}/\varepsilon)\frac{\partial u_m^{\varepsilon}}{\partial x_n}. \tag{9.2}$$

The coefficients h_{ijmn} are a known combination of the tensor of elasticity constants $\varepsilon^{-4}c_{ijmn}$ and initial stresses $\sigma{*}_{ij}$, which, in the case under consideration, is taken in the form,

$$h_{ijmn}(x_1, \mathbf{x}/\varepsilon) = \varepsilon^{-4} c_{ijmn}(\mathbf{x}/\varepsilon) + \varepsilon^{-4} b^{(-4)}_{ijmn}(x_1, \mathbf{x}/\varepsilon) \tag{9.3}$$

$$+ \varepsilon^{-3} b^{(-3)}_{ijmn}(x_1, \mathbf{x}/\varepsilon) + \varepsilon^{-2} b^{(-2)}_{ijmn}(x_1, \mathbf{x}/\varepsilon),$$

where

$$b^{(p)}_{ijmn}(x_1, \mathbf{x}/\varepsilon) = \sigma^{*(p)}_{jn}(x_1, \mathbf{x}/\varepsilon)\delta_{im} \; (p = -4, -3, -2). \tag{9.4}$$

Formula (9.2) may be considered as a local governing equation for the body under consideration.

The boundary conditions can be written in the form

$$\sigma^{\varepsilon}_{ij} n_j = h^{\varepsilon}_i(\mathbf{x}) \text{ on } S_\varepsilon, \tag{9.5}$$

$$\mathbf{u}^{\varepsilon}(\mathbf{x}) = 0 \text{ on } S_u.$$

The mass force is taken in the form $\mathbf{f}^{\varepsilon}(\mathbf{x}) = (\varepsilon^{-2} f_1, \varepsilon^{-2} f_2, \varepsilon^{-3} f_3)$ and the surface force is taken in the form $\mathbf{h}^{\varepsilon}(\mathbf{x}) = (\varepsilon^{-1} h_1, \varepsilon^{-1} h_2, \varepsilon^{-2} h_3)$.

Here, S_ε is the lateral surface of the beam; now it are not a cylinder. Generally, all coordinates of the normal $\mathbf{n}$ to S_ε are non-zero. The beam is fastened along the surface S_u (Fig. 9.1).

Using the two-scale method, in the versions for beams, a function of arguments $x_1, \mathbf{x}/\varepsilon$ is considered as a function of the "slow" variable x_1 and the "fast" variables $\mathbf{y} = \mathbf{x}/\varepsilon$. In accordance with this note, the functions will be written in the form $c_{ijmn}(\mathbf{y})$ and $\sigma^{*(p)}_{jn}(x_1, \mathbf{y})$, $h_{ijmn}(x_1, \mathbf{y})$. The functions are periodic in y_1 with period m, where m is the projection of the periodicity cell Y in the "fast" variables $\mathbf{y}$ on the Oy_1-axis, see Fig. 9.1.

In (9.3), the multiplier ε^{-4} at c_{ijmn} guarantees that the bending and torsion stiffnesses of the beam will be nonzero, as $\varepsilon \to 0$. The term $\varepsilon^{-2} b^{(-2)}_{ijmn} = \varepsilon^{-2} \sigma^{*(-2)}_{jn} \delta_{im}$ corresponds to the axial resultant initial forces independent on ε. The term $\varepsilon^{-3} b^{(-3)}_{ijmn} = \varepsilon^{-3} \sigma^{*(-3)}_{jn} \delta_{im}$ corresponds to the resultant moments independent on ε. The term $\varepsilon^{-4} b^{(-4)}_{ijmn} = \varepsilon^{-3} \sigma^{*(-3)}_{jn} \delta_{im}$ corresponds to the initial stresses of order of elastic constants.

9.2 Asymptotic Expansion and Homogenized Equations of Equilibrium

We will study global deformations of the beams, as $\varepsilon \to 0$. To do this, we use the following asymptotic expansion:

asymptotic expansion for displacements

$$\mathbf{u}^{\varepsilon} = \mathbf{u}^{(0)}(x_1) + \sum_{k=1}^{\infty} \varepsilon^k \mathbf{u}^{(k)}(x_1, \mathbf{y}), \tag{9.6}$$

asymptotic expansion for stresses

$$\sigma_{ij}^{\varepsilon} = \sum_{p=-4}^{\infty} \varepsilon^p \sigma_{ij}^{(p)}(x_1, \mathbf{y}). \tag{9.7}$$

The functions on the right-hand side of (9.6) and (9.7) are assumed to be periodic in y_1 with periodicity cell T.

The leading terms in the series (9.6) depends only on "slow" variable, thus, the deflection of beam have limit value. The leading term in the series (9.7) is highly oscillating and do not have a strong limit, $\varepsilon \to 0$, but the average values of the stresses, as well as moments of stresses, have limit values.

The difference of the asymptotic expansions (9.6) and (9.6) from the asymptotic expansions (8.7) and (8.8) is that the functions on the right-hand parts (9.6) and (9.6) depend on all components of the "fast" variable $\mathbf{y} = (y_1, y_2, y_3)$, not only y_1, y_2. It gives rise to some differences in mathematical analysis of the problem. However, the most significant differences arise not in mathematical, but in mechanical analysis of the problem.

The analysis of problem (9.1)–(9.4) breaks down into two stages. The first entails obtaining the equations of equilibrium for the beam considered as a one-dimensional structure. As in the case of plates, see Chap. 6, this stage does not involve local governing equations and it is the same stage for any governing equations.

For function $Z(x_1, \mathbf{y})$ of the arguments x_1 and $\mathbf{y} = (y_1, y_2, y_3)$, like on the right-hand side of (9.6) and (9.7), the differential operators are represented in the form

$$\frac{\partial Z}{\partial x_\alpha} \to Z_{,1x} + \varepsilon^{-1} Z_{,1y}\,, \quad \frac{\partial Z}{\partial x_\alpha} \to \varepsilon^{-1} Z_{,\alpha y} \quad (\alpha = 2, 3). \tag{9.8}$$

Here and below in this chapter, the Latin indexes take the values 1, 2, 3; and Greek indexes take the values 2, 3; subscript,αx means $\partial/\partial x_\alpha$ and , $1y$ means $\partial/\partial y_1$.

We introduce

$$\langle\rangle = \frac{1}{T} \int_Y d\mathbf{y}, \quad \langle\rangle_S = \frac{1}{T} \int_S d\mathbf{y} \tag{9.9}$$

the average values of function over periodic cell Y and its lateral surface S in "fast" variables $\mathbf{y}$ (Fig. 9.1).

Proposition 9.1 *Let function $Z(x_1, \mathbf{y})$ have continuous derivatives with respect to variable x_1 and it is periodic with respect to the variable y_1 with period m. Then,*

$$\varepsilon^{-1}\int_{G_\varepsilon} Z(x_1, \mathbf{x}/\varepsilon)d\mathbf{x} \to \int_{-a}^{a} \langle Z\rangle(x_1)dx_1;$$

$$\int_{S_\varepsilon} Z(x_1, \mathbf{x}/\varepsilon)d\mathbf{x} \to \int_{-a}^{a} \langle Z\rangle_S(x_1)dx_1, \ as\ \varepsilon \to 0.$$

The proof of Proposition 9.1 is similar to the proof of Proposition 3.1. See note after Proposition 3.1. Another proof may be found in Trabucho and Viano (1996).

To obtain the homogenized equilibrium equations, consider a weak form of problem (9.1)

$$-\int_{G_\varepsilon} \sigma_{ij}^{\varepsilon}\frac{\partial v_i}{\partial x_j}d\mathbf{x} + \int_{S_\varepsilon} \mathbf{h}^{\varepsilon}\mathbf{v}d\mathbf{x} + \int_{G_\varepsilon} \mathbf{f}^{\varepsilon}\mathbf{v}d\mathbf{x} = 0 \text{ for any } \mathbf{v} \in V_\varepsilon, \tag{9.10}$$

where

$$V_\varepsilon = \{\mathbf{v} \in \{H^1(G_\varepsilon)\}^3 : \mathbf{v}(\mathbf{x}) = 0 \text{ on } S_u\}$$

is the set of admissible displacements of the beam considered as three-dimensional elastic body.

Substituting (9.7) into (9.10), we obtain, with allowance for the differentiation rule (9.8)

$$\sum_{p=-4}^{\infty}\int_{G_\varepsilon} \varepsilon^p\sigma_{ij}^{(p)}(v_{i,1x}\delta_{j1} + \varepsilon^{-1}v_{i,jy})d\mathbf{x} \tag{9.11}$$

$$= \int_{S_\varepsilon} \mathbf{h}^{\varepsilon}\mathbf{v}d\mathbf{x} + \int_{G_\varepsilon} \mathbf{f}^{\varepsilon}\mathbf{v}d\mathbf{x} \quad \text{for any } \mathbf{v} \in U_\varepsilon.$$

We use functional space,

$$H = \{H^{01}([-a,a])\}^3 = \{\mathbf{w}(x_1) \in \{H^1([-a,a])\}^3 : \mathbf{w}(\pm a) = 0\} \tag{9.12}$$

is the set of admissible displacements of one-dimensional beam. It is obvious that $H \subset U_\varepsilon$.

Taking in (9.11) trial function $\mathbf{v} = \mathbf{w}(x_1) \in H$, we arrive at the equality

$$\sum_{p=-4}^{\infty}\int_{G_\varepsilon} \varepsilon^p\sigma_{i1}^{(p)}v_{i,1x}\delta_{j1}d\mathbf{x} \tag{9.13}$$

$$= \int_{S_\varepsilon} \mathbf{h}^{\varepsilon}\mathbf{v}d\mathbf{x} + \int_{G_\varepsilon} \mathbf{f}^{\varepsilon}\mathbf{v}d\mathbf{x} \quad \text{for any } \mathbf{v} \in H.$$

By virtue of Proposition 9.1, Eq. (9.13) may be rewritten as

$$\sum_{p=-4}^{\infty}\int_{-a}^{a}\varepsilon^p\langle\sigma_{i1}^{(p)}\rangle v_{i,1x}dx_1 \tag{9.14}$$

$$=\int_{-a}^{a}\langle\mathbf{h}^{\varepsilon}\rangle_S\mathbf{v}dx_1+\int_{-a}^{a}\langle\mathbf{f}^{\varepsilon}\rangle\mathbf{v}dx_1 \quad \text{for any } \mathbf{v}\in H.$$

The equality (9.14) suggests to us that in the case under consideration it would be reasonable to introduce the resultant stresses by the formula

$$N_{ij}^{(p)}(x_1)=\langle\sigma_{ij}^{(p)}(x_1,\mathbf{y})\rangle \ (i,j=1,2,3;\ p=-4,-3,...). \tag{9.15}$$

In contrast to the classical theory, the definition of the resultant forces assumes the integration over the periodicity cell Y, not over a cross section of the beam.

Changing in (9.14) $\langle\sigma_{ij}^{(p)}\rangle$ for $N_{ij}^{(p)}$, we obtain

$$\sum_{p=-4}^{\infty}\int_{-a}^{a}\varepsilon^p N_{i1}^{(p)}v_{i,1x}\delta_{j1}dx_1 \tag{9.16}$$

$$=\int_{-a}^{a}\langle\mathbf{h}^{\varepsilon}\rangle_S\mathbf{v}dx_1+\int_{-a}^{a}\langle\mathbf{f}^{\varepsilon}\rangle\mathbf{v}dx_1 \quad \text{for any } \mathbf{v}\in H.$$

Integrating in (9.16) by parts and equating the terms with identical power of ε, we obtain the following set of equilibrium equations with respect to the resultant stresses (we present only the equations which will be used below):

$$N_{11,1x}^{(-4)}=0; \tag{9.17}$$

$$N_{11,1x}^{(-3)}+\langle f_1^{\varepsilon}\rangle+\langle h_1^{\varepsilon}\rangle_S=0; \tag{9.18}$$

$$N_{\beta 1,1x}^{(-2)}+\langle f_\beta^{\varepsilon}\rangle+\langle h_\beta^{\varepsilon}\rangle_S=0 \tag{9.19}$$

Now, we take the trial function in (9.11) in the form $\mathbf{v}=\mathbf{v}(x_1)y_\alpha$. As a result, we arrive at the equality

$$\sum_{p=-4}^{\infty}\int_{G_\varepsilon}(\varepsilon^p\sigma_{i1}^{(p)}y_\alpha v_{i,1x}+\varepsilon^{p-1}\sigma_{ij}^{(p)}v_i\delta_{j\alpha})d\mathbf{x} \tag{9.20}$$

$$=\int_{S_\varepsilon}\mathbf{h}^{\varepsilon}\mathbf{v}d\mathbf{x}+\int_{G_\varepsilon}\mathbf{f}^{\varepsilon}\mathbf{v}d\mathbf{x} \quad \text{for any } \mathbf{v}\in H.$$

By virtue of Proposition 9.1, Eq. (9.20) may be rewritten as

$$\sum_{p=-4}^{\infty} \int_{-a}^{a} (\varepsilon^{p} \langle \sigma_{i1}^{(p)} y_\alpha \rangle v_{i,1x} + \varepsilon^{p-1} \langle \sigma_{i\alpha}^{(p)} \rangle v_i) dx_1 \tag{9.21}$$

$$= \int_{a}^{a} \langle \mathbf{h}^{\varepsilon} y_\alpha \rangle_S \mathbf{v} dx_1 + \int_{-a}^{a} \langle \mathbf{f}^{\varepsilon} y_\alpha \rangle \mathbf{v} dx_1 \quad \text{for any } \mathbf{v} \in H.$$

The equality (9.21) suggests to us that in the case under consideration the resultant moments must be introduced by the integration over the periodicity cell Y:

$$M_{i\beta}^{(p)} = -\langle \sigma_{i1}^{(p)} y_\beta \rangle \ (i = 1, 2, 3; \ \beta = 2, 3 \text{ and } p = -4, -3, ...), \tag{9.22}$$

not over a cross-section of the beam.

Changing in (9.14) $\langle \sigma_{i1}^{(p)} y_\alpha \rangle$ for $M_{i\alpha}^{(p)}$ and $\langle \sigma_{i\alpha}^{(p)} \rangle$ for $N_{i\alpha}^{(p)}$, we obtain

$$\sum_{p=-4}^{\infty} \int_{-a}^{a} (\varepsilon^{p} M_{i\alpha}^{(p)} v_{i,1x} + \varepsilon^{p-1} N_{i\alpha}^{(p)} v_i) dx_1 \tag{9.23}$$

$$= \int_{-a}^{a} \langle \mathbf{h}^{\varepsilon} y_\alpha \rangle_S \mathbf{v} dx_1 + \int_{-a}^{a} \langle \mathbf{f}^{\varepsilon} y_\alpha \rangle \mathbf{v} dx_1 \quad \text{for any } \mathbf{v} \in H.$$

Integrating in (9.23) by parts and equating the terms with identical power of ε, we obtain the following set of equilibrium equations with respect to the resultant moments (we present only the equations which will be used in the following):

$$M_{i\beta,1x}^{(-4)} - N_{i\beta}^{(-3)} = 0; \tag{9.24}$$

$$M_{i\beta,1x}^{(-3)} - N_{i\beta}^{(-2)} + \delta_{i1}(\langle f_1^{\varepsilon} y_\beta \rangle + \langle h_1^{\varepsilon} y_\beta \rangle_S) = 0; \tag{9.25}$$

$$N_{i\alpha}^{(-4)} = \langle \sigma_{i\alpha}^{(-4)} \rangle = 0. \tag{9.26}$$

Equations (3.2.18) can be broken down into the following two groups: equilibrium equations for the bending moments $M_{1\beta}^{-3}$ ($\beta = 2, 3$),

$$M_{1\beta,1x}^{(-3)} - N_{1\beta}^{(-2)} + \langle f_1^{\varepsilon} y_\beta \rangle + \langle h_1^{\varepsilon} y_\beta \rangle_S = 0, \tag{9.27}$$

and equilibrium equation for the torsion moment $M = M_{32}^{-3} - M_{23}^{-3}$,

$$M_{,1x} + (N_{32}^{(-2)} - N_{23}^{(-2)}) = 0. \tag{9.28}$$

Equations (9.17), (9.18), (9.27), and (9.28) look like the classical ones. But in contrast to the classical equations in (9.18), (9.27),

$$N_{\beta 1}^{(-2)} \neq N_{1\beta}^{(-2)}.$$

and in (9.28)

$$N_{32}^{(-2)} \neq N_{23}^{(-2)}$$

because the local stresses $\sigma_{ij}^{\varepsilon}$ (9.2) are not symmetric in i and j, see (4.18).

Substituting expansion (9.7) into the equilibrium equations (9.1), we obtain, with allowance for (9.8),

$$\sum_{p=-4}^{\infty} \varepsilon^p \sigma_{i1,1x}^{(p)} + \varepsilon^{p-1} \sigma_{ij,jy}^{(p)} + f_i^{\varepsilon}(\mathbf{x}) = 0 \text{ in } G_{\varepsilon}; \tag{9.29}$$

$$\sum_{p=-4}^{\infty} \varepsilon^p \sigma_{ij}^{(p)} n_j = h_i^{\varepsilon}(\mathbf{x}) \text{ on } S_{\varepsilon}.$$

Equating in (9.27) the terms with identical negative power of ε, we obtain in particular,

$$\sigma_{ij,jy}^{(-4)} = 0 \text{ in } Y \ (p = -4); \tag{9.30}$$

$$\sigma_{ij,jy}^{(-3)} + \sigma_{i1,1x}^{(-4)} = 0 \text{ in } Y \ (p = -3);$$

$$\sigma_{ij}^{(p)} n_j = 0 \text{ on } S \ (p = -4, -3, \ldots).$$

Here S means the lateral surface of periodicity cell Y in "fast" variables; $\mathbf{n}$ is the normal to S, see Fig. 9.1.

9.3 Homogenized Governing Equations

The second stage of the analysis of the problem consists of obtaining the governing equations for the beam as a one-dimensional structure and excluding unknown quantities from the equilibrium equations. In contrast to the first stage, this stage does involve local governing equations.

In this section, we analyze the case,

$$\sigma_{ij}^{*(-3)} = 0 \text{ and } \sigma_{11}^{*(-2)} \neq 0$$

corresponding to nonzero axial force.

Substituting (9.6), (9.7) in local governing equation (9.2), with allowance for the differentiation rule (9.8), and equating the terms with identical power of ε, we obtain in particular,

$$\sigma_{ij}^{(p)} = c_{ijm1}(\mathbf{y})u_{m,1x}^{(p+4)} + c_{ijmn}(\mathbf{y})u_{m,ny}^{(p+5)} \ (p = -4, -3); \tag{9.31}$$

$$\sigma_{ij}^{(-2)} = c_{ijm1}(\mathbf{y})u_{m,1x}^{(2)} + b_{ijm1}(x_1, \mathbf{y})u_{m,1x}^{(3)} + \tag{9.32}$$

$$+ c_{ijmn}(\mathbf{y})u_{m,ny}^{(3)} + b_{ijmn}(x_1, \mathbf{y})u_{m,nx}^{(1)}.$$

As was assumed above,

$$\mathbf{u}^{(k)}(\mathbf{y}) \text{ is periodic in } y_1 \text{ with period } m \quad (k = 1, 2, ...). \tag{9.33}$$

Let us consider the problem (9.30, $p = -4$), (9.31, $p = -4$), (9.33, $k = 1$). It can be written in the form

$$\begin{cases} (c_{ijmn}(\mathbf{y})u_{m,ny}^{(1)} + c_{ijm1}(\mathbf{y})u_{m,1x}^{(0)}(x_1))_{,jy} = 0 \text{ in } Y; \\ (c_{ijmn}(\mathbf{y})u_{m,ny}^{(1)} + c_{ijm1}(\mathbf{y})u_{m,1x}^{(0)}(x_1))n_j = 0 \text{ on } S. \end{cases} \tag{9.34}$$

In order to solve problems of this kind, periodicity cell problems are introduced. In the case under consideration, we introduce the functions $\mathbf{N}^{0p}(\mathbf{y})$ $(p = 1, 2, 3)$, which are determined by solutions to periodicity cell problems of the beam theory (corresponding to axial deformation):

$$\begin{cases} (c_{ijmn}(\mathbf{y})N_{m,ny}^{0p} + c_{ijp1}(\mathbf{y}))_{,jy} = 0 \text{ in } Y; \\ (c_{ijmn}(\mathbf{y})N_{m,ny}^{0p} + c_{ijp1}(\mathbf{y}))n_j = 0 \text{ on } S; \\ \mathbf{N}^{0p}(\mathbf{y}) \text{ is periodic in } y_1 \text{ with period } m. \end{cases} \tag{9.35}$$

Proposition 9.2 *For $m = \alpha = 2, 3$ the solution to periodicity cell problem (9.37) is given by the following formula:*

$$\mathbf{N}^{0\alpha}(\mathbf{y}) = -y_\alpha \mathbf{e}_1, \tag{9.36}$$

where $\{\mathbf{e}_i\}$ are basis vectors of the coordinate system.

The proof may be found in Kolpakov (2004).

By virtue of Proposition 9.2, only one from the problems (9.35) is non-trivial. It is the problem for $p = 1$, which corresponds to axial tension of the beam:

$$\begin{cases} (c_{ijmn}(\mathbf{y})N_{m,ny}^{01} + c_{ij11}(\mathbf{y}))_{,jy} = 0 \text{ in } Y; \\ (c_{ijmn}(\mathbf{y})N_{m,ny}^{01} + c_{ij11}(\mathbf{y}))n_j = 0 \text{ on } S; \\ \mathbf{N}^{01}(\mathbf{y}) \text{ is periodic in } y_1 \text{ with period } m. \end{cases} \tag{9.37}$$

Proposition 9.3 *Periodicity cell problem (9.37) is degenerate. The kernel of problem (9.37) is generated by the function* **const** $= (C_1, C_2, C_3)$ *and the function*

$$\mathbf{U}(\mathbf{y}) = y_\Gamma s_\gamma \mathbf{e}_\gamma, \tag{9.38}$$

where $\{\mathbf{e}_i\}$ *are basis vectors of the coordinate system.*

Following (8.38), we use the notations

$$s_1 = 0, s_2 = -1, s_3 = 1; \tag{9.39}$$

$$\Gamma = 2 \text{ if } \gamma = 3 \text{ and } \Gamma = 3 \text{ if } \gamma = 2.$$

The summation in γ, Γ is assumed (if otherwise not indicated) in accordance with the rule $x_\Gamma y_\gamma = x_3 y_2 + x_2 y_3$.

The Propositions 9.2 and 9.3 are not equivalent to Propositions 8.2 and 8.3 because $bf N^{0p}$ and c_{ijmn} depend on the all three variables y_1, y_2, y_3. The proofs of the propositions may be found in Kolpakov (2004).

Allowing for the fact that the function $\mathbf{u}^{(0)}(x_1)$ of the argument x_1 plays the role of a parameter in the problems in the "fast" variable $\mathbf{y}$, solution to the problem (9.33, $k = 1$) and (9.34) can be found in the form,

$$\mathbf{u}^{(1)} = \mathbf{N}^{01}(\mathbf{y}) u^{(0)}_{1,1x}(x_1) - y_\alpha \mathbf{e}_1 u^{(0)}_{\alpha,1x}(x_1) \tag{9.40}$$

$$+ y_\Gamma s_\gamma \mathbf{e}_\gamma \phi(x_1) + \mathbf{V}(x_1).$$

Here, $\phi(x_1)$ and $\mathbf{V}(x_1)$ are arbitrary functions of the argument x_1, which will be determined later.

From (9.31, $p = -4$) and (9.40), we have

$$\sigma^{(-4)}_{ij} = (c_{ijmn}(\mathbf{y}) N^{01}_{m,ny}(\mathbf{y}) + c_{ij11}(\mathbf{y})) u^{(0)}_{1,1x}(x_1). \tag{9.41}$$

Averaging (9.41) with $ij = 11$, we obtain

$$N^{(-4)}_{11} = A u^{(0)}_{1,1x}(x_1), \tag{9.42}$$

where

$$A = \langle c_{1111}(\mathbf{y}) + c_{11mn}(\mathbf{y}) N^{01}_{m,ny}(\mathbf{y}) \rangle. \tag{9.43}$$

In accordance with the homogenized equilibrium equation (9.17) from (9.36), we obtain

$$(A u^{(0)}_{1,1x})_{,1x} = 0. \tag{9.44}$$

The boundary condition for $u_1^{(0)}$ follows from (9.5) and expansion (9.6), and has the form

$$u_1^{(0)}(-a) = u_1^{(0)}(a). \tag{9.45}$$

The solution to problem (9.44), (9.45) is $u_1^{(0)} = 0$. Then (9.40) takes a form

$$\mathbf{u}^{(1)} = -y_\alpha \mathbf{e}_1 u^{(0)}_{\alpha,1x}(x_1) + y_\Gamma s_\gamma \mathbf{e}_\gamma \phi(x_1) + \mathbf{V}(x_1), \tag{9.46}$$

Substituting (9.46) in (9.40) gives the following equations:

$$\sigma_{ij}^{(-4)} = 0; \tag{9.47}$$

$$\sigma_{ij}^{(-3)} = c_{ijmn}(\mathbf{y}) u^{(2)}_{m,ny} - c_{ij11}(\mathbf{y}) y_\alpha u^{(0)}_{\alpha,1x1x}(x_1) \tag{9.48}$$

$$+ c_{ij11}(\mathbf{y}) V_{1,1x}(x_1).$$

Let us examine problem (9.30, $p = -3$), (9.48), and (9.33, $k = 2$). It has the form

$$\begin{cases} (c_{ijmn}(\mathbf{y}) u^{(2)}_{m,ny} - c_{ij11}(\mathbf{y}) y_\alpha u^{(0)}_{\alpha,1x1x}(x_1) \\ + c_{ij11}(\mathbf{y}) V_{1,1x}(x_1))_{,jy} = 0 \text{ in } G_\varepsilon; \\ (c_{ijmn}(\mathbf{y}) u^{(2)}_{m,ny} - c_{ij11}(\mathbf{y}) y_\alpha u^{(0)}_{\alpha,1x1x}(x_1) \\ + c_{ij11}(\mathbf{y}) V_{1,1x}(x_1)) n_j = 0 \text{ on } S_\varepsilon. \end{cases} \tag{9.49}$$

In order to solve the problem (9.49), we introduce functions $\mathbf{X}^{1\alpha}(\mathbf{y})$ ($\alpha = 2, 3$) and $\mathbf{X}^3(\mathbf{y})$ which are determined by solutions of the following periodicity cell problems (corresponding to bending and torsion of beam):

$$\begin{cases} (c_{ijmn}(\mathbf{y}) N^{1\alpha}_{m,ny} - c_{ij11}(\mathbf{y}) y_\alpha)_{,jy} = 0 \text{ in } Y; \\ (c_{ijmn}(\mathbf{y}) N^{1\alpha}_{m,ny} - c_{ij11}(\mathbf{y}) y_\alpha) n_j = 0 \text{ on } S; \\ \mathbf{N}^{1\alpha}(\mathbf{y}) \text{ is periodic in } y_1 \text{ with period } m, \end{cases} \tag{9.50}$$

and

$$\begin{cases} (c_{ijmn}(\mathbf{y}) X^3_{m,ny} + c_{ij\gamma 1}(\mathbf{y}) s_\gamma y_\Gamma)_{,jy} = 0 \text{ in } Y; \\ (c_{ijmn}(\mathbf{y}) X^3_{m,ny} + c_{ij\gamma 1}(\mathbf{y}) s_\gamma y_\Gamma) n_j = 0 \text{ on } S; \\ \mathbf{X}^3(\mathbf{y}) \text{ is periodic in } y_1 \text{ with period } m. \end{cases} \tag{9.51}$$

The solution to problem (9.30, $p = -3$), (9.48) and (9.33, $k = 2$) can be expressed through the functions $\mathbf{N}^{01}(\mathbf{y})$, $\mathbf{N}^{1\alpha}(\mathbf{y})$, $\mathbf{X}^3(\mathbf{y})$, as follows:

$$\mathbf{u}^{(2)} = \mathbf{N}^{1\alpha}(\mathbf{y})u^{(0)}_{\alpha,1x1x}(x_1) - y_\alpha V_{\alpha,1x}(x_1)\mathbf{e}_1 + \mathbf{N}^{01}(\mathbf{y})V_{1,1x}(x_1)$$

$$+ \mathbf{X}^3(\mathbf{y})\phi_{,1x}(x_1) + \mathbf{U}(x_1) + s_\gamma y_\Gamma \mathbf{e}_\gamma \psi(x_1). \tag{9.52}$$

Writing (9.52), we use the equality $c_{ij\gamma\Gamma s_\gamma} = 0$ following from the symmetry of the elastic constants (see condition C1, Sect. 3.1). In (9.52), $\psi(x_1)$ is a new function (that will not influence the problem).

Substituting (9.52) in (9.47), we obtain

$$\sigma^{(-3)}_{ij} = (c_{ijmn}(\mathbf{y})N^{01}_{m,ny} - c_{ij11}(\mathbf{y}))V_{1,1x}(x_1) \tag{9.53}$$

$$+ (c_{ijmn}(\mathbf{y})N^{1\alpha}_{m,ny} + c_{ij11}(\mathbf{y})y_\alpha)u^{(0)}_{\alpha,1x1x}(x_1)$$

$$+ (c_{ijmn}(\mathbf{y})X^3_{m,ny} + c_{ij\gamma 1}(\mathbf{y})s_\gamma y_\Gamma)\phi_{,1x}(x_1).$$

Averaging (9.53) with $ij = 11$ over periodicity cell Y, we obtain the homogenized governing equations of the beam

$$N^{(-3)}_{11} = AV_{1,1x} + A^1_{\alpha\beta}u^{(0)}_{\alpha,1x1x} + b\phi(x_1), \tag{9.54}$$

Multiplying (9.53) with $j = 1$ by y_β and averaging over periodicity cell Y, we obtain equations, which can be broken down into two groups: the homogenized equations for bending moments,

$$M^{(-3)}_{1\beta} = \langle \sigma^{-3}_{11} y_\beta \rangle;$$

and the homogenized equations for torsion moments,

$$M = M^{(-3)}_{\Gamma\gamma} s_\gamma = M^{(-3)}_{32} - M^{(-3)}_{23} = \langle \sigma^{-3}_{\Gamma 1} y_\gamma s_\gamma \rangle = \langle \sigma^{-3}_{31} y_2 - \sigma^{-3}_{21} y_3 \rangle.$$

The homogenized governing equations are the following:

$$M^{(-3)}_{1\beta} = A_\beta V_{1,1x} + A^2_{\beta\alpha}u^{(0)}_{\alpha,1x1x} + B_\beta\phi(x_1); \tag{9.55}$$

$$M = A^1 V_{1,1x} + A^2_\alpha u^{(0)}_{\alpha,1x1x} + B\phi(x_1), \tag{9.56}$$

The coefficients on the right-hand parts of (9.54) and (9.55) are

$$A = \langle c_{1111}(\mathbf{y}) + c_{11\delta mn}(\mathbf{y}) N^{01}_{m,ny}(\mathbf{y}) \rangle; \tag{9.57}$$

$$A^1_\alpha = \langle c_{1111}(\mathbf{y}) y_\alpha + c_{11mn}(\mathbf{y}) N^{1\alpha}_{m,ny}(\mathbf{y}) \rangle;$$

$$b = \langle c_{11\gamma 1}(\mathbf{y})_\gamma y_\Gamma + c_{11mn}(\mathbf{y}) X^3_{m,ny}(\mathbf{y}) \rangle;$$

$${}^1A_\beta = \langle (c_{1111}(\mathbf{y}) + c_{11mn}(\mathbf{y}) N^{01}_{m,ny}(\mathbf{y})) y_\beta \rangle;$$

$$A^2_{\alpha\beta} = \langle (c_{1111}(\mathbf{y})) y_\alpha + c_{11mn}(\mathbf{y}) N^{1\alpha}_{m,ny}(\mathbf{y})) y_\beta \rangle;$$

$$B_\beta = \langle (c_{11\gamma 1}(\mathbf{y}) s_\gamma y_\Gamma + c_{11mn}(\mathbf{y}) X^3_{m,ny}(\mathbf{y})) y_\beta \rangle s_\beta;$$

$$A^1 = \langle (c_{3111}(\mathbf{y}) + c_{31mn}(\mathbf{y}) N^{01}_{m,ny}(\mathbf{y})) y_2$$

$$- (c_{2111}(\mathbf{y}) + c_{21mn}(\mathbf{y}) N^{01}_{m,ny}(\mathbf{y})) y_3 \rangle;$$

$$A^2_\alpha = \langle (c_{3111}(\mathbf{y})) y_\alpha + c_{31mn}(\mathbf{y}) N^{1\alpha}_{m,ny}(\mathbf{y})) y_2$$

$$- (c_{2111}(\mathbf{y}) y_\alpha + c_{21mn}(\mathbf{y}) N^{1\alpha}_{m,ny}(\mathbf{y})) y_3 \rangle$$

$(\alpha, \beta, \gamma, \delta = 2, 3)$.

The values (9.57) are the homogenized stiffnesses of the beam. In particular, A is the homogenized tension stiffness, $A^2_{\alpha\beta}$ are the homogenized bending stiffnesses, and A^1 is the homogenized torsional stiffness. Other stiffnesses in (9.57) are the the asymmetric homogenized stiffnesses. Note that the homogenized governing equations derived coincide with the classical ones in form, but the homogenized stiffnesses are calculated in anther way. At the same time, all the homogenized stiffnesses are expressed through solutions of periodicity cell problems (9.37), (9.50), and (9.51). In general, none of the homogenized rigidities is expressed explicitly.

Let us denote the right-hand parts of the governing equations (9.54)–(9.56) by

$$N(V_{1,1x}, u^{(0)}_{\alpha,1x1x}, \phi_{,1x}) = AV_{1,1x} + A^1_{\alpha\beta} u^{(0)}_{\alpha,1x1x} + b\phi(x_1)_{,1x};$$

$$M_\beta(V_{1,1x}, u^{(0)}_{\alpha,1x1x}, \phi_{,1x}) = {}^1A_\beta V_{1,1x} + A^2_{\beta\alpha} u^{(0)}_{\alpha,1x1x} + B_\beta \phi(x_1);$$

$$M(V_{1,1x}, u^{(0)}_{\alpha,1x1x}, \phi_{,1x}) = A^1 V_{1,1x} + A^2_\alpha u^{(0)}_{\alpha,1x1x} + B\phi(x_1)_{,1x}.$$

9.4 Elimination of Shear Forces

We collect the non-trivial homogenized equilibrium equations from (9.18), (9.27), and (9.28). They are the equations for the resultant forces

$$N^{(-3)}_{11,1x} + \langle f^{\varepsilon}_1 \rangle + \langle h^{\varepsilon}_1 \rangle_S = 0; \tag{9.58}$$

$$N^{(-2)}_{\beta 1,1x} + \langle f^{\varepsilon}_\beta \rangle + \langle h^{\varepsilon}_\beta \rangle_S = 0; \tag{9.59}$$

and equations for the resultant moments

$$M^{(-3)}_{1\beta,1x} - N^{(-2)}_{1\beta} + \langle f^{\varepsilon}_1 y_\beta \rangle + \langle g^{\varepsilon}_1 y_\beta \rangle_S = 0; \tag{9.60}$$

$$M_{,1x} + (N^{(-2)}_{32} - N^{(-2)}_{23}) = 0. \tag{9.61}$$

In the absence of initial stresses $\sigma^{(-2)}_{ij} = \sigma^{(-2)}_{ji}$ due to $c_{ijkl} = c_{jikl}$, see (9.29). Then $N^{(-2)}_{ij} = N^{(-2)}_{ji}$ and we can eliminate $N^{(-2)}_{1\beta}$ and $N^{(-2)}_{\beta 1}$ from Eqs. (9.59) and (9.60) using this symmetry, only, as done in both the classical theory of beams Timoshenko and Woinowsky-Krieger (1959) and in the asymptotic theory of beams no initial stresses (Trabucho and Viano 1987; Kolpakov 1991; Trabucho and Viano 1996).

In our case $N^{(-2)}_{ij}$ *is not symmetrical* in i and j because b_{ijmn} (9.4) are not symmetric in i and j. We use the method proposed in Chap. 8 to eliminate the shear forces form the homogenized equilibrium equations.

Let us insert $\mathbf{u}^{(1)}$ into (3.2.26) in accordance with (3.2.38). Then, we obtain

$$\sigma^{(-2)}_{ij} = c_{ijm1}(\mathbf{y}) u^{(2)}_{m,1x} + c_{ijmn}(\mathbf{y}) u^{(3)}_{m,ny} \tag{9.62}$$

$$+ b_{ij\alpha 1}(x_1, \mathbf{y}) u^{(0)}_{\alpha,1x} + b_{ij\gamma\Gamma m}(x_1, \mathbf{y}) s_\gamma \phi(x_1).$$

The first and second terms on the right-hand side of equation (9.62) are symmetrical with respect to i and j by virtue of the symmetry of the elastic constants. The antisymmetric part of the tensor $\sigma^{(-2)}_{ij}$ is

$$\sigma^{(-2)}_{ij} - \sigma^{(-2)}_{ji} = b_{ij\alpha 1}(x_1, \mathbf{y}) u^{(0)}_{3,\beta x}(x_1) - b_{ji\alpha 1}(x_1, \mathbf{y}) u^{(0)}_{\alpha,1x}(x_1)$$

$$+ b_{ij\gamma\Gamma}(x_1, \mathbf{y}) s_\gamma \phi(x_1) - b_{ji\gamma\Gamma}(x_1, \mathbf{y}) s_\gamma \phi(x_1).$$

Averaging (9.63) over periodicity cell Y, we obtain

$$N^{(-2)}_{ij} - N^{(-2)}_{ji} = \langle b_{ij\alpha 1}(x_1, \mathbf{y}) \rangle u^{(0)}_{3,\beta x}(x_1) - \langle b_{ji\alpha 1}(x_1, \mathbf{y}) \rangle u^{(0)}_{\alpha,1x}(x_1)$$

$$+ \langle b_{ij\gamma\Gamma}(x_1, \mathbf{y}) \rangle s_\gamma \phi(x_1) - \langle b_{ji\gamma\Gamma}(x_1, \mathbf{y}) \rangle s_\gamma \phi(x_1).$$

To continue our investigation, we need some facts about the average values of the initial stresses.

Proposition 9.4 *Let initial stresses* σ^*_{ij} , *which can be represented as* $\sigma^*_{ij} = \varepsilon^{-3}\sigma^{*(-3)}_{ij} + \varepsilon^{-2}\sigma^{*(-2)}_{ij} + \cdots$, *satisfy the equilibrium equations,*

$$\begin{cases} \dfrac{\partial \sigma^*_{ij}}{\partial x_j} + \varepsilon^a F_i = 0 \text{ in } G_\varepsilon; \\ \\ \sigma^*_{ij} n_j = \varepsilon^b H_i(x_1, \mathbf{y}) \text{ on } S_\varepsilon. \end{cases} \tag{9.63}$$

Then, if $a < -4, b < -3$, $\langle b^{*(p)}_{ijmn}(x_1, \mathbf{y}) Z_{,ny} \rangle = 0$, $p = -3, -2$ *for any differentiable function* $Z(\mathbf{y})$ *periodic in* y_1 *with period* m.

Substituting the expansion

$$\sigma^*_{ij} = \varepsilon^{-3}\sigma^{*(-3)}_{ij} + \varepsilon^{-2}\sigma^{*(-2)}_{ij} + \ldots$$

into (9.63), we obtain

$$\begin{cases} \sigma^{*(-3)}_{ij,jy} = 0 \text{ in } Y; \\ \\ \sigma^{*(-3)}_{ij} n_j = 0 \text{ on } S. \end{cases} \tag{9.64}$$

Taking into account the definition of $b^{(-3)}_{ijmn}$, it is sufficient to verify that $\langle \sigma^{*(-3)}_{jn} Z_{,ny} \rangle = 0$. Let us consider $\langle \sigma^*_{jn} Z_{,ny} \rangle$. Accounting the definition of the average value over the periodicity cell Y, integrating by parts, we find that this value is equal to

$$\int_Y \sigma^{*(-3)}_{jn,ny} Z(\mathbf{y}) d\mathbf{y} + \int_S \sigma^{*(-3)}_{jn} n_n Z(\mathbf{y}) d\mathbf{y} + \int_\omega \sigma^*_{j1} n_l Z(\mathbf{y}) d\mathbf{y}.$$

Here ω denote the opposite faces of the periodicity cell Y perpendicular to the Oy_1 axis (Fig. 9.1). The integrals over Y and S are equal to zero due to (9.64). The integral over ω is equal to zero as a consequence of the periodicity of the functions σ^*_{jn} and Y and the antiperiodicity of the vector—the normal with respect to y_1 (see Fig. 9.1).

Proposition 9.5 *Under the conditions of Proposition 8.4,* $\langle \sigma^{*(-3)}_{i\alpha} \rangle = 0$ *and* $\langle \sigma^{*(-3)}_{1\alpha} y_\alpha \rangle = 0$, *and* $\langle \sigma^{*(-2)}_{i\alpha} \rangle = 0$ *when* $\langle \sigma^{*(-3)}_{ij} \rangle = 0$.

Proposition 9.5 is the consequence of Proposition 9.4. It is sufficient to put $Z(\mathbf{y}) = y_\alpha$ and $Z(\mathbf{y}) = y^2_\alpha/2$, and take into account that y_α and $y^2_\alpha/2$ ($\alpha = 2, 3$) are periodic in y_1.

Now, we can exclude shear forces $N^{(-2)}_{\beta 1}$ and $N^{(-2)}_{1\beta}$ from the equilibrium equations.

9.5 Bending

Consider the following equilibrium equations (9.59) and (9.60). To eliminate $N^{(-2)}_{\beta 1}$ and $N^{(-2)}_{1\beta}$, we use the following relationship, which follows from (9.63) and the definition of $b^{(-2)}_{ijmn}$ (9.4):

$$N^{(-2)}_{1\beta} - N^{(-2)}_{\beta 1} = K_\beta, \tag{9.65}$$

where $K_\beta(y_1)$ is defined by the formula

$$\begin{aligned} K_\beta &= \langle b_{\beta 1\alpha}(x_1, \mathbf{y})\rangle u^{(0)}_{\alpha,1x}(x_1) - \langle b_{1\beta\alpha}(x_1, \mathbf{y})\rangle u^{(0)}_{\alpha,1x}(x_1) \\ &= \langle b_{\beta 1\gamma\Gamma}(x_1, \mathbf{y})\rangle s_\gamma \phi(x_1) - \langle b_{1\beta\gamma\Gamma}(x_1, \mathbf{y})\rangle s_\gamma \phi(x_1) \\ &= \langle \sigma^*_{11}(x_1, \mathbf{y})\rangle \delta_{\beta\alpha} u^{(0)}_{\alpha,1x}(x_1) + \langle \sigma_{1\Gamma}(x_1, \mathbf{y})\rangle \delta_{\beta\gamma} s_\gamma \phi(x_1). \end{aligned} \tag{9.66}$$

Differentiating (9.59) and using (9.60) and (9.65), we obtain

$$(M^{(-3)}_{1\beta,1x} + \langle f^\varepsilon_1 y_\beta\rangle + \langle h^\varepsilon_1 y_\beta\rangle)_{,1x} = K_{\beta,1x} - \langle f^\varepsilon_\beta\rangle - \langle h^\varepsilon_\beta\rangle_S. \tag{9.67}$$

Under the conditions of Proposition 9.4

$$K_\beta = \langle \sigma^*_{11}(x_1, \mathbf{y})\rangle u^{(0)}_{\beta,1x}(x_1) = N^*_{11}(x_1) u^{(0)}_{\beta,1x}(x_1),$$

where $N^*_{11} = \langle \sigma^*_{11}(x_1, \mathbf{y})\rangle$ is the resultant initial axial force. Then (3.2.60) takes the classical form,

$$\begin{aligned} &M^{(-3)}_{1\beta,1x1x} + \langle f^\varepsilon_\beta\rangle + \langle h^\varepsilon_\beta\rangle_S + (\langle f^\varepsilon_1 y_\beta\rangle + \langle h^\varepsilon_1 y_\beta\rangle_S)_{,1x} \\ &= (N^*_{11}(x_1) u^{(0)}_{\beta,1x}(x_1))_{,1x}. \end{aligned} \tag{9.68}$$

9.6 Torsion

Here, as above, we encounter the asymmetry of $N^{(-2)}_{ij}$. Due to the asymmetry, term $N^{(-2)}_{32} - N^{(-2)}_{23}$ in (9.61) is an unknown (and non-zero). In accordance with (9.63) and the definition of $b^{(-2)}_{ijmn}$ (9.4), we can write

$$N^{(-2)}_{23} = N^{(-2)}_{32} + K, \tag{9.69}$$

where

$$K = N_{32}^{(-2)} - N_{23}^{(-2)} = \langle b_{32\alpha 1}(x_1, \mathbf{y})\rangle u_{3,\beta x}^{(0)}(x_1) - \langle b_{23\alpha 1}(x_1, \mathbf{y})\rangle u_{\alpha,1x}^{(0)}(x_1)$$

$$+ \langle b_{32\gamma \Gamma}(x_1, \mathbf{y})\rangle s_\gamma \phi(x_1) - \langle b_{23\gamma \Gamma}(x_1, \mathbf{y})\rangle s_\gamma \phi(x_1) \tag{9.70}$$

$$= \langle \sigma_{31}^*(x_1, \mathbf{y})\rangle \delta_{2,\alpha} u_{\alpha,1x}^{(0)}(x_1) - \langle \sigma_{3\Gamma}^*(x_1, \mathbf{y})\rangle \delta_{2\gamma} \phi(x_1).$$

Then, (9.61) can be rewritten in the form,

$$M_{1,1x} - K = 0. \tag{9.71}$$

Under the conditions of Proposition 9.4, $K = 0$. This means, that the initial stresses of the order ε^{-2} do not influence the torsion of the beam.

9.7 Homogenized One-Dimensional Problem

By virtue of the homogenized governing equations (9.54)–(9.56), we can write the homogenized equations of equilibrium (9.58), (9.60), (9.71) in the form,

$$(AV_{1,1x} + A_{\alpha\beta}^1 u_{\alpha,1x1x}^{(0)} + b\phi_{,1x})_{,1x} + \langle f_1^\varepsilon\rangle + \langle h_1^\varepsilon\rangle_S = 0; \tag{9.72}$$

$$({}^1A_\beta V_{1,1x} + A_{\beta\alpha}^2 u_{\alpha,1x1x}^{(0)} + B_\beta \phi_{,1x})_{,1x1x} \tag{9.73}$$

$$+ \langle f_\beta^\varepsilon\rangle + \langle h_\beta^\varepsilon\rangle_S + (\langle f_1^\varepsilon y_\beta\rangle + \langle h_1^\varepsilon y_\beta\rangle_S)_{,1x} = (N_{11}^* u_{\alpha,1x}^{(0)})_{,1x};$$

$$(A^1 V_{1,1x} + A_\alpha^2 u_{\alpha,1x1x}^{(0)} + B\phi_{,1x})_{,1x} = 0. \tag{9.74}$$

Here, $N_{11}^* = \langle \sigma_{11}\rangle$.

The boundary conditions are the following:

$$V_1(-a) = V_1(a) = 0; \tag{9.75}$$

$$u_\alpha^{(0)}(-a) = u_\alpha^{(0)}(a) = u_{\alpha,1x}^{(0)}(-a) = u_{\alpha,1x}^{(0)}(a) = 0;$$

$$\phi(-a) = \phi(a) = 0.$$

Problem (9.72)–(9.74) with boundary conditions (9.75) is a one-dimensional model of a beam with respect to the functions $u_\alpha^{(0)}(x_1)$ $(\alpha = 2, 3)$, $V_1(x_1)$, and $\phi(x_1)$. Here $V_1(x_1)$ has the meaning of the axial displacement, $u_\alpha^{(0)}(x_1)$ $(\alpha = 2, 3)$ are normal deflections, and $\phi(x_1)$ is the torsion angle. This model coincides with the classical model of beam in form. As was noted, the definitions of the resultant forces and the resultant moments are qualitatively different from the classical definitions of the resultant forces and the resultant moments.

9.8 Local Stresses

In accordance with the general theory (Washizu 1992), the total stresses $\sigma_{ij}^{total}(\mathbf{x})$ in a beam with initial stresses are the sum of the initial stresses $\sigma_{ij}^{*}(\mathbf{x})$ and additional stresses $\sigma_{ij}^{\varepsilon}(\mathbf{x})$ (Washizu 1992):

$$\sigma_{ij}^{total}(\mathbf{x}) = \sigma_{ij}^{*}(\mathbf{x}) + \sigma_{ij}^{\varepsilon}(\mathbf{x}). \tag{9.76}$$

In accordance with (9.7), the additional stresses

$$\sigma_{ij}^{\varepsilon} = \varepsilon^{-3}\sigma_{ij}^{(-3)}(x_1, \mathbf{y}) + \varepsilon^{-2}\sigma_{ij}^{(-2)}(x_1, \mathbf{y}) + \cdots . \tag{9.77}$$

In accordance with (9.47), $\sigma_{ij}^{(-4)} = 0$, thus, the local stresses in the beam are determined by the term $\sigma_{ij}^{(-3)}$, which is given by (9.53), and we arrive at the following formula for the computation of the total stresses σ_{ij}^{total} in the uniform beam with initial stresses σ_{ij}^{*}:

$$\begin{aligned}\sigma_{ij}^{total}(\mathbf{x}) &= \sigma_{ij}^{*}(\mathbf{x}) + \sigma_{ij}^{(-3)} \\ &= \sigma_{ij}^{*}(\mathbf{x}) + (c_{ijmn}(\mathbf{y})N_{m,ny}^{01} - c_{ij11}(\mathbf{y}))V_{1,1x}(x_1) \\ &+ (c_{ijmn}(\mathbf{y})N_{m,ny}^{1\alpha} + c_{ij11}(\mathbf{y})y_\alpha)u_{\alpha,1x1x}^{(0)}(x_1) \\ &+ (c_{ijmn}(\mathbf{y})X_{m,ny}^{3} + c_{ij\gamma 1}(\mathbf{y})s_\gamma y_\Gamma)\phi_{,1x}(x_1).\end{aligned} \tag{9.78}$$

References

Agarwal BD, Broutman LJ, Chandrashekhara K (2006) Analysis and performance of fiber composites. Wiley India, New Dehli

Andrianov IV, Kolpakov AA, Faella L (2024) Asymptotic model of a piezoelectric composite beam. J Appl Mech Tech Phys 65(2):345–353

Araujo JAF, Rodrigues MC, de Pontes Pires RB (2023) Multiscale structural analysis of oil rig mast using mean fields and finite element method. J Braz Soc Mech Sci Eng 45(20)

Buannic N, Cartraud P (2001) Higher-order effective modeling of periodic heterogeneous beams. I. asymptotic expansion method. Int J Solids Struct 38:7139–7161

Cartraud P, Messager T (2006) Computational homogenization of periodic beam-like structures. Int J Solids Struct 43(3–4):686–696

Gesualdo A (2018) A direct technique for the homogenization of periodic beam-like structures by transfer matrix eigen-analysis. Latin Am J Solids Struct 15(5):44–57

Groen JP, Stutz FC, Aage N, Baentzen JA, Sigmund O (2020) De-homogenization of optimal multi-scale 3d topologies. Comput Meth Appl Mech Eng 364:112979

He Z, Liu J, Chen Q (2023) Higher-order asymptotic homogenization for piezoelectric composites. Int J Solids Struct 264:7139–7161
Hebiba A, Bouferguene A, Moon S, Han S (2022) Wind-wise automated stability analysis for selection of tower crane and location. J Const Eng Manag 148(11)
Jiang JP, Li DX (2007) A new finite element model for piezothermoelastic composite beam. J Sound Vib 306(3–5):849–864
Johnson RP (2004) Composite structures of steel and concrete: beams, slabs, columns, and frames for buildings. Wiley-Blackwell, Hoboken, NJ
Kalamkarov AL, Kolpakov AG (1997) Analysis, design and optimization of composite structures. Wiley, Chichester
Kapuria S, Dumir PC, Jain NK (2004) Assessment of zigzag theory for static loading, buckling, free and forced response of composite and sandwich beams. Comput Struct 64(3–4):317–327
Kollar LP, Springer GS (2009) Mechanics of composite structures. Cambridge University Press, Cambridge
Kolpakov AG (2004) Stressed composite structures: homogenized models for thin-walled nonhomogeneous structures with initial stresses. Springer, Berlin
Kolpakov AG (1991) Calculation of the characteristics of thin elastic rods of periodic structure. J Appl Math Mech 57:440–448
Kolpakov AG (1998) Variational principles for stiffnesses of a non-homogeneous beam. J Mech Phys Solids 46(6):1039–1053
Kolpakov AA, Kolpakov AG (2006) Design of smart beam—an integrated design procedure. Struct Multidisc Opt 31(13):76–80
Kondaiah P, Shankar K, Ganesan N (2012) Studies on magneto-electro-elastic cantilever beam under thermal environment. Coupled Syst Mech 1(2):205–217
Kuczma M (2010) Composite beams with embedded shape memory alloy. In: Albers B (ed) Continuous media with microstructure. Springer, Berlin, pp 301–315
Levinski T (1995) Effective stiffness of transversally non-homogeneous plates with unidirectional periodic structure. Int J Solids Struct 32(19):3261–3287
Panasenko GP (1993) Asymptotic solution of the elasticity theory system of equations for lattice and skeletal structures. Russian Math Surv (Matem. Zb.) 75(1):85–110
Pshenichnov GI (1993) Theory of latticed plates and shells. World Scientific, Singapore
Qiao WT, Wang D, Zhao MS (2017) Mechanical features of cable-supported ribbed beam composite slab structure. Steel Comput Struct 25(5):523–534
Timoshenko S, Woinowsky-Krieger K (1959) Theory of plates and shells. McGraw-Hill, New York
Trabucho L, Viano JM (1987) Derivation of generalized models for linear elastic beams by asymptotic expansion method. In: Ciarlet PG, Lions J-L (eds) Application of multiple scaling in mechanics. Masson, Paris, pp 129–148
Trabucho L, Viano JM (1996) Mathematical modeling of rods. In: Ciarlet PG, Lions J-L (eds) Handbook of numerical analysis, Vol IV. Elsevier, Amsterdam, pp 487–974
Treyssede F, Cartraud p (2022) A two-dimensional formulation for the homogenization of helical beam-like structures under bending loads. Int J Solids Struct 234–235
Vinot P, Cogan S, Piranda J (2001) Shape optimization of thin-walled beam-like structures. Thin-Walled Struct 39:611–630
Washizu K (1992) Variational methods in elasticity and plasticity. Pergamon Press, Oxford
Yadav AK (2023) Magnet-thermo-piezo-elastic wave in an initially stressed rotating monoclinic crystal in a two-temperature theory. Int J Appl Mech Eng 28(3):127–158
Zhen W, Wanji C (2008) An assessment of several displacement based theories for the vibration and stability analysis of laminated composite and sandwich beams. Comput Struct 84(4):337–349

Chapter 10
Initial Stresses Comparable with the Elastic Constants of the Material of the Beam

In this chapter, we consider the case where the initial stresses are of the same order as the elastic constants of the material of the beam. As was noted above, this is not a rare, but usual case in engineering. A typical case is thermal stresses in non-uniform structural elements. Note that the use of thermal stresses is widely used in the production of pre-stressed structural elements. Thermal stresses also play the role of initial stresses in continuous rails, see Chap. 2. Of course, the initial stresses are small as comparable with the elastic constants in the absolute value. We will make a transition from a three-dimensional problem of elasticity theory with initial stresses in a small-diameter body to a problem of beam theory.

10.1 Statement of the Problem

We start with the three-dimensional elasticity theory problem for a body with initial stresses described in Sect. 9. According to this model, the equilibrium equations of a beam considered as a small diameter three-dimensional elastic body G_ε can be written as

$$\frac{\partial \sigma_{ij}^\varepsilon}{\partial x_j} + f_i^\varepsilon(\mathbf{x}) = 0 \text{ in } G_\varepsilon, \tag{10.1}$$

with the boundary conditions

$$\begin{aligned} \sigma_{ij}^\varepsilon n_j &= h_i^\varepsilon(\mathbf{x}) \text{ on } S_\varepsilon; \\ \mathbf{u}^\varepsilon &= 0 \text{ on } S_u. \end{aligned} \tag{10.2}$$

In this chapter, we consider the case when the initial stresses are of the same order in ε as elastic constants. In this case, the relationship between the additional stresses σ_{ij}^ε, the additional displacements $\mathbf{u}^\varepsilon$ and the initial stresses σ_{ij}^* is the following:

A. Kolpakov, *Thin-Walled Structures with Initial Stresses*, Mathematical Engineering,
https://doi.org/10.1007/978-3-032-07247-4_10

$$\sigma_{ij}^{\varepsilon} = \varepsilon^{-4} h_{ijmn}(x_1, \mathbf{x}/\varepsilon) \frac{\partial u_k^{\varepsilon}}{\partial x_l}, \tag{10.3}$$

where

$$h_{ijmn}(x_1, \mathbf{x}/\varepsilon) = c_{ijmn}(\mathbf{x}/\varepsilon) + b_{ijmn}(x_1, \mathbf{x}/\varepsilon);$$
$$b_{ijmn}(x_1, \mathbf{x}/\varepsilon) = \sigma_{jn}^{*\varepsilon}(x_1, \mathbf{x}/\varepsilon)\delta_{im}. \tag{10.4}$$

For small ε we have the following hierarchy of powers: $\varepsilon^{-2} \ll \varepsilon^{-3} \ll \varepsilon^{-4}$. This means that the initial axial force of order ε^{-2} is the smallest in magnitude and manifests itself first (in the form of loss of stability of the beam). The moments of initial stresses correspond to terms of order ε^{-3}. They are greater in magnitude than the initial stresses generating the axial force. Their influence manifests itself only if the resultant axial force is zero, otherwise the beam loses stability before the influence of the resultant moments of initial stresses is revealed. The manifestation of effects associated with initial stresses of the order of ε^{-4} (ε^{-4} can be observed only if both the resultant initial axial force and the resultant initial moments are equal to zero:

$$\langle \sigma_{1i}^{*\varepsilon} \rangle = 0, \langle \sigma_{1i}^{*\varepsilon} y_{\alpha} \rangle = 0 (i = 1, 2, 3; \alpha = 2, 3). \tag{10.5}$$

10.2 Asymptotic Expansions and Equilibrium Equations

The following expansions are used to analyze the problem:

asymptotic expansion for displacement

$$\mathbf{u}^{\varepsilon} = \mathbf{u}^{(0)}(x_1) + \sum_{k=1}^{\infty} \varepsilon^k \mathbf{u}^{(k)}(x_1, \mathbf{y}); \tag{10.6}$$

asymptotic expansion for stresses

$$\sigma_{ij}^{\varepsilon} = \sum_{p=-4}^{\infty} \varepsilon^p \sigma_{ij}^{(p)}(x_1, \mathbf{y}) \tag{10.7}$$

and we use the representation of differentiation operators in the form (9.8).

The functions in the right-hand sides (10.6), (10.7) are assumed to be periodic in y_1 with period m, see Fig. 9.1. The expansion (10.7) starts with a term of order ε^{-4} according to the governing equations (10.3) and the expansion (10.6) for the displacement.

Substituting (10.6), (10.7) into (10.3), taking into account the differentiation rule for replacing derivatives (9.8), we obtain

$$\sum_{m=-4}^{\infty} \varepsilon^m \sigma_{ij}^{(m)} = \sum_{n=0}^{\infty} \varepsilon^{n-4} h_{ijkl}(x_1, \mathbf{y})(u_{k,1x}^{(n)} \delta_{1l} + \varepsilon^{-1} u_{k,ly}^{(n)}). \tag{10.8}$$

Equating in (10.8) the terms corresponding to the identical orders of ε, we obtain

$$\sigma_{ij}^{(m)} = h_{ijkl}(x_1, \mathbf{y})(u_{k,1x}^{(m+4)} \delta_{l1} + u_{k,ly}^{(m+5)}) \tag{10.9}$$

$(m = -4, -3, ...)$.

Substituting the expansion (10.7) into the equilibrium equations, as in Chap. 8, leads to equalities (see (9.30))

$$\sigma_{ij,jy}^{(-4)} = 0, \tag{10.10}$$

$$\sigma_{i1,1x}^{(-4)} + \sigma_{ij,jy}^{(-3)} = 0.$$

The analysis of problem (10.1)–(10.5) is carried out in two stages. The first is the obtaining of the equilibrium equations for the beam, considered as a one-dimensional structure. This stage does not involve the local governing equations (in this case, Eq. (10.3)) and is the same for any governing equations.

The homogenized equilibrium equations do not depend on the local governing equations, see Sect. 8.2.1, and they consist of the equations for the resultant forces

$$N_{11,1x}^{(p)} = 0, \; (p = -4, -3) \tag{10.11}$$

$$N_{\beta 1,1x}^{(-2)} + \langle f_\beta^\varepsilon \rangle - \langle h_\beta^\varepsilon \rangle_S = 0, \tag{10.12}$$

and equations for the resultant moments

$$M_{i\beta,1x}^{(-4)} - N_{i\beta}^{(-3)} = 0 \;\; (p = -4); \tag{10.13}$$

$$M_{1\beta,1x}^{(-3)} - N_{1\beta}^{(-2)} + \langle f_1^\varepsilon y_\beta \rangle + \langle g_1^\varepsilon y_\beta \rangle_S = 0 \;\; (p = -3)$$

$(i, j$=1, 2, 3 and $\alpha, \beta = 2, 3)$. The equations corresponding to $p > -3$ are not used further, and therefore are not given.

10.2.1 Homogenized Constitutive Equations

The second stage of the problem analysis is obtaining the governing equations for the beam, considered as a one-dimensional structure. This stage makes significant use of local governing equations.

According to (10.10)

$$\sigma^{(-4)}_{ij,jy} = 0 \text{ in } Y, \sigma^{(-4)}_{ij} n_j = 0 \text{ on } S. \tag{10.14}$$

According to (10.9, $p = -4$)

$$\sigma^{(-4)}_{ij} = h_{ijk1}(x_1, \mathbf{y}) u^{(1)}_{k,1x} + h_{ijk1}(x_1, \mathbf{y}) u^{(0)}_{k,1x}(x_1). \tag{10.15}$$

Substituting (10.15) into (10.14), we arrive at the boundary-value problem

$$\begin{cases} (h_{ijk1}(x_1, \mathbf{y}) u^{(1)}_{k,1x} + h_{ijk1}(x_1, \mathbf{y}) u^{(0)}_{k,1x}(x_1))_{,jy} = 0 \text{ in } G_\varepsilon; \\ (h_{ijk1}(x_1, \mathbf{y}) u^{(1)}_{k,1x} + h_{ijk1}(x_1, \mathbf{y}) u^{(0)}_{k,1x}(x_1)) n_j = 0 \text{ on } S_\varepsilon. \end{cases} \tag{10.16}$$

Introduce $\mathbf{K}^{1p}(\mathbf{y})$ as solution to the following periodicity cell problem:

$$\begin{cases} (h_{ijkl}(x_1, \mathbf{y}) K^{1p}_{k,ly} + h_{ijp1}(\mathbf{y})(x_1, \mathbf{y}))_{,jy} = 0 \text{ in } Y; \\ (h_{ijkl}(x_1, \mathbf{y}) K^{1p}_{k,ly} + h_{ijp1}(\mathbf{y})(x_1, \mathbf{y})) n_j = 0 \text{ on } S; \\ \mathbf{K}^{1p}(\mathbf{y}) \text{ periodic in } y_1 \text{ with period } m. \end{cases} \tag{10.17}$$

It should be noted that the periodicity cell problem (10.17) differs from the periodicity cell problem (9.37). Coefficients $c_{ijmn}(\mathbf{y})$ in the periodicity cell problem (9.37) are elastic constants of the material of the beam. In the periodicity cell problem (10.17), coefficients $h_{ijmn}(\mathbf{y})$ are combination of the elastic characteristics of the material of the beam and initial stresses in the beam, see (10.4).

Proposition 10.1 *If the initial stresses $\sigma^{*\varepsilon}_{ij}$ satisfy the conditions of Proposition 9.4, then the periodicity cell problem (10.17) has particular solutions*

$$\mathbf{K}^{0\alpha}(\mathbf{y}) = -y_\alpha \mathbf{e}_1, \tag{10.18}$$

and the kernel of the homogeneous problem corresponding to (10.17) is generated by the functions **const** $= (C_1, C_2, C_3)$ *and*

$$\mathbf{X}(\mathbf{y}) = y_B s_\beta \mathbf{e}_\beta,$$

where B, β are determined similarly to the definition of Γ, γ in (8.38).

Proposition 10.1 is similar but not equivalent to Proposition 9.3 because coefficients in the periodicity cell problems (9.37) and (10.17) are different. Proposition 10.1 states that the partial solutions to the periodicity cell problem for a beam with initial stresses are the same as for a beam without initial stresses if the initial stresses are self-equilibrated.

To check that $-y_\alpha \mathbf{e}_1$ is indeed a solution to the periodicity cell problem (10.17), we substitute $-y_\alpha \mathbf{e}_1$ into the left-hand side of the equations from (10.17). Taking into account the definitions of h_{ijkl} and b_{ijkl}, see (10.4), we obtain:

$$-\left(c_{ij1\alpha}(\mathbf{y}) + b_{ij1\alpha}(x_1, \mathbf{y})\right) + c_{ij\alpha 1}(\mathbf{y}) = -c_{ij1\alpha}(\mathbf{y}) + c_{ij\alpha 1}(\mathbf{y}) - b_{ij1\alpha}(x_1, \mathbf{y}).$$

In the last expression, $-c_{ij1\alpha}(\mathbf{y}) + c_{ij\alpha 1} = 0$ due to the symmetry of the elastic constants c_{ijkl} (see condition C1, Sect. 3.1), and $b_{ij1\alpha,jy} = \delta_{i1}\sigma^{*\varepsilon}_{\alpha j,jy} = 0$ in Y due to the condition of Proposition 10.1. Here we use symmetry of the initial stresses $\sigma^{*\varepsilon}_{\alpha j}$ in the indices i and α.

Similarly, it is verified that due to the symmetry of the elastic constants c_{ijkl} and $\sigma^{*\varepsilon}_{\alpha j}(x_1, \mathbf{y})n_j = 0$ on S

$$\begin{aligned}&(-(c_{ij1\alpha}(\mathbf{y}) + b_{ij1\alpha}(x_1, \mathbf{y})) + c_{ij\alpha 1}(\mathbf{y}))n_j \\ &= -b_{ij1\alpha}(x_1, \mathbf{y})n_j = -\delta_{i1}\sigma^{*\varepsilon}_{\alpha j}(x_1, \mathbf{y})n_j = 0.\end{aligned}$$

Here we use symmetry of the initial stresses $\sigma^{*\varepsilon}_{\alpha j}$ in the indices i and α.

We verify that the function $\mathbf{X}(\mathbf{y}) = y_B s_\beta \mathbf{e}_\beta$ is a solution to the homogeneous problem (10.17). Substituting $\mathbf{X}(\mathbf{y}) = y_B s_\beta \mathbf{e}_\beta$ into the homogeneous equation corresponding to the equation from (10.17), taking into account the definition of h_{ijkl} (10.4), we obtain:

$$(c_{ij\beta B}(\mathbf{y}) + b_{ij\beta B}(x_1, \mathbf{y}))s_\beta = c_{ij23}(\mathbf{y}) - c_{ij32}(\mathbf{y}) + \sigma^{*\varepsilon}_{23}(x_1, \mathbf{y}) - \sigma^{*\varepsilon}_{32}(x_1, \mathbf{y}) = 0$$

due to the symmetry of the elastic constants c_{ijkl} and initial stresses $\sigma^{*\varepsilon}_{ij}$. Hence, we have proved the proposition.

Function **const** is solution to the homogenized problem, it is evident.

As a result, the following problems remain from (10.17):

$$\begin{cases} (h_{ijkl}(x_1, \mathbf{y})K^{01}_{k,ly} + h_{ij11}(\mathbf{y})(x_1, \mathbf{y}))_{,jy} = 0 \text{ in } Y; \\ (h_{ijkl}(x_1, \mathbf{y})K^{01}_{k,ly} + h_{ij11}(\mathbf{y})(x_1, \mathbf{y}))n_j = 0 \text{ on } S; \\ \mathbf{K}^{01}(\mathbf{y}) \text{ periodic in } y_1 \text{ with period } m, \end{cases} \tag{10.19}$$

corresponding to the homogenized (macroscopic) axial tension of the beam. Taking into account Proposition 10.1 and using solution of the periodicity cell problem (10.19), we can write solution to the problem (10.16) in the form

$$\mathbf{u}^{(1)} = \mathbf{K}^{01}(\mathbf{y})u^{(0)}_{1,1x} - y_\alpha \mathbf{e}_1 u^{(0)}_{\alpha,1x}(x_1) + y_B s_\beta \mathbf{e}_\beta \phi(x_1) + \mathbf{V}(x_1). \tag{10.20}$$

Note that in the case under consideration, the solution to the periodicity cell problem depends on the initial stresses, since $\mathbf{K}^{01}$ is the solution to the periodicity cell problem (10.19), which coefficients h_{ijkl} (10.4) depend on the initial stresses.

Substituting (10.20) into (10.15) taking into account the definition of h_{ijkl} (10.4), we obtain:

$$\begin{aligned}\sigma_{ij}^{(-4)} &= (c_{ij\alpha 1}(\mathbf{y}) + \sigma_{j1}^{*\varepsilon}(x_1, \mathbf{y})\delta_{i\alpha} - c_{ij1\alpha}(\mathbf{y}) - \sigma_{j\alpha}^{*\varepsilon}(x_1, \mathbf{y})\delta_{i1})u_{\alpha,1x}^{(0)}(x_1) \\ &+(h_{ij11}(x_1, \mathbf{y}) + h_{ijkl}(x_1, \mathbf{y})K_{k,ly}^{01}(\mathbf{y}))u_{1,1x}^{(0)}(x_1) \\ &+((c_{ij\beta B}(\mathbf{y})s_\beta + \sigma_{jB}^{*\varepsilon}(x_1, \mathbf{y})s_\beta\delta_{i\beta})\phi(x_1).\end{aligned} \tag{10.21}$$

In (10.21), $c_{ij\alpha 1} - c_{ij1\alpha} = 0$ and $c_{ij\beta B}s_\beta = c_{ij23} - c_{ij32} = 0$ due to the symmetry of the elastic constants c_{ijkl}. Averaging the remaining terms in (10.21) over the periodicity cell Y, and taking into account Proposition 10.1, we get

$$\begin{aligned}N_{ij}^{(-4)} &= \langle\sigma_{ij}^{(-4)}\rangle \\ &= (\langle\sigma_{j1}^{*\varepsilon}(x_1, \mathbf{y})\rangle\delta_{i\alpha} - \langle\sigma_{j\alpha}^{*\varepsilon}(x_1, \mathbf{y})\rangle\delta_{i1})u_{\alpha,1x}^{(0)}(x_1) \\ &+\langle h_{ij11}(x_1, \mathbf{y}) + h_{ijkl}(x_1, \mathbf{y})K_{k,ly}^{01}(\mathbf{y})\rangle u_{1,1x}^{(0)}(x_1) + \langle\sigma_{jB}^{*\varepsilon}(x_1, \mathbf{y})\rangle s_\beta\delta_{i\beta}\phi(x_1).\end{aligned} \tag{10.22}$$

For $ij = 11$, by (10.5) and conditions (10.5), formula (10.22) takes the form

$$N_{11}^{(-4)} = \langle\sigma_{11}^{(-4)}\rangle = A(\sigma)u_{1,1x}^{(0)}(x_1),$$

where

$$A(\sigma) = \langle h_{1111}(x_1, \mathbf{y}) + h_{11kl}(x_1, \mathbf{y})K_{k,ly}^{01}\rangle. \tag{10.23}$$

As discussed in Sect. 4.5, it cannot be claimed that $A(\sigma) > 0$ for arbitrary $\sigma_{ij}^{*\varepsilon}$. However, the initial stresses $\sigma_{ij}^{*\varepsilon}$ are small as compared to the elastic constants c_{ijkl} in absolute value: $|\sigma_{ij}^{*\varepsilon}| \ll |c_{ijkl}|$, see Sect. 4.5. The small perturbation of the elastic constant preserves the positive-definiteness of $A(\sigma)$ (it is known that $A(0) > 0$ Kolpakov 1991; Trabucho and Viano 1996). Then from the equilibrium equations in (10.12) and the boundary condition

$$u_1^{(0)}(-a) = u_1^{(0)}(a) = 0,$$

which follows from (10.1) and the expansion (10.6), we obtain that $u_1^{(0)}(x_1) = 0$.

Then the equality (10.20) takes the form

$$\mathbf{u}^{(1)} = -y_\alpha\mathbf{e}_1u_{\alpha,1x}^{(0)}(x_1) + y_Bs_\beta\mathbf{e}_\beta\phi(x_1) + \mathbf{V}(x_1). \tag{10.24}$$

Proposition 10.2 *The equilibrium equations (10.13, $p = -4$) are satisfied identically under conditions (10.5) and the conditions of Proposition 9.4.*

To verify this proposition, let us consider the quantities included in (10.13). The equality (10.21) when taking into account (10.24) yields

$$\begin{aligned}\sigma_{ij}^{(-4)} &= (\sigma_{j1}^{*\varepsilon}(x_1, \mathbf{y})\delta_{i\alpha} - \sigma_{j\alpha}^{*\varepsilon}(x_1, \mathbf{y})\delta_{i1})u_{\alpha,1x}^{(0)}(x_1) \\ &+\sigma_{jB}^{*\varepsilon}(x_1, \mathbf{y})s_\beta\delta_{i\beta}\phi(x_1).\end{aligned} \tag{10.25}$$

We integrate (10.25) over the periodicity cell Y. Then we multiply (10.25) by y_β, and also integrate over the periodicity cell Y at $j = 1$. As a result, taking into account (10.5), we obtain

$$\begin{aligned} N_{ij}^{(-4)} &= (\langle\sigma_{j1}^{*\varepsilon}(x_1,\mathbf{y})\rangle\delta_{i\alpha} - \langle\sigma_{j\alpha}^{*\varepsilon}(x_1,\mathbf{y})\rangle\delta_{i1})u_{\alpha,1x}^{(0)}(x_1) \\ &\quad+\langle\sigma_{jB}^{*\varepsilon}(x_1,\mathbf{y})\rangle s_\beta\delta_{i\beta}\phi(x_1); \\ M_{i\beta}^{(-4)} &= (\langle\sigma_{11}^{*\varepsilon}(x_1,\mathbf{y})y_\beta\rangle\delta_{i\alpha} - \langle\sigma_{1\alpha}^{*\varepsilon}(x_1,\mathbf{y})y_\beta\rangle\delta_{i1})u_{\alpha,1x}^{(0)}(x_1) \\ &\quad+\langle\sigma_{1B}^{*\varepsilon}(x_1,\mathbf{y})y_\beta\rangle s_\beta\delta_{i\beta}\phi(x_1). \end{aligned} \tag{10.26}$$

Formula (10.9) for $m = -3$ yields

$$\sigma_{ij}^{(-3)} = h_{ijk1}(x_1,\mathbf{y})u_{k,1x}^{(1)} + h_{ijkl}(x_1,\mathbf{y})u_{k,ly}^{(2)}. \tag{10.27}$$

Substituting expression (10.24) into (10.27), we obtain

$$\begin{aligned} \sigma_{ij}^{(-3)} &= -h_{ij11}(x_1,\mathbf{y})y_\alpha u_{\alpha,1x1x}^{(0)}(x_1) + h_{ij\beta1}(x_1,\mathbf{y})y_B s_\beta\phi_{,1x}(x_1) \\ &\quad+h_{ijk1}(x_1,\mathbf{y})V_{k,1x} + h_{ijkl}(x_1,\mathbf{y})u_{k,ly}^{(2)}. \end{aligned} \tag{10.28}$$

The equilibrium equations (10.13) for $i = 1$ contain the quantities $N_{1\alpha}^{(-3)}$—the shear forces. To exclude them from (10.13), we will use the method proposed in Sect. 8.2.3. Let us consider the difference $N_{1\alpha}^{(-3)} - N_{\alpha1}^{(-3)}$, which, using (10.28) and the definition of the quantities h_{ijkl}, can be written as

$$\begin{aligned} N_{1\alpha}^{(-3)} - N_{\alpha1}^{(-3)} &= \langle\sigma_{1\alpha}^{(-3)} - \sigma_{\alpha1}^{(-3)}\rangle \\ &= \langle(\sigma_{\alpha1}^{*\varepsilon}(x_1,\mathbf{y}) - \sigma_{11}^{*\varepsilon}(x_1,\mathbf{y})\delta_{\alpha1})y_\beta\rangle u_{\alpha,1x1x}^{(0)}(x_1) \\ &\quad+\langle(\sigma_{\alpha1}^{*\varepsilon}(x_1,\mathbf{y}) - \sigma_{11}^{*\varepsilon}(x_1,\mathbf{y})\delta_{\alpha1})y_B\rangle s_\beta\phi_{,1x}(x_1) \\ &\quad+\langle(\sigma_{\alpha1}^{*\varepsilon}(x_1,\mathbf{y})\delta_{1k} - \sigma_{11}^{*\varepsilon}(x_1,\mathbf{y})\delta_{\alpha1})y_B\rangle V_{k,1x}(x_1). \end{aligned} \tag{10.29}$$

The coefficients on the right-hand sides of the formulas for $N_{ij}^{(-4)}$, $M_{i\beta}^{(-4)}$ from (10.26) and in (10.29) are equal to zero due to conditions (10.5) and equality (10.5). Consequently, Eqs. (10.13) for $i = 1$ are satisfied identically. The proposition is proved.

Let us study of the next terms of the asymptotic expansion (10.6). For this, we consider the problem (10.10, $p = -3$)

$$\sigma_{ij,jy}^{(-3)} + \sigma_{i1,1x}^{(-4)} = 0 \text{ in } Y;\ \sigma_{ij}^{(-3)}n_j = 0 \text{ on } S \tag{10.30}$$

with $\sigma_{ij}^{(-3)}$ given by (10.28) and $\sigma_{ij}^{(-4)}$ given by (10.25).

For $j = 1$, (10.25) takes the form

$$\begin{aligned} \sigma_{i1,1x}^{(-4)} &= (\sigma_{11}^{*\varepsilon}(x_1,\mathbf{y})\delta_{i\alpha} - \sigma_{1\alpha}^{*\varepsilon}(x_1,\mathbf{y})\delta_{i1})u_{\alpha,1x1x}^{(0)}(x_1) \\ &\quad+\sigma_{1B}^{*\varepsilon}(x_1,\mathbf{y})s_\beta\delta_{i\beta}\phi(x_1)_{,1x}. \end{aligned} \tag{10.31}$$

Substituting (10.28) and (10.31) into (10.30), we arrive at the problem ($\alpha, \beta = 2, 3$)

$$\begin{cases} (h_{ijkl}(x_1, \mathbf{y})u^{(2)}_{k,ly} - h_{ij11}(x_1, \mathbf{y})y_\alpha u^{(0)}_{\alpha,1x1x}(x_1) \\ +h_{ij\beta1}(x_1, \mathbf{y})y_B s_\beta \phi_{,1x}(x_1) + h_{ijk1}(x_1, \mathbf{y})V_{k,1x})_{,jy}(x_1) \\ +(\sigma^{*\varepsilon}_{11}(x_1, \mathbf{y})\delta_{i\alpha} - \sigma^{*\varepsilon}_{1\alpha}(x_1, \mathbf{y})\delta_{i1})u^{(0)}_{\alpha,1x1x}(x_1) \\ +\sigma^{*\varepsilon}_{1B}(x_1, \mathbf{y})s_\beta\delta_{i\beta}\phi_{,1x}(x_1) = 0 \text{ in } G_\varepsilon; \\ (h_{ijkl}(x_1, \mathbf{y})u^{(2)}_{k,ly} - h_{ij11}(x_1, \mathbf{y})y_\alpha u^{(0)}_{\alpha,1x1x}(x_1) \\ +h_{ij\beta1}(x_1, \mathbf{y})y_B s_\beta \phi_{,1x}(x_1) + h_{ijk1}(x_1, \mathbf{y})V_{k,1x}(x_1))n_j = 0 \text{ on } S_\varepsilon. \end{cases} \tag{10.32}$$

To solve problems of this type, we introduce the following periodicity cell problems: periodicity cell problem for the function $\mathbf{K}^{1\alpha}(\mathbf{y})$, corresponding to the beam bending:

$$\begin{cases} (h_{ijmn}(x_1, \mathbf{y})K^{1\alpha}_{m,ny} - h_{ij11}(x_1, \mathbf{y})y_\alpha)_{,jy} = 0 \text{ in } Y; \\ (h_{ijmn}(x_1, \mathbf{y})K^{1\alpha}_{m,ny} - h_{ij11}(x_1, \mathbf{y})y_\alpha)n_j = 0 \text{ on } S; \\ \mathbf{K}^{1\alpha}(\mathbf{y}) \text{ periodic in } y_1 \text{ with period } m \quad (\alpha = 2, 3). \end{cases} \tag{10.33}$$

and the periodicity cell problem for the function $\mathbf{L}^3(\mathbf{y})$, corresponding to the beam torsion

$$\begin{cases} (h_{ijmn}(x_1, \mathbf{y})L^3_{m,ny} + h_{ij\beta1}(x_1, \mathbf{y})y_B s_\beta)_{,jy} = 0 \text{ in } Y; \\ (h_{ijmn}(x_1, \mathbf{y})L^3_{m,ny} + h_{ij\beta1}(x_1, \mathbf{y})y_B s_\beta)n_j = 0 \text{ on } S; \\ \mathbf{L}^3(\mathbf{y}) \text{ periodic in } y_1 \text{ with period } m. \end{cases} \tag{10.34}$$

We also introduce the periodicity cell problems for the functions $\mathbf{Z}^3(\mathbf{y})$ and $\mathbf{Y}(\mathbf{y})$, corresponding to initial stresses (10.25) in the beam:

$$\begin{cases} (h_{ijmn}(x_1, \mathbf{y})Z^\alpha_{m,ny})_{,jy} + \sigma^{*\varepsilon}_{11}(x_1, \mathbf{y})\delta_{i\alpha} - \sigma^{*\varepsilon}_{1\alpha}(x_1, \mathbf{y})\delta_{i1} = 0 \text{ in } Y; \\ h_{ijmn}(x_1, \mathbf{y})Z^\alpha_{m,ny}n_j = 0 \text{ on } S; \\ \mathbf{Z}^\alpha(\mathbf{y}) \text{ periodic in } y_1 \text{ with period } m \quad (\alpha = 2, 3), \end{cases} \tag{10.35}$$

and

$$\begin{cases} (h_{ijmn}(x_1,\mathbf{y})Y_{m,ny})_{,jy} + \sigma^{*\varepsilon}_{1B}(x_1,\mathbf{y})s_\beta\delta_{i\beta} = 0 \text{ in } Y; \\ h_{ijmn}(x_1,\mathbf{y})Y_{m,ny}n_j = 0 \text{ on } S; \\ \mathbf{Y}(\mathbf{y}) \text{ periodic in } y_1 \text{ with period } m(\alpha = 2,3) \end{cases} \tag{10.36}$$

Taking into account the equalities (10.18) and using the functions $\mathbf{K}^{01}$, $\mathbf{K}^{1\alpha}$, $\mathbf{L}^3$, $\mathbf{Z}^\alpha$, $\mathbf{Y}$, we can express the solution to the problem (10.32) in the form

$$\begin{aligned} \mathbf{u}^{(2)} = (\mathbf{K}^{1\alpha}(\mathbf{y}) + \mathbf{Z}^\alpha(\mathbf{y}))u^{(0)}_{\alpha,1x1x}(x_1) + (\mathbf{L}^3(\mathbf{y}) + \mathbf{Y}(\mathbf{y}))\phi_{,1x}(x_1) \\ +\mathbf{K}^{01}(\mathbf{y})V_{1,1x}(x_1) - y_\alpha\mathbf{e}_1V_{\alpha,1x}(x_1), \end{aligned} \tag{10.37}$$

and substituting (10.37) into (10.28) yields

$$\begin{aligned} \sigma^{(-3)}_{ij} &= (h_{ij11}(x_1,\mathbf{y})y_\alpha + h_{ijmn}(x_1,\mathbf{y})K^{1\alpha}_{m,ny} \\ &\quad +h_{ijmn}(x_1,\mathbf{y})Z^\alpha_{m,ny})u^{(0)}_{\alpha,1x1x}(x_1)+ \\ &\quad +(h_{ij\beta1}(x_1,\mathbf{y})y_Bs_\beta + h_{ijmn}(x_1,\mathbf{y})L^3_{m,ny} + h_{ijmn}(x_1,\mathbf{y})Y_{m,ny})\phi_{,1x}(x_1) \\ &\quad +(h_{ij11}(x_1,\mathbf{y}) + h_{ijmn}(x_1,\mathbf{y})K^{01}_{m,ny})V_{1,1x}(x_1)+ \\ &\quad +(h_{ij\alpha1}(\mathbf{y}) - h_{ij1\alpha}(\mathbf{y}))V_{\alpha,1x}(x_1) \\ &= (h_{ij11}(x_1,\mathbf{y})y_\alpha + h_{ijmn}(x_1,\mathbf{y})K^{1\alpha}_{m,ny} + h_{ijmn}(x_1,\mathbf{y})Z^\alpha_{m,ny})u^{(0)}_{\alpha,1x1x}(x_1)+ \\ &\quad +(h_{ij\beta1}(x_1,\mathbf{y})y_Bs_\beta + h_{ijmn}(x_1,\mathbf{y})L^3_{m,ny} + h_{ijmn}(x_1,\mathbf{y})Y_{m,ny})\phi_{,1x}(x_1) \\ &\quad +(h_{ij11}(x_1,\mathbf{y}) + h_{ijmn}(x_1,\mathbf{y})K^{01}_{m,ny})V_{1,1x}(x_1). \end{aligned} \tag{10.38}$$

We integrate (10.38) over the periodicity cell Y at $ij = 11$. Taking into account the definition of h_{ijkl} and b_{ijkl} (10.4), and the equality (10.5), we obtain

$$N^{(-3)}_{11} = A^0(\sigma)V_{1,1x} + A^1_\alpha(\sigma)u^{(0)}_{\alpha,1x1x}(x_1) + B^0(\sigma)\phi_{,1x}. \tag{10.39}$$

We multiply (10.38) at $j = 1$ by y_β and integrate the result over the periodicity cell Y. Taking into account the second condition from (10.5), we obtain

$$M^{(-3)}_{i\beta} =^1 A_{i\beta}(\sigma)V_{1,1x} + A^2_{i\beta\alpha}(\sigma)u^{(0)}_{\alpha,1x1x}(x_1) + B^1_{i\beta}(\sigma)\phi_{,1x}. \tag{10.40}$$

We separate the homogenized governing equations (10.40) for the resultant moments into two groups: equations for the bending moments $M^{(-3)}_{1\beta}$ ($i = 1$; $\beta = 2,3$)

$$M^{(-3)}_{1\beta} =^1 A_{1\beta}(\sigma)V_{1,1x} + A^2_{\beta\alpha}(\sigma)u^{(0)}_{\alpha,1x1x}(x_1) + B^1_{1\beta}(\sigma)\phi_{,1x}, \tag{10.41}$$

and equations for the torsion moment

$$\begin{aligned} M^{(-3)} &= M_{32}^{(-3)} - M_{23}^{(-3)} = M_{\Theta\theta}^{(-3)} s_\Theta \\ &=^1 A(\sigma) V_{1,1x} + A_\alpha^2(\sigma) u_{\alpha,1x1x}^{(0)}(x_1) + B^1(\sigma)\phi_{,1x}. \end{aligned} \tag{10.42}$$

Here

$$s_1 = 0, s_2 = -1, s_3 = 1; \text{ and } \Theta = 2 \text{ if } \theta = 3 \text{ and } \Theta = 3 \text{ if } \theta = 2.$$

In (10.39)–(10.42),

$$\begin{aligned} A^0(\sigma) &= \langle h_{1111}(\mathbf{y}) + h_{11mn}(x_1, \mathbf{y}) K_{m,ny}^{01} \rangle; \\ A_\alpha^1(\sigma) &= \langle h_{1111}(x_1, \mathbf{y}) y_\alpha + h_{11mn}(x_1, \mathbf{y}) K_{m,ny}^{1\alpha} \\ &\quad + h_{11mn}(x_1, \mathbf{y}) Z_{m,ny}^\alpha \rangle; \\ B^0(\sigma) &= \langle h_{11\beta 1}(x_1, \mathbf{y}) y_B s_\beta + h_{11mn}(x_1, \mathbf{y}) L_{m,ny}^3 \\ &\quad + h_{11mn}(x_1, \mathbf{y}) Y_{m,ny} \rangle; \\ {}^1A_\beta(\sigma) &= \langle (h_{1111}(x_1, \mathbf{y}) + h_{11mn}(x_1, \mathbf{y}) K_{m,ny}^{01}) y_\beta \rangle; \\ A_{\beta\alpha}^2(\sigma) &= \langle (h_{1111}(x_1, \mathbf{y}) y_\alpha + h_{11mn}(x_1, \mathbf{y}) K_{m,ny}^{1\alpha} \\ &\quad + h_{i1mn}(x_1, \mathbf{y}) Z_{m,ny}^\alpha) y_\beta \rangle; \\ B_\beta^1(\sigma) &= \langle (h_{11\gamma 1}(x_1, \mathbf{y}) y_\Gamma s_\gamma + h_{11mn}(x_1, \mathbf{y}) L_{m,ny}^3 \\ &\quad + h_{11mn}(x_1, \mathbf{y}) Y_{m,ny}) y_\beta \rangle, \\ {}^1A(\sigma) &= \langle (h_{\Theta 111}(\mathbf{y}) + h_{\Theta 1mn}(x_1, \mathbf{y}) K_{m,ny}^{01}) y_\theta \rangle s_\Theta; \\ A_\alpha^2 &= \langle (h_{\Theta 111}(x_1, \mathbf{y}) y_\theta + h_{\Theta 1mn}(x_1, \mathbf{y}) K_{m,ny}^{1\alpha} \\ &\quad + h_{\Theta 1mn}(x_1, \mathbf{y}) Z_{m,ny}^\alpha) y_\theta \rangle s_\Theta; \\ B^1 = B_{\Theta\theta}^1(\sigma) s_\Theta &= \langle (h_{\Theta 1\beta 1}(x_1, \mathbf{y}) y_B s_\beta + h_{\Theta 1mn}(x_1, \mathbf{y}) L_{m,ny}^3 \\ &\quad + h_{\Theta 1mn}(x_1, \mathbf{y}) Y_{m,ny}) y_\theta \rangle s_\Theta. \end{aligned} \tag{10.43}$$

Formulas (10.39) and (10.40) are the governing equations of a beam with initial stresses considered as on-dimensional object. The quantities (10.43) are the beam stiffnesses in tension, bending and torsion. Although the formulas (10.43) are similar in appearance to the corresponding formulas for a beam without initial stresses from Sect. 9.3, there is a significant difference between them. Namely, all the stiffnesses in (10.43) depend on the initial stresses.

10.2.2 Shear Forces in a Beam with Initial Stresses

Let's discuss the problem of shear forces for the problem under consideration. The equilibrium equations (9.19) and (9.25) are

$$N_{\alpha 1,1x}^{(-2)} + \langle f_\alpha^\varepsilon \rangle + \langle h_\alpha^\varepsilon \rangle_S = 0; \tag{10.44}$$

$$M_{1\alpha,1x}^{(-3)} - N_{1\alpha}^{(-2)} + \langle f_1^\varepsilon y_\alpha \rangle + \langle h_1^\varepsilon y_\alpha \rangle_S = 0.$$

The shear forces $N_{\alpha 1}^{(-2)}$ and $N_{1\alpha}^{(-2)}$ included in (10.44) are asymmetric and we cannot use the classical method of Timoshenko and Woinowsky-Krieger (1959) to eliminate them. To eliminate $N_{1\alpha}^{(-2)}$, we use the method proposed in Sect. 8.2.3. Taking into account that in the expression (10.9, $m = -2$) some terms are symmetric with respect to i, j, we write the following equality:

$$\begin{gathered} K_{ij} = N_{ij}^{(-2)} - N_{ji}^{(-2)} \\ = \langle (b_{ijk1}(x_1, \mathbf{y}) - b_{jik1}(x_1, \mathbf{y}))u_{k,1x}^{(2)} \rangle + \langle (b_{ijkl}(x_1, \mathbf{y}) - b_{jikl}(x_1, \mathbf{y}))u_{k,ly}^{(3)} \rangle. \end{gathered} \tag{10.45}$$

By virtue of Proposition 9.4, the last term on the right-hand side of (10.45) is zero. Substituting $\mathbf{u}^{(2)}$ into the remaining term on the right-hand side of (10.45) according to (10.37), we obtain

$$\begin{gathered} K_{ij} = N_{ij}^{(-2)} - N_{ji}^{(-2)} \\ = \langle (b_{ijk1}(x_1, \mathbf{y}) - b_{jik1}(x_1, \mathbf{y}))K_k^{01} \rangle V_{1,1x1x}(x_1) \\ - \langle (b_{ij11}(x_1, \mathbf{y}) - b_{ji11}(x_1, \mathbf{y})y_\alpha) \rangle V_{\alpha,1x1x}(x_1) \\ + \langle (b_{ijk1}(x_1, \mathbf{y}) - b_{jik1}(x_1, \mathbf{y}))(K_k^{1\alpha} + Z_k^\alpha) \rangle u_{\alpha,1x1x1x}^{(0)}(x_1) \\ + \langle (b_{ijk1}(x_1, \mathbf{y}) - b_{jik1}(x_1, \mathbf{y}))(L_k^3 + Y_k) \rangle \phi_{,1x1x}(x_1). \end{gathered} \tag{10.46}$$

By virtue of the second condition from (10.5)

$$\langle (b_{ij11}(x_1, \mathbf{y}) - b_{ji11}(x_1, \mathbf{y}))y_\alpha \rangle = \langle \sigma_{j1}^{*\varepsilon}(x_1, \mathbf{y})y_\alpha \rangle \delta_{i1} - \langle \sigma_{j1}^{*\varepsilon}(x_1, \mathbf{y})y_\alpha \rangle \delta_{j1} = 0.$$

For $i = 1$, $j = \beta$, we have

$$K_{1\beta} = k_\beta V_{1,1x1x} + k_{\beta\alpha} u_{\alpha,1x1x1x}^{(0)} + l_\beta \phi_{,1x1x}. \tag{10.47}$$

The coefficients in (10.47) are

$$\begin{aligned} k_\beta &= \langle (b_{1\beta k1}(x_1, \mathbf{y}) - b_{1\beta k1}(x_1, \mathbf{y}))K_k^{01} \rangle \\ &\quad - \langle (b_{1\beta 11}(x_1, \mathbf{y}) - b_{1\beta 11}(x_1, \mathbf{y})y_\alpha) \rangle; \\ k_{\beta\alpha} &= \langle (b_{1\beta k1}(x_1, \mathbf{y}) - b_{1\beta k1}(x_1, \mathbf{y}))(K_k^{1\alpha} + Z_k^\alpha) \rangle; \\ l_\beta &= \langle (b_{1\beta k1}(x_1, \mathbf{y}) - b_{1\beta k1}(x_1, \mathbf{y}))(L_k^3 + Y_k) \rangle. \end{aligned} \tag{10.48}$$

Substituting b_{jikl} (10.4), we obtain from (10.48)

$$\begin{aligned} k_\beta &= \langle \sigma_{\beta 1}^{*\varepsilon}(x_1, \mathbf{y})K_k^{01} \rangle - \langle \sigma_{11}^{*\varepsilon}(x_1, \mathbf{y})K_k^{01} \rangle \delta_{\beta k}; \\ k_{\beta\alpha} &= \langle (\sigma_{\beta 1}^{*\varepsilon}(x_1, \mathbf{y}))\delta_{ik} - \sigma_{11}^{*\varepsilon}(x_1, \mathbf{y})) \rangle \delta_{\beta k})(K_k^{1\alpha} + Z_k^\alpha) \rangle; \\ l_\beta &= \langle (\sigma_{\beta 1}^{*\varepsilon}(x_1, \mathbf{y}))\delta_{1k} - \sigma_{11}^{*\varepsilon}(x_1, \mathbf{y}))\delta_{\beta k})(L_k^3 + Y_k) \rangle. \end{aligned} \tag{10.49}$$

We can separate the homogenized equilibrium equations (10.44) into two groups: equation for bending moments and equations for torsion moments. Obtain the equation for bending moments. First, as in the classical theory, we differentiate the first equation from (10.44), after which we use the equality

$$N_{1\beta}^{(-2)} = N_{\beta 1}^{(-2)} + K_{1\beta}$$

for the antisymmetric part of the shear forces and the second equation from (10.44). As a result, we obtain

$$M_{1\beta,1x1x}^{(-3)} = K_{1\beta,1x}. \tag{10.50}$$

Note. The formula for beam bending with initial stresses (10.50) actually includes not all shear forces $N_{ij}^{(-2)}$, but only their antisymmetric part, given by formula (10.47). If one want to compute the shear forces $N_{ij}^{(-2)}$, it is necessary to use the subsequent terms of the asymptotic expansion. For calculating the antisymmetric part of the shear forces of this calculation, the subsequent terms of the asymptotic expansion are not required.

For the torsion moment $M = M_{32}^{(-3)} - M_{23}^{(-3)}$, we obtain from (10.13, $p = -3$) equation

$$M_{,1x} = N_{32}^{(-3)} - N_{23}^{(-2)}.$$

Let us denote

$$N_{32}^{(-2)} - N_{23}^{(-2)} = K.$$

According to (10.46), we can write

$$K_{32} = kV_{1,1x1x} + k_\alpha u_{\alpha,1x1x1x}^{(0)} + l\phi_{,1x1x}, \tag{10.51}$$

The coefficients in (10.51) taking into account the definition of b_{jikl} (10.4) are

$$k = \langle \sigma_{21}^{*\varepsilon}(x_1, \mathbf{y}) K_k^{11} \rangle \delta_{3k} - \langle \sigma_{31}^{*\varepsilon}(x_1, \mathbf{y}) K_k^{11} \rangle \delta_{2k};$$

$$k_\alpha = \langle \sigma_{21}^{*\varepsilon}(x_1, \mathbf{y})(K_k^{2\alpha} + Z_k^\alpha) \rangle \delta_{3k} - \langle \sigma_{31}^{*\varepsilon}(x_1, \mathbf{y})(K_k^{2\alpha} + Z_k^\alpha) \rangle \delta_{2k}; \tag{10.52}$$

$$l = \langle \sigma_{21}^{*\varepsilon}(x_1, \mathbf{y})(L_k^3 + Y_k) \rangle \delta_{3k} - \langle \sigma_{31}^{*\varepsilon}(x_1, \mathbf{y})(L_k^3 + Y_k) \rangle \delta_{2k}.$$

As a result, we obtain the equation for the torsion moment

$$-M_{,1x} + K = 0. \tag{10.53}$$

The solutions to the periodicity cell problem are determined, as noted above, with an accuracy of up to an arbitrary constant. In addition, the solution to the periodicity cell problem (10.34) is determined with an accuracy of up to the function $y_\beta s_\beta \mathbf{e}_\beta \phi(x_1)$. The indicated arbitrariness does not lead to the appearance in (10.49) and (10.52) of undefined terms of the form $\langle \sigma_{i1}^{*\varepsilon} \rangle$ and $\langle \sigma_{\alpha 1}^{*\varepsilon} y_\beta \rangle$, since such terms vanish due to the condition (10.5) and the equality (10.5).

10.2.3 Boundary Conditions

When the beam is fixed as three-dimensional body over the surface S_u, from the initial boundary condition $\mathbf{u}^\varepsilon = 0$ for $x_1 = \pm a$, the expansion (10.6), and the representation (10.24), we obtain the classical clamped boundary conditions for one-dimensional beam

$$\begin{gathered} V_1(-a) = V_1(a) = 0; \\ u_\alpha^{(0)}(-a) = u_\alpha^{(0)}(a) = u_{\alpha,1x}^{(0)}(-a) = u_{\alpha,1x}^{(0)}(a) = 0; \\ \phi(-a) = \phi(a) = 0. \end{gathered} \tag{10.54}$$

For other types of beam support, such as a hinge or roller support, or free edge, classical boundary conditions are take place.

10.3 Cylindrical Beams. Dependence of the Homogenized Stiffnesses on Initial Stresses

Let the initial stresses in the beam be of the form

$$\sigma_{11}^* = \sigma_{11}^*(x_2, x_3), \sigma_{ij}^* = 0 \text{ when } ij \neq 11.$$

In this case

$$\langle \sigma_{11}^* \rangle = 0.$$

Stresses of this type can arise in a inhomogeneous cylindrical beam of coaxial structure.

In the case under consideration, the functions c_{ijkl}, σ_{ij}^* do not depend on and y_1, and solutions to the cell problem can be sought as functions of the variables y_2, y_3. We have ($\mathbf{M}$ means one of the functions $\mathbf{K}^{01}$, $\mathbf{K}^{1\alpha}$ or $\mathbf{L}^3$):

$$h_{ijkl} M_{k,ly} = h_{ijk\alpha} M_{k,\alpha y} = c_{ijk\alpha} M_{k,\alpha y}$$

due to the fact that $h_{ijk\alpha} = c_{ijk\alpha} + \sigma_{\alpha j}^* \delta_{ik}$ and $\sigma_{\alpha 1}^* = 0$.

The free terms from the equations in the periodicity cell problem (10.17), (10.34) and (10.37) are zero due to the fact that $\sigma_{ij}^* = 0$ for $ij \neq 11$ (this is verified by performing differentiation in them). In the same periodicity cell problems (10.17), (10.34) and (10.37), by Proposition 9.4 and the fact that $n_1 = 0$ on S and $\sigma_{\alpha 1}^* = 0$, we have $\sigma_{\alpha\beta}^* n_\beta = 0$ on S.

By virtue of this,the periodicity cell problems (10.17), (10.34) and (10.37) in the case under consideration coincide with the periodicity cell problems for a beam without initial stresses. Then $\mathbf{K}^{01}$, $\mathbf{K}^{1\alpha}$ and $\mathbf{L}^3$ in (10.43) can be replaced, respectively, by $\mathbf{N}^{(01)}$, $\mathbf{N}^{(1\alpha)}$ and $\mathbf{X}^3$—solutions to periodicity cell problems (9.37), (9.48), (9.50) for beams without initial stresses. Taking into account the above, the formulas (10.43), in the case under consideration, can be given the following form:

$$\begin{aligned} A^0(\sigma) &= A^0(0), \\ A^1_\alpha(\sigma) &= A^1_\alpha(0), \\ B^0(\sigma) &= B^0(0), \\ {}^1A_\beta &= {}^1A_\beta(0), \\ A^2_{\beta\alpha}(\sigma) &= A^2_{\beta\alpha}(0) + \langle \sigma^{*(-2)}_{11}(x_1, \mathbf{y}) y_\alpha y_\beta \rangle, \\ B^1_\beta(\sigma) &= B^1_\beta(0) + \langle \sigma^{*(-2)}_{11}(x_1, \mathbf{y}) y_\Gamma y_\beta \rangle s_\gamma, \\ A^2_\alpha(\sigma) &= A^2_\alpha(0) + \langle \sigma^{*(-2)}_{11}(x_1, \mathbf{y}) y^2_\theta \rangle s_\Theta, \\ A_1(\sigma) &= A_1(0), \\ B^1 &= B^1(0) + \langle \sigma^{*(-2)}_{11}(x_1, \mathbf{y})(y_2^2 + y_3^2) \rangle. \end{aligned} \tag{10.55}$$

The value of Γ is determined by γ in (8.38). The stiffness with argument 0 in (10.55) denote the homogenized stiffnesses of the beam with no initial stresses. In particular, $A^2_{i\beta\alpha}(0)$ and $B^1_{i\beta}(0)$ bending and torsional homogenized stiffnesses of the beam with no initial stresses. They are determined in in (9.57) .

According to (10.55), the axial initial stresses affect only the beam torsional and bending stiffnesses.

Let's consider the torsional stiffness of uniform beam, which is introduced as $B(\sigma) = B^1_{32}(\sigma) - B^1_{23}(\sigma)$. For it we obtain from (10.55) the following formula:

$$B(\sigma) = B(0) + \langle \sigma^*_{11}(x_1, \mathbf{y})(y_2^2 + y_3^2) \rangle.$$

In the resulting formula, the last term coincides with the term taking into account the initial stresses in the problem of torsion of a homogeneous cylindrical beam with initial stresses, the solution of which is given in Washizu (1992). This coincidence is not complete. In Washizu (1992), a homogeneous thick rod is considered, to which a non-zero resultant axial force is applied. We consider thin beam under condition $\langle \sigma^*_{11}(x_1, \mathbf{y}) \rangle = 0$. If this condition is not met, the beam will lose stability before we notice the effect of the initial stresses on the beam stiffness.

10.4 Case of Small Initial Stresses

In this section, we consider the case when the initial stresses has the order of the elastic constants but are essentially smaller the elastic constants in absolute value.

We made a remark about the term "small initial stresses" in Sect. 6.2.1 devoted to the plates with the initial stresses of the order of elastic constants. In the case of beams, the elastic constants have the order of ε^{-4}. This guaranty that the stiffness of the plate has the order of unity, as $\varepsilon \to 0$. In this section, we consider the initial stresses of the order of ε^{-4} - the order of elastic constants. Despite the elastic constants and initial stresses are of the same order in ε, the initial stresses are considerable smaller the elastic constants in absolute value, see Sect. 4.5, in particular, Table 4.1. Since the ratio of the strength limit to Young's modulus less 0.02 for the most materials, the linear (the first order) corrector may be sufficient to obtain accurate approximate formulas for calculating beam stiffnesses. Below, we construct the linear correctors.

The periodicity cell problems (10.19), (10.33), (10.34), (10.35), (10.36) may be written in the following general forms:

$$\begin{cases} (h_{ijmn}(x_1, \mathbf{y}) H^{\alpha}_{m,ny} + f^{\alpha}_{ij}(x_1, \mathbf{y}))_{,jy} = 0 \text{ in } Y; \\ (h_{ijmn}(x_1, \mathbf{y}) H^{\alpha}_{m,ny} + f^{\alpha}_{ij}(x_1, \mathbf{y})) n_j = 0 \text{ on } S; \\ \mathbf{H}^{\alpha}(\mathbf{y}) \text{ is periodic in } y_1 \text{ with period } m; \end{cases} \tag{10.56}$$

and

$$\begin{cases} (h_{ijmn}(x_1, \mathbf{y}) H_{m,ny})_{,jy} + g^{\alpha}_i = 0 \text{ in } Y; \\ h_{ijmn}(x_1, \mathbf{y}) H_{m,ny} n_j = 0 \text{ on } S; \\ \mathbf{H}(\mathbf{y}) \text{ is periodic in } y_1 \text{ with period } m. \end{cases} \tag{10.57}$$

In (10.56),

$$f^{0}_{ij}(x_1, \mathbf{y}) = h_{ij11}(x_1, \mathbf{y}) \text{ for (10.19)}, \quad H = K^{01};$$

$$f^{\alpha}_{ij}(x_1, \mathbf{y}) = h_{ij11}(x_1, \mathbf{y}) y_{\alpha} \text{ for (10.33)}, \quad H = K^{1\alpha};$$

$$f^{1}_{ij}(x_1, \mathbf{y}) = h_{ij\beta 1}(x_1, \mathbf{y}) y_B s_{\beta} \text{ for (10.34)}, \quad H = L^3;$$

and in (10.57),

$$g^{\alpha}_i(x_1, \mathbf{y}) = \sigma^{*\varepsilon}_{11}(x_1, \mathbf{y}) \delta_{i\alpha} - \sigma^{*\varepsilon}_{1\alpha}(x_1, \mathbf{y}) \delta_{i1} \text{ for (10.35)}, \quad H = Z^{\alpha};$$

$$g^{1}_i(x_1, \mathbf{y}) = \sigma^{*\varepsilon}_{j\beta}(x_1, \mathbf{y}) s_{\beta} \delta_{i\beta} \text{ for (10.36)}, \quad H = Y.$$

The periodicity cell problems (10.56) and (10.57) may be written in the following general form:

$$\begin{cases} (h_{ijmn}(x_1, \mathbf{y}) J^{K}_{m,ny} + f^{K}_{ij}(x_1, \mathbf{y}))_{,jy} + g^{K}_i = 0 \text{ in } Y; \\ (h_{ijmn}(x_1, \mathbf{y}) J^{K}_{m,ny} + f^{K}_{ij}(x_1, \mathbf{y})) n_j = 0 \text{ on } S; \\ \mathbf{J}^{K}(\mathbf{y}) \text{ is periodic in } y_1 \text{ with period } m; \end{cases} \tag{10.58}$$

where $K = 0, 1, 2, 3$ and functions $\mathbf{J}^K$, f^K_{ij} and g^K_i are determined as follows:

$$K = 0: \tag{10.59}$$
$$f_{ij}^{0}(x_1, \mathbf{y}) = h_{ij11}(x_1, \mathbf{y}), \;\; g_i^{0}(x_1, \mathbf{y}) = 0, \;\; \mathbf{J}^0 = \mathbf{K}^{01} \text{ for (10.19)};$$

$$K = \alpha = 2, 3: \tag{10.60}$$
$$f_{ij}^{\alpha}(x_1, \mathbf{y}) = h_{ij11}(x_1, \mathbf{y}) y_\alpha, \;\; g_i^{\alpha}(x_1, \mathbf{y}) = \sigma_{11}^{*\varepsilon}(x_1, \mathbf{y})\delta_{i\alpha} - \sigma_{1\alpha}^{*\varepsilon}(x_1, \mathbf{y})\delta_{i1},$$
$$\mathbf{J}^\alpha = \mathbf{K}^{1\alpha} + \mathbf{Z}^\alpha \text{ for (10.33) and (10.35)};$$

$$K = 1: \tag{10.61}$$
$$f_{ij}^{1}(x_1, \mathbf{y}) = h_{ij\beta 1}(x_1, \mathbf{y}) y_B s_\beta, \;\; g_i^{1}(x_1, \mathbf{y}) = \sigma_{1B}^{*\varepsilon}(x_1, \mathbf{y}) s_\beta \delta_{i\beta},$$
$$\mathbf{J}^1 = \mathbf{L}^3 + \mathbf{Y} \text{ for (10.34) and (10.36)}.$$

The functions (10.59) in (10.58) have the form

$$\begin{aligned} h_{ijmn}(x_1, \mathbf{y}) &= c_{ijmn}(\mathbf{y}) + \mu b_{ijmn}(x_1, \mathbf{y}); \\ f_{ij}^{K}(x_1, \mathbf{y}) &= f_{ij}^{K0}(x_1, \mathbf{y}) + \mu f_{ij}^{K1}(\mathbf{y}); \\ g_i^{K}(x_1, \mathbf{y}) &= \mu g_i^{K1}(\mathbf{y}), \quad K = 0, 1, 2, 3, \end{aligned} \tag{10.62}$$

where μ is a small parameter which has the meaning of ratio of the characteristic value of the initial tresses to the characteristic value of the elastic constants. Note that the functions h_{ijmn}, f_{ij}^{K}, and g_i^{K} have zero-order part (they are not equal to zero when $\mu = 0$), and g_i^{K} do not have zero-order part (they are equal to zero when $\mu = 0$).

10.4.1 Expansions of the Solutions to the Periodicity Cell Problems in a Series in a Small Parameter μ

We will seek solution to the periodicity cell problem (10.58) in the form

$$\mathbf{J}^\alpha(x_1, \mathbf{y}) = \sum_{s=0}^{\infty} \mu^s \mathbf{J}^{\alpha s}(x_1, \mathbf{y}), \tag{10.63}$$

where functions $\mathbf{J}^{\alpha s}(\mathbf{y})$ are periodic in y_1 with the period m.

Substituting (10.63), (10.62) into (10.56) yields

$$\begin{cases} \Big((c_{ijmn}(\mathbf{y}) + \mu b_{ijmn}(x_1, \mathbf{y}) \sum_{s=0}^{\infty} \mu^s J^{Ks}_{m,ny} + f^{K0}_{ij}(\mathbf{y}) \\ +\mu f^{K1}_{ij}(x_1, \mathbf{y})\Big)_{,jy} + \mu g^K_i = 0 \text{ in } Y; \\ \Big(c_{ijmn}(\mathbf{y}) + \mu b_{ijmn}(x_1, \mathbf{y}) \sum_{s=0}^{\infty} \mu^s J^{Ks}_{m,ny} + f^{K0}_{ij}(x_1, \mathbf{y}) \\ +\mu f^{K1}_{ij}(x_1, \mathbf{y})\Big) n_j = 0 \text{ on } S; \\ \mathbf{J}^{Ks}(\mathbf{y}) \text{ periodic in } y_1 \text{ with period } m. \end{cases} \tag{10.64}$$

By equating the expressions corresponding to the identical powers of μ in (10.64), we obtain an infinite sequence of equations, the first two of which have the form:

$$\begin{cases} (c_{ijmn}(\mathbf{y}) J^{K0}_{m,ny} + f^{K0}_{ij}(x_1, \mathbf{y}))_{,jy} = 0 \text{ in } Y; \\ (c_{ijmn}(\mathbf{y}) J^{K0}_{m,ny} + f^{K0}_{ij}(x_1, \mathbf{y})) n_j = 0 \text{ on } S; \\ \mathbf{J}^{K0}(\mathbf{y}) \text{ is periodic in } y_1 \text{ with period } m; \end{cases} \tag{10.65}$$

and

$$\begin{cases} (c_{ijmn}(\mathbf{y}) J^{K1}_{m,ny} + b_{ijmn}(x_1, \mathbf{y}) J^{K0}_{m,ny} \\ + f^{K1}_{ij}(x_1, \mathbf{y}))_{,jy} + g^{K1}_i = 0 \text{ in } Y; \\ (c_{ijmn}(\mathbf{y}) J^{K1}_{m,ny} + b_{ijmn}(x_1, \mathbf{y}) J^{K0}_{m,ny} \\ + f^{K1}_{ij}(x_1, \mathbf{y})) n_j = 0 \text{ on } S; \\ \mathbf{J}^{K1}(\mathbf{y}) \text{ is periodic in } y_1 \text{ with period } m. \end{cases} \tag{10.66}$$

Note. In the periodicity cell problem (10.65), corresponding to zero-order terms,

$$\begin{gathered} K = 0: \ f^{00}_{ij}(x_1, \mathbf{y}) = c_{ij11}(x_1, \mathbf{y}), \ g^0_i(x_1, \mathbf{y}) = 0, \\ \mathbf{J}^0 = \mathbf{K}^{00} = \mathbf{N}^0 \text{ for (10.19)}; \\ K = \alpha = 2, 3: \ f^{\alpha 0}_{ij}(x_1, \mathbf{y}) = c_{ij11}(x_1, \mathbf{y}) y_\alpha, \ g^{\alpha 0}_i(x_1, \mathbf{y}) = 0, \\ \mathbf{Z}^{\alpha 0} = 0, \ \mathbf{J}^\alpha = \mathbf{K}^{1\alpha 0} = \mathbf{N}^{1\alpha} \ (\alpha = 2, 3) \text{ for (10.33) and (10.35)}; \\ K = 1: \ f^{10}_{ij}(x_1, \mathbf{y}) = c_{ij\beta 1}(x_1, \mathbf{y}) y_B s_\beta, \ g^1_i(x_1, \mathbf{y}) = 0, \\ \mathbf{Y} = 0, \ \mathbf{J}^{10} = \mathbf{X}^3 \text{ for (10.34) and (10.36)}. \end{gathered}$$

Multiplying equation in (10.65) by $J^{L1}_i(\mathbf{y})$ and integrating by parts with account of the periodicity $J^{L1}_i(\mathbf{y})$, we obtain

$$\langle c_{ijmn}(\mathbf{y})J^{K0}_{m,ny}J^{L1}_{i,jy} + f^{K0}_{ij}(x_1,\mathbf{y})J^{L1}_i\rangle = 0, \tag{10.67}$$

or

$$\langle c_{ijmn}(\mathbf{y})J^{L0}_{m,ny}J^{K1}_{i,jy} + f^{L0}_{ij}(x_1,\mathbf{y})J^{K1}_i\rangle = 0. \tag{10.68}$$

Multiplying equation in (10.66) by $J^{L0}_i(\mathbf{y})$ and integrating by parts with account of the periodicity $J^{L0}_i(\mathbf{y})$, we obtain

$$\begin{aligned}\langle c_{ijmn}(\mathbf{y})J^{K1}_{m,ny}J^{L0}_{i,jy} + b_{ijmn}(x_1,\mathbf{y})J^{K0}_{m,ny}J^{L0}_{i,jy}\\ +f^{K1}_{ij}(x_1,\mathbf{y})J^{L0}_{i,jy} + g^{K1}_i J^{L0}_i\rangle = 0.\end{aligned} \tag{10.69}$$

Taking into account symmetry of the elastic constants $c_{ijmn}(\mathbf{y})$, see C1 (3.5), we conclude that the quadratic terms in (10.68) and (10.69) are equal, and obtain the following equality:

$$\begin{aligned}\langle b_{ijmn}(x_1,\mathbf{y})J^{K0}_{m,ny}J^{L0}_{i,jy} + f^{K1}_{ij}(x_1,\mathbf{y})J^{L0}_{i,jy} + g^K_i J^{L0}_i\rangle\\ = \langle f^{L0}_{ij}(x_1,\mathbf{y})J^{K1}_i\rangle.\end{aligned} \tag{10.70}$$

10.4.2 Elimination of the First-Order Terms

The formulas (10.43) for calculating the homogenized stiffnesses can be rewritten as

$$\begin{aligned}
A^0(\sigma) - \langle h_{1111}(\mathbf{y})\rangle &= \langle h_{11mn}(x_1,\mathbf{y})K^{01}_{m,ny}\rangle;\\
A^1_\alpha(\sigma) - \langle h_{1111}(x_1,\mathbf{y})y_\alpha\rangle &= \langle h_{11mn}(x_1,\mathbf{y})[K^{1\alpha}_{m,ny} + Z^\alpha_{m,ny}]\rangle;\\
B^0(\sigma) - \langle h_{11\beta1}(x_1,\mathbf{y})y_B s_\beta\rangle &= \langle h_{11mn}(x_1,\mathbf{y})[L^3_{m,ny} + Y_{m,ny}]\rangle;\\
{}^1A_\beta(\sigma) - \langle h_{1111}(\mathbf{y})y_\beta\rangle &= \langle h_{11mn}(x_1,\mathbf{y})K^{01}_{m,ny}y_\beta\rangle;\\
A^2_{\beta\alpha}(\sigma) - \langle h_{1111}(x_1,\mathbf{y})y_\alpha\rangle &= \langle h_{11mn}(x_1,\mathbf{y})[K^{1\alpha}_{m,ny} + Z^\alpha_{m,ny}]y_\beta\rangle;\\
B^1_\beta(\sigma) - \langle h_{11\beta1}(x_1,\mathbf{y})y_\Theta s_\theta y_\beta\rangle &= \langle h_{11mn}(x_1,\mathbf{y})[L^3_{m,ny} + Y_{m,ny}]y_\beta\rangle,\\
{}^1A(\sigma) - \langle h_{\Theta111}(\mathbf{y})\rangle &= \langle h_{\Theta1mn}(x_1,\mathbf{y})K^{01}_{m,ny}y_\theta\rangle s_\Theta;\\
A^2_\alpha - \langle h_{\Theta111}(x_1,\mathbf{y})y_\theta\rangle &= \langle h_{\Theta1mn}(x_1,\mathbf{y})[K^{1\alpha}_{m,ny} + Z^\alpha_{m,ny}]y_\theta\rangle s_\Theta;\\
B^1 - \langle h_{\Theta1\beta1}(x_1,\mathbf{y})y_B s_\beta\rangle &= \langle h_{\Theta1mn}(x_1,\mathbf{y})[L^3_{m,ny} + Y_{m,ny}]y_\theta\rangle s_\Theta.
\end{aligned} \tag{10.71}$$

The right-hand parts of equalities (10.71) have the form

$$\begin{aligned}
&\langle h_{11mn}(x_1,\mathbf{y})J^K_{m,ny}\rangle\\
&= \langle (c_{11mn}(x_1,\mathbf{y}) + \mu b_{i1mn}(x_1,\mathbf{y}))(J^{K0}_{m,ny} + \mu J^{K1}_{m,ny})\rangle\\
&= \langle c_{11mn}(x_1,\mathbf{y})J^{K0}_{m,ny}\rangle\\
&+\mu[(b_{11mn}(x_1,\mathbf{y}))J^{K0}_{m,ny} + c_{11mn}(x_1,\mathbf{y})J^{K1}_{m,ny}))] + \cdots;
\end{aligned} \tag{10.72}$$

$$\begin{gathered}\langle h_{11mn}(x_1,\mathbf{y})J^K_{m,ny}y_\beta\rangle\\ =\langle(c_{11mn}(x_1,\mathbf{y})+\mu b_{i1mn}(x_1,\mathbf{y}))(J^{K0}_{m,ny}+\mu J^{K1}_{m,ny})y_\beta\rangle\\ =\langle c_{11mn}(x_1,\mathbf{y})J^{K0}_{m,ny}y_\beta\\ +\mu[(b_{11mn}(x_1,\mathbf{y}))J^{K0}_{m,ny}+c_{11mn}(x_1,\mathbf{y})J^{K1}_{m,ny})y_\beta\rangle]+\cdots;\end{gathered}\tag{10.73}$$

and

$$\begin{gathered}\langle h_{\Theta 1mn}(x_1,\mathbf{y})J^K_{m,ny}y_\theta\rangle s_\Theta\\ =\langle(c_{\Theta 1mn}(x_1,\mathbf{y})+\mu b_{i1mn}(x_1,\mathbf{y}))(J^{K0}_{m,ny}+\mu J^{K1}_{m,ny})y_\theta\rangle s_\Theta\\ =\langle c_{\Theta 1mn}(x_1,\mathbf{y})J^{K0}_{m,ny}y_\theta\rangle s_\Theta\\ +\mu\langle(b_{\Theta 1mn}(x_1,\mathbf{y})J^{K0}_{m,ny}+c_{\Theta 1mn}(x_1,\mathbf{y})J^{K1}_{m,ny})y_\theta\rangle s_\Theta+\cdots;\end{gathered}\tag{10.74}$$

In (10.72), (10.73), and (10.74), "..." means the terms of the order higher than μ.

The non-trivial terms in (10.73) and (10.74) are

$$\langle c_{11mn}(x_1,\mathbf{y})J^{K1}_{m,ny}\rangle=\langle c_{mn11}(x_1,\mathbf{y})J^{K1}_{m,ny}\rangle;\tag{10.75}$$

$$\langle c_{11mn}(x_1,\mathbf{y})J^{K1}_{m,ny}y_\beta\rangle=\langle c_{mn11}(x_1,\mathbf{y})J^{K1}_{m,ny}y_\beta\rangle;\tag{10.76}$$

and

$$\langle c_{\Theta 1mn}(x_1,\mathbf{y})J^{K1}_{m,ny})y_\theta\rangle s_\Theta=\langle c_{mn\Theta 1}(x_1,\mathbf{y})J^{K1}_{m,ny}y_\theta\rangle s_\Theta\tag{10.77}$$

which contains the functions J^{K1}.

Comparing expressions (10.76) and (10.77) with the definitions

$$f^{00}_{ij}(x_1,\mathbf{y})=c_{ij11}(x_1,\mathbf{y})\text{ for (10.19)},\ \mathbf{J}^0=\mathbf{K}^{01};$$

$$f^{\alpha 0}_{ij}(x_1,\mathbf{y})=c_{ij11}(x_1,\mathbf{y})y_\alpha\text{ for (10.33)},\ \mathbf{J}^\alpha=\mathbf{K}^{1\alpha}+\mathbf{Z}^\alpha,\ \alpha=2,3;$$

and

$$f^{10}_{ij}(x_1,\mathbf{y})=c_{ij\beta 1}(x_1,\mathbf{y})y_B s_\beta\text{ for (10.34)},\ \mathbf{J}=\mathbf{L}^3+\mathbf{Y};$$

we conclude that the non-trivial terms (10.76) and (10.77) in (10.73) and (10.74) may be rewritten as follows:

$$\langle c_{mn11}(x_1,\mathbf{y})J^{K1}_{m,ny}\rangle=\langle f^{00}_{ij}(x_1,\mathbf{y})J^{K1}_{m,ny}\rangle;\tag{10.78}$$

$$\langle c_{mn11}(x_1,\mathbf{y})y_\beta J^{K1}_{m,ny}\rangle=\langle f^{\beta 0}_{ij}(x_1,\mathbf{y})J^{K1}_{m,ny}\rangle,\ \beta=2,3;\tag{10.79}$$

and (we can change Θ and θ for β and B)

$$\langle c_{mn\Theta 1}(x_1, \mathbf{y}) y_\theta s_\Theta J^{K1}_{m,ny}\rangle = \langle f^{10}_{ij}(x_1, \mathbf{y}) J^{K1}_{m,ny}\rangle. \tag{10.80}$$

Let's rewrite (10.70) as

$$\begin{aligned}&\langle f^{L0}_{ij}(x_1, \mathbf{y}) J^{K1}_i\rangle \\ &= \langle b_{ijmn}(x_1, \mathbf{y}) J^{K0}_{m,ny} J^{L0}_{i,jy} + f^{K1}_{ij}(x_1, \mathbf{y}) J^{L0}_{i,jy} + g^{K1}_i J^{L0}_i\rangle;\end{aligned} \tag{10.81}$$

and substitute (10.81) with proper indices K, L into (10.78), (10.79), and (10.80). As a result, we obtain

$$\begin{aligned} K, L = 0:\ &\langle f^{00}_{ij}(x_1, \mathbf{y}) J^{K1}_{i,jy}\rangle = \langle b_{ijmn}(x_1, \mathbf{y}) J^{K0}_{m,ny} J^{00}_{i,jy} \\ &+ f^{K1}_{ij}(x_1, \mathbf{y}) J^{00}_{i,jy} + g^{K1}_i J^{00}_i\rangle; \\ K, L = \beta:\ &\langle f^{\beta 0}_{ij}(x_1, \mathbf{y}) J^{K1}_{m,ny}\rangle = \langle b_{ijmn}(x_1, \mathbf{y}) J^{K0}_{m,ny} J^{\beta 0}_{i,jy} \\ &+ f^{\alpha 1}_{ij}(x_1, \mathbf{y}) J^{\beta 0}_{i,jy} + g^{\alpha 1}_i J^{\beta 0}_i\rangle;\end{aligned}$$

and

$$\begin{aligned}&K, L = 1:\ \langle f^{10}_{ij}(x_1, \mathbf{y}) J^{K1}_i\rangle \\ &= \langle b_{ijmn}(x_1, \mathbf{y}) J^{K0}_{m,ny} J^{10}_{i,jy} + f^{K1}_{ij}(x_1, \mathbf{y}) J^{10}_{i,jy} + g^{K1}_i J^{10}_i\rangle.\end{aligned}$$

The correctors (the multipliers by μ) in the right-hand parts of the equalities (10.72), (10.73), and (10.74) are given by the formulas

$$\begin{aligned}&\langle (b_{11mn}(x_1, \mathbf{y})) J^{K0}_{m,ny} + c_{11mn}(x_1, \mathbf{y}) J^{K1}_{m,ny})\rangle \\ &= (b_{11mn}(x_1, \mathbf{y})) J^{K0}_{m,ny} + f^{00}_{mn}(x_1, \mathbf{y}) J^{K1}_{m,ny})\rangle \\ &= \langle (b_{11mn}(x_1, \mathbf{y})) J^{\alpha 0}_{m,ny}\rangle + \langle b_{ijmn}(x_1, \mathbf{y}) J^{K0}_{m,ny} J^{00}_{i,jy}\rangle \\ &+ \langle f^{K1}_{ij}(x_1, \mathbf{y}) J^{00}_{i,jy} + g^{K1}_i J^{00}_i\rangle;\end{aligned} \tag{10.82}$$

$$\begin{aligned}&\langle (b_{11mn}(x_1, \mathbf{y})) J^{K0}_{m,ny} + c_{11mn}(x_1, \mathbf{y}) J^{K1}_{m,ny}) y_\beta\rangle \\ &= \langle (b_{11mn}(x_1, \mathbf{y})) J^{\alpha 0}_{m,ny} + f^{\beta 0}_{mn}(x_1, \mathbf{y}) J^{\alpha 1}_{m,ny})\rangle \\ &= \langle (b_{11mn}(x_1, \mathbf{y})) J^{\alpha 0}_{m,ny}\rangle \langle b_{ijmn}(x_1, \mathbf{y}) J^{K0}_{m,ny} J^{\beta 0}_{i,jy}\rangle \\ &+ \langle f^{K1}_{ij}(x_1, \mathbf{y}) J^{\beta 0}_{i,jy} + g^{K1}_i J^{\beta 0}_i\rangle;\end{aligned} \tag{10.83}$$

and

$$\begin{aligned}&\langle (b_{\Theta 1mn}(x_1, \mathbf{y}) J^{K0}_{m,ny} + c_{\Theta 1mn}(x_1, \mathbf{y}) J^{\alpha 1}_{m,ny}) y_\theta s_\Theta\rangle \\ &= \langle (b_{\Theta 1mn}(x_1, \mathbf{y}) J^{K0}_{m,ny} + f^{10}_{mn}(x_1, \mathbf{y}) J^{K1}_{m,ny}) y_\theta s_\Theta\rangle \\ &= \langle b_{\Theta 1mn}(x_1, \mathbf{y}) J^{K0}_{m,ny}\rangle s_\Theta + \langle b_{ijmn}(x_1, \mathbf{y}) J^{K0}_{m,ny} J^{10}_{i,jy}\rangle \\ &+ \langle f^{K1}_{ij}(x_1, \mathbf{y}) J^{10}_{i,jy} + g^{K1}_i J^{10}_i\rangle.\end{aligned} \tag{10.84}$$

Substituting (10.88), (10.89), and $b_{ijmn}(x_1, \mathbf{x}/\varepsilon) = \sigma^{*\varepsilon}_{jn}(x_1, \mathbf{x}/\varepsilon)\delta_{im}$ in accordance with (10.4), we obtain that the correctors (10.82), (10.83), and (10.84) are (the correctors are marked by the symbol Δ)

$$\begin{aligned}
\Delta A^0 &= \langle b_{11mn}(x_1,\mathbf{y}) J^{00}_{m,ny}\rangle + \langle b_{ijmn}(x_1,\mathbf{y}) J^{00}_{m,ny} J^{00}_{i,jy}\rangle \\
&\quad + \langle f^{01}_{ij}(x_1,\mathbf{y}) J^{00}_{i,jy} + g^{01}_i J^{00}_i\rangle; \\
\Delta A^1_\alpha &= \langle b_{11mn}(x_1,\mathbf{y}) J^{\alpha 0}_{m,ny}\rangle + \langle b_{ijmn}(x_1,\mathbf{y}) J^{\alpha 0}_{m,ny} J^{00}_{i,jy}\rangle \\
&\quad + \langle f^{\alpha 1}_{ij}(x_1,\mathbf{y}) J^{00}_{i,jy} + g^{\alpha 1}_i J^{00}_i\rangle; \\
\Delta B^0 &= \langle b_{11mn}(x_1,\mathbf{y}) J^{10}_{m,ny}\rangle + \langle b_{ijmn}(x_1,\mathbf{y}) J^{10}_{m,ny} J^{00}_{i,jy}\rangle \\
&\quad + \langle f^{11}_{ij}(x_1,\mathbf{y}) J^{00}_{i,jy} + g^{11}_i J^{00}_i\rangle;
\end{aligned} \tag{10.85}$$

$$\begin{aligned}
\Delta^1 A_\beta &= \langle b_{11mn}(x_1,\mathbf{y}) J^{00}_{m,ny} y_\beta\rangle + \langle b_{ijmn}(x_1,\mathbf{y}) J^{00}_{m,ny} J^{\beta 0}_{i,jy}\rangle \\
&\quad + \langle f^{01}_{ij}(x_1,\mathbf{y}) J^{\beta 0}_{i,jy} + g^{01}_i J^{\beta 0}_i\rangle; \\
\Delta A^2_{\beta\alpha} &= \langle b_{11mn}(x_1,\mathbf{y}) J^{\alpha 0}_{m,ny} y_\beta\rangle + \langle b_{ijmn}(x_1,\mathbf{y}) J^{\alpha 0}_{m,ny} J^{\beta 0}_{i,jy}\rangle \\
&\quad + \langle f^{\alpha 1}_{ij}(x_1,\mathbf{y}) J^{\beta 0}_{i,jy} + g^{\alpha 1}_i J^{\beta 0}_i\rangle; \\
\Delta B^1_\beta(\sigma) &= \langle (b_{11mn}(x_1,\mathbf{y})) J^{\alpha 0}_{m,ny}\rangle + \langle b_{ijmn}(x_1,\mathbf{y}) J^{K0}_{m,ny} J^{\beta 0}_{i,jy}\rangle \\
&\quad + \langle f^{K1}_{ij}(x_1,\mathbf{y}) J^{\beta 0}_{i,jy} + g^{K1}_i J^{\beta 0}_i\rangle;
\end{aligned} \tag{10.86}$$

and

$$\begin{aligned}
\Delta^1 A &= \langle b_{\Theta 1mn}(x_1,\mathbf{y}) J^{00}_{m,ny} s_\Theta + \langle b_{ijmn}(x_1,\mathbf{y}) J^{00}_{m,ny} J^{10}_{i,jy}\rangle \\
&\quad + \langle f^{01}_{ij}(x_1,\mathbf{y}) J^{10}_{i,jy} + g^{01}_i J^{10}_i\rangle; \\
\Delta A^2_\alpha &= \langle b_{\Theta 1mn}(x_1,\mathbf{y}) J^{\alpha 0}_{m,ny}\rangle s_\Theta + \langle b_{ijmn}(x_1,\mathbf{y}) J^{\alpha 0}_{m,ny} J^{10}_{i,jy}\rangle \\
&\quad + \langle f^{\alpha 1}_{ij}(x_1,\mathbf{y}) J^{10}_{i,jy} + g^{\alpha 1}_i J^{10}_i\rangle; \\
\Delta B^1 &= \langle b_{\Theta 1mn}(x_1,\mathbf{y}) J^{10}_{m,ny}\rangle s_\Theta + \langle b_{ijmn}(x_1,\mathbf{y}) J^{10}_{m,ny} J^{10}_{i,jy}\rangle \\
&\quad + \langle f^{11}_{ij}(x_1,\mathbf{y}) J^{10}_{i,jy} + g^{11}_i J^{10}_i\rangle.
\end{aligned} \tag{10.87}$$

Remind that $\alpha = 2, 3$ and

$$\begin{gathered}
g^{K0}_i(x_1,\mathbf{y}) = 0; \\
f^{01}_{ij}(x_1,\mathbf{y}) = b_{ij11}(x_1,\mathbf{y}),\ \ g^{01}_i(x_1,\mathbf{y}) = 0 \text{ for (10.19)}; \\
f^{\alpha 1}_{ij}(x_1,\mathbf{y}) = b_{ij11}(x_1,\mathbf{y}) y_\alpha, \\
g^{\alpha 1}_i(x_1,\mathbf{y}) = \sigma^{*\varepsilon}_{11}(x_1,\mathbf{y})\delta_{i\alpha} - \sigma^{*\varepsilon}_{1\alpha}(x_1,\mathbf{y})\delta_{i1} \text{ for (10.33) and (10.35)}; \\
f^{11}_{ij}(x_1,\mathbf{y}) = b_{ij\beta 1}(x_1,\mathbf{y}) y_B s_\beta, \\
g^{11}_i(x_1,\mathbf{y}) = \sigma^{*\varepsilon}_{1B}(x_1,\mathbf{y}) s_\beta \delta_{i\beta} \text{ for (10.34) and (10.36)};
\end{gathered} \tag{10.88}$$

and

$$\mathbf{J}^{00} = \mathbf{N}^{01},\ \ \mathbf{J}^{\alpha 0} = \mathbf{N}^{1\alpha},\ \ \mathbf{J}^{10} = \mathbf{X}^3; \tag{10.89}$$

where $\mathbf{N}^0$, $\mathbf{N}^{1\alpha}$, $\mathbf{X}^3$ are solutions to the periodicity cell problems (9.37), (9.50), (9.51) with no initial stresses.

Substituting (10.88) and (10.89) into (10.85), (10.86), and (10.87), we have ($\alpha = 2, 3$)

$$\begin{aligned} \Delta A^0 &= \langle b_{11mn}(x_1, \mathbf{y}))N^{01}_{m,ny}\rangle + \langle b_{ijmn}(x_1, \mathbf{y})N^{01}_{m,ny}N^{01}_{i,jy}\rangle \\ &+\langle b_{ij11}(x_1, \mathbf{y})N^{01}_{i,jy} + g^{K1}_i N^{01}_i\rangle; \\ \Delta A^1_\alpha &= \langle b_{11mn}(x_1, \mathbf{y}))N^{1\alpha}_{m,ny}\rangle + \langle b_{ijmn}(x_1, \mathbf{y})N^{1\alpha}_{m,ny}N^{01}_{i,jy}\rangle \\ &+\langle b_{ij11}(x_1, \mathbf{y})y_\alpha N^{01}_{i,jy} + g^{\alpha 1}_i N^{01}_i\rangle; \\ &+\langle b_{ijmn}(x_1, \mathbf{y})X^3_{m,ny}N^{01}_{i,jy}\rangle \\ &+\langle b_{ij\beta 1}(x_1, \mathbf{y})y_B s_\beta N^{01}_{i,jy} + g^{11}_i N^{01}_i\rangle; \end{aligned} \tag{10.90}$$

$$\begin{aligned} \Delta^1 A_\beta &= \langle b_{11mn}(x_1, \mathbf{y})N^{01}_{m,ny}y_\beta\rangle \\ &+\langle b_{ijmn}(x_1, \mathbf{y})N^{01}_{m,ny}N^{1\beta}_{i,jy}\rangle \\ &+\langle f^{01}_{ij}(x_1, \mathbf{y})N^{1\beta}_{i,jy} + g^{01}_i N^{1\beta}_i\rangle; \\ \Delta A^2_{\beta\alpha} &= \langle b_{11mn}(x_1, \mathbf{y}))N^{1\alpha}_{m,ny}y_\beta\rangle + \langle b_{ijmn}(x_1, \mathbf{y})N^{1\alpha}_{m,ny}N^{1\beta}_{i,jy}\rangle \\ &+\langle b_{ij11}(x_1, \mathbf{y})y_\alpha N^{1\beta}_{i,jy} + g^{\alpha 1}_i N^{1\beta}_i\rangle; \\ \Delta B^1_\beta &= \langle b_{11mn}(x_1, \mathbf{y}))X^3_{m,ny}y_\beta\rangle + \langle b_{ijmn}(x_1, \mathbf{y})X^3_{m,ny}N^{1\beta}_{i,jy}\rangle \\ &+\langle b_{ij\beta 1}(x_1, \mathbf{y})y_B s_\beta N^{1\beta}_{i,jy} + g^{11}_i N^{1\beta}_i\rangle; \end{aligned} \tag{10.91}$$

and

$$\begin{aligned} \Delta^1 A &= \langle b_{\Theta 1mn}(x_1, \mathbf{y})N^{01}_{m,ny}s_\Theta\rangle + \langle b_{ijmn}(x_1, \mathbf{y})N^{01}_{m,ny}J^{10}_{i,jy}\rangle \\ &+\langle b_{ij11}(x_1, \mathbf{y})X^3_{i,jy} + g^{01}_i X^3_i\rangle; \\ \Delta A^2_\alpha &= \langle b_{\Theta 1mn}(x_1, \mathbf{y}))N^{1\alpha}_{m,ny}s_\Theta\rangle + \langle b_{ijmn}(x_1, \mathbf{y})N^{1\alpha}_{m,ny}X^3_{i,jy}\rangle \\ &+\langle b_{ij11}(x_1, \mathbf{y})y_\alpha X^3_{i,jy} + g^{\alpha 1}_i X^3_i\rangle; \\ \Delta B^1 &= \langle b_{\Theta 1mn}(x_1, \mathbf{y}))X^3_{m,ny}s_\Theta\rangle + \langle b_{ijmn}(x_1, \mathbf{y})X^3_{m,ny}X^3_{i,jy}\rangle \\ &+\langle b_{ij\beta 1}(x_1, \mathbf{y})y_B s_\beta X^3_{i,jy} + g^{11}_i X^3_i\rangle. \end{aligned} \tag{10.92}$$

As a result, the stiffnesses of a beam with initial stresses are the sum of the stiffnesses of the beam with no initial stresses and correctors (10.90), (10.91), and (10.92) multiplied by μ:

$$\begin{aligned} A^0(\sigma) &= A^0(0) + \mu\Delta A^0; \\ A^1_\alpha(\sigma) &= A^1_\alpha(0) + \mu\Delta A^1_\alpha; \\ B^0(\sigma) &= B^0(0) + \mu\Delta B^0; \\ {}^1A_\beta(\sigma) &= {}^1A_\beta(0) + \mu\Delta C^1 A_\beta; \\ A^2_{\beta\alpha}(\sigma) &= A^2_{\beta\alpha}(0) + \mu\Delta A^2_{\beta\alpha}; \\ B_\beta(\sigma) &= B_\beta(0) + \mu\Delta B_\beta. \\ A(\sigma) &= A(0) + \mu\Delta A; \\ A^2_\alpha(\sigma) &= A^2_\alpha(0) + \mu\Delta A^1_\alpha; \\ B^1(\sigma) &= B^1(0) + \mu\Delta B^1; \end{aligned} \tag{10.93}$$

where $A^0(0)$, $A^1_\alpha(0)$, $B^0(0)$, ${}^1A_\beta(0)$, $A^2_{\beta\alpha}(0)$, $B_\beta(0)$, $A_\alpha(0)$, $A^2_\alpha(0)$, $B^1(0)$ are the stiffnesses of the beam with no initial stresses.

10.5 Homogenized Model Parameters of the Order of Initial Stresses

The homogenized governing equations (10.48) and (10.52) involve the parameters (10.48) that are of the order of the initial stresses

$$\begin{aligned} k_\beta(\sigma) &= \langle \sigma^{*\varepsilon}_{\beta 1}(x_1, \mathbf{y}) K^{01}_1 \rangle - \langle \sigma^{*\varepsilon}_{11}(x_1, \mathbf{y}) K^{01}_\beta \rangle; \\ k_{\beta\alpha}(\sigma) &= \langle (\sigma^{*\varepsilon}_{\beta 1}(x_1, \mathbf{y}) \delta_{ik} - \sigma^{*\varepsilon}_{11}(x_1, \mathbf{y}) \delta_{\beta k})(K^{1\alpha}_k + Z^\alpha_k) \rangle; \\ l_\beta(\sigma) &= \langle (\sigma^{*\varepsilon}_{\beta 1}(x_1, \mathbf{y})) \delta_{1k} - \sigma^{*\varepsilon}_{11}(x_1, \mathbf{y})) \delta_{\beta k})(L^3_k + Y_k) \rangle; \end{aligned} \tag{10.94}$$

and

$$\begin{aligned} k &= \langle \sigma^{*\varepsilon}_{21}(x_1, \mathbf{y}) K^{01}_k \rangle \delta_{3k} - \langle \sigma^{*\varepsilon}_{31}(x_1, \mathbf{y}) K^{01}_k \rangle \delta_{2k}; \\ k_\alpha &= \langle \sigma^{*\varepsilon}_{21}(x_1, \mathbf{y})(K^{1\alpha}_k + Z^\alpha_k) \rangle \delta_{3k} - \langle \sigma^{*\varepsilon}_{31}(x_1, \mathbf{y})(K^{1\alpha}_k + Z^\alpha_k) \rangle \delta_{2k}; \\ l &= \langle \sigma^{*\varepsilon}_{21}(x_1, \mathbf{y})(L^3_k + Y_k) \rangle \delta_{3k} - \langle \sigma^{*\varepsilon}_{31}(x_1, \mathbf{y})(L^3_k + Y_k) \rangle \delta_{2k}. \end{aligned} \tag{10.95}$$

The functions $\mathbf{K}^{01}$, $\mathbf{K}^{1\alpha}$, and $\mathbf{L}^3$ have the form $\mathbf{K}^{01} = \mathbf{N}^{01} + \mu ...$, $\mathbf{K}^{1\alpha} = \mathbf{N}^{1\alpha} + \mu ...$, and $\mathbf{L}^3 = \mathbf{X}^3 + \mu ...$ where $\mathbf{N}^{01}$, $\mathbf{N}^{1\alpha}$, and $\mathbf{X}^3$ are solutions of the periodicity cell problems (10.19), (10.33), and (10.34) for $\sigma^{*\varepsilon}_{ij} = 0$ (i.e. solutions to the periodicity cell problems (9.37), (9.50) for a beam with no initial stresses); "$\mu ...$" mean the terms of order of μ. The functions $\mathbf{Z}^\alpha$ and $\mathbf{Y}$ are of the order of μ.

Saving in (10.94) and (10.95) the terms linear in μ, we have (we keep in mind that in the case under consideration the initial stresses $\sigma^{*\varepsilon}_{ij}$ are of the order of μ)

$$\begin{aligned} k_\beta(\sigma) &= \langle \sigma^{*\varepsilon}_{\beta 1}(x_1, \mathbf{y}) N^{01}_k \rangle - \langle \sigma^{*\varepsilon}_{11}(x_1, \mathbf{y}) N^{01}_\beta \rangle; \\ k_{\beta\alpha}(\sigma) &= \langle \sigma^{*\varepsilon}_{\beta 1}(x_1, \mathbf{y}) N^{1\alpha}_i \rangle - \langle \sigma^{*\varepsilon}_{11}(x_1, \mathbf{y}) N^{1\alpha}_\beta \rangle; \\ l_\beta(\sigma) &= \langle \sigma^{*\varepsilon}_{\beta 1}(x_1, \mathbf{y}) X^3_1 \rangle - \langle \sigma^{*\varepsilon}_{11}(x_1, \mathbf{y}) X^3_\beta \rangle; \end{aligned} \tag{10.96}$$

and

$$\begin{aligned} k &= \langle \sigma^{*\varepsilon}_{21}(x_1, \mathbf{y}) N^{01}_k \rangle \delta_{3k} - \langle \sigma^{*\varepsilon}_{31}(x_1, \mathbf{y}) N^{01}_k \rangle \delta_{2k}; \\ k_\alpha &= \langle \sigma^{*\varepsilon}_{21}(x_1, \mathbf{y}) N^{1\alpha}_k \rangle \delta_{3k} - \langle \sigma^{*\varepsilon}_{31}(x_1, \mathbf{y}) N^{1\alpha}_k \rangle \delta_{2k}; \\ l &= \langle \sigma^{*\varepsilon}_{21}(x_1, \mathbf{y}) L^3_k \rangle \delta_{3k} - \langle \sigma^{*\varepsilon}_{31}(x_1, \mathbf{y}) L^3_k \rangle \delta_{2k}. \end{aligned} \tag{10.97}$$

10.6 Local Stresses in the Beam

The total stresses $\sigma^{total}_{ij}(\mathbf{x})$ in the beam with initial stresses are the sum of the initial stresses $\sigma^{*\varepsilon}_{ij}(\mathbf{x})$ and additional stresses $\sigma^{\varepsilon}_{ij}(\mathbf{x})$. In the case under consideration, the additional stresses are the stresses $\sigma^{(-3)}_{ij}$ (10.38), and we have

$$\begin{gathered}\sigma_{ij}^{total} = \sigma_{ij}^{*\varepsilon} + \sigma_{ij}^{(-3)} = \sigma_{ij}^{*\varepsilon} \\ +(h_{ij11}(x_1, \mathbf{y})y_\alpha + h_{ijmn}(x_1, \mathbf{y})K_{m,ny}^{1\alpha} + h_{ijmn}(x_1, \mathbf{y})Z_{m,ny}^{\alpha})u_{\alpha,1x1x}^{(0)}(x_1) \\ +(h_{ij\beta1}(x_1, \mathbf{y})y_B s_\beta + h_{ijmn}(x_1, \mathbf{y})L_{m,ny}^{3} + h_{ijmn}(x_1, \mathbf{y})Y_{m,ny})\phi_{,1x}(x_1) \\ +(h_{ij11}(x_1, \mathbf{y}) + h_{ijmn}(x_1, \mathbf{y})K_{m,ny}^{01})V_{1,1x}(x_1).\end{gathered} \tag{10.98}$$

For small initial stresses

$$\mathbf{J}^\alpha = \mathbf{J}^{\alpha 0} + \mu \mathbf{J}^{\alpha 1}, \tag{10.99}$$

where, see (10.89)

$$\mathbf{J}^{00} = \mathbf{N}^{01},\ \mathbf{J}^{\alpha 0} = \mathbf{N}^{1\alpha},\ \mathbf{J}^{10} = \mathbf{X}^3,$$

where $\mathbf{N}^0$, $\mathbf{N}^{1\alpha}$, $\mathbf{X}^3$ are solutions to the periodicity cell problems (9.37), (9.50), (9.51) with no initial stresses.

Correspondingly,

$$\mathbf{K}^0 = \mathbf{N}^{01} + \mu \mathbf{K}^{011},\ \mathbf{K}^\alpha = \mathbf{N}^{1\alpha} + \mu \mathbf{K}^{1\alpha 1},\ \mathbf{L}^3 = \mathbf{X}^3 + \mu \mathbf{L}^1,$$

where $\mathbf{K}^{011}$, $\mathbf{K}^{1\alpha 1}$, and $\mathbf{L}^1$ are solutions to the problem (10.66) for proper functions.

Substituting (10.99) and $h_{ijmn}(x_1, \mathbf{y}) = c_{ijmn}(\mathbf{y}) + \mu b_{ijmn}(x_1, \mathbf{y})$ (10.62) into (10.98), we obtain

$$\sigma_{ij}^{total} = \sigma_{ij}^{*\varepsilon} + \sigma_{ij}^{(-3)} + \mu S_{ij}, \tag{10.100}$$

where

$$\begin{gathered}\sigma_{ij}^{(-3)} = (c_{ij11}(x_1, \mathbf{y})y_\alpha + c_{ijmn}(x_1, \mathbf{y})N_{m,ny}^{1\alpha})u_{\alpha,1x1x}^{(0)}(x_1) \\ +(c_{ij\beta1}(x_1, \mathbf{y})y_B s_\beta + c_{ijmn}(x_1, \mathbf{y})X_{m,ny}^{3})\phi_{,1x}(x_1) \\ +(c_{ij11}(x_1, \mathbf{y}) + c_{ijmn}(x_1, \mathbf{y})N_{m,ny}^{01})V_{1,1x}(x_1),\end{gathered} \tag{10.101}$$

and corrector

$$\begin{gathered}S_{ij} = (b_{ij11}(x_1, \mathbf{y})y_\alpha + b_{ijmn}(x_1, \mathbf{y})N_{m,ny}^{1\alpha})u_{\alpha,1x1x}^{(0)}(x_1) \\ +(b_{ij\beta1}(x_1, \mathbf{y})y_B s_\beta + b_{ijmn}(x_1, \mathbf{y})X_{m,ny}^{3})\phi_{,1x}(x_1) \\ +(b_{ij11}(x_1, \mathbf{y}) + b_{ijmn}(x_1, \mathbf{y})N_{m,ny}^{01})V_{1,1x}(x_1) \\ +(b_{ij\alpha1}(\mathbf{y}) - b_{ij1\alpha}(\mathbf{y}))V_{\alpha,1x}(x_1) \\ +(c_{ij11}(x_1, \mathbf{y})y_\alpha + c_{ijmn}(x_1, \mathbf{y})K_{m,ny}^{1\alpha 1})u_{\alpha,1x1x}^{(0)}(x_1) \\ +(c_{ij\beta1}(x_1, \mathbf{y})y_B s_\beta + c_{ijmn}(x_1, \mathbf{y})L_{m,ny}^{31})\phi_{,1x}(x_1) \\ +(c_{ij11}(x_1, \mathbf{y}) + c_{ijmn}(x_1, \mathbf{y})K_{m,ny}^{011})V_{1,1x}(x_1) \\ +(c_{ij\alpha1}(\mathbf{y}) - c_{ij1\alpha}(\mathbf{y}))V_{\alpha,1x}(x_1) \\ = (c_{ij11}(x_1, \mathbf{y})y_\alpha + c_{ijmn}(x_1, \mathbf{y})K_{m,ny}^{1\alpha 1})u_{\alpha,1x1x}^{(0)}(x_1) \\ +(c_{ij\beta1}(x_1, \mathbf{y})y_B s_\beta + c_{ijmn}(x_1, \mathbf{y})L_{m,ny}^{31})\phi_{,1x}(x_1) \\ +(c_{ij11}(x_1, \mathbf{y}) + c_{ijmn}(x_1, \mathbf{y})K_{m,ny}^{010})V_{1,1x}(x_1).\end{gathered} \tag{10.102}$$

From the final formulas for local stresses, it is not possible to exclude the first-order correctors $\mathbf{K}^{011}$, $\mathbf{K}^{1\alpha 1}$, $\mathbf{L}^{31}$ determined from the problem (10.66) for the corresponding values of the index K in functions f_{ij}^{K1} and g_i^{K1}.

References

Kolpakov AG (1991) Calculation of the characteristics of thin elastic rods of periodic structure. J Appl Math Mech 57:440–448
Timoshenko S, Woinowsky-Krieger K (1959) Theory of plates and shells. McGraw-Hill, New York
Trabucho L, Viano JM (1996) Mathematical modeling of rods. In: Ciarlet PG, Lions J-L (eds) Handbook of numerical analysis, vol IV. Elsevier, Amsterdam, pp 487–974
Washizu K (1992) Variational methods in elasticity and plasticity. Pergamon Press, Oxford

Chapter 11
Comparison of the Classical Hypothesis and Terms of the Asymptotic Expansions

The Euler-Bernoulli beam theory (also known as engineering beam theory or classical beam theory) several centuries history. The age of the Kirchhoff-Love theory is not so large, but it is also greater than a hundred years old. In the common literature and textbooks, the Euler-Bernoulli beam theory praised as providing engineers with a fairly easy-to-understand means of calculating the load-bearing properties of beams. It is noted that the simplicity of beam theory makes it an important tool in the sciences, especially in structural and mechanical engineering. The Kirchhoff-Love plate theory is praised as an extension of the Euler-Bernoulli ideas to plates. Indeed, at first glance, the similarities between the hypotheses underlying these theories are obvious. Even the imperfection—the neglecting the normal deformation(s)—is the same for both the theories.

It should be noted that the beam and plate stiffness formulas derived from the Euler-Bernoulli beam theory and the Kirchhoff-Love plate theory have been confirmed by centuries of intensive engineering practice. Indeed, numerous engineering structures have served as experimental tests of the Euler-Bernoulli stiffness formulas for beams and the Kirchhoff-Love formulas for plates and have validated these formulas. After such verification, these formulas are beyond doubt.

Finalizing the book, the author would like to discuss the relationship between the Euler-Bernoulli beam theory and Kirchhoff-Love plate theory with the asymptotic theory.

Considering the formula (5.49) for the first-order asymptotic corrector of plate displacements

$$\mathbf{u}^{(1)} = -y_3 \mathbf{e}_\alpha u^{(0)}_{3,\alpha x}(x_1, x_2) + \mathbf{V}(x_1, x_2), \tag{11.1}$$

we find that it coincides with the Kirchhoff-Love hypothesis. Considering the formula (8.45) for the first-order asymptotic corrector of beam displacements

$$\mathbf{u}^{(1)} = -y_\alpha \mathbf{e}_1 u^{(0)}_{\alpha,1x}(x_1) + y_\Gamma s_\gamma \mathbf{e}_\gamma \phi(x_1) + \mathbf{V}(x_1), \tag{11.2}$$

A. Kolpakov, *Thin-Walled Structures with Initial Stresses*, Mathematical Engineering,
https://doi.org/10.1007/978-3-032-07247-4_11

we find that it coincides with the Euler-Bernoulli hypothesis. The indicated formulas are valid for uniform plates and beams.

A look at Propositions 9.2 and 9.3 show that the ideas of Euler-Bernoulli and Coulomb are directly related to the deformation of non-uniform beams and beams of variable thickness. Proposition 6.2 shows that the ideas of Kirchhoff-Love are also directly related to the deformation of inhomogeneous plates and plates of variable thickness.

For example, consider formula (9.40)

$$\mathbf{u}^{(1)} = \mathbf{N}^{01}(\mathbf{y})u^{(0)}_{1,1x}(x_1) - y_\alpha \mathbf{e}_1 u^{(0)}_{\alpha,1x}(x_1) \tag{11.3}$$

$$+ y_\Gamma s_\gamma \mathbf{e}_\gamma \phi(x_1) + \mathbf{V}(x_1),$$

which corresponds to the axial tension—torsion of the beam. In formula (11.3), one see the sum

$$-y_\alpha \mathbf{e}_1 u^{(0)}_{\alpha,1x}(x_1) + y_\Gamma s_\gamma \mathbf{e}_\gamma \phi(x_1) + \mathbf{V}(x_1) \tag{11.4}$$

exactly coinciding with the classical hypothesis concerning the beam displacements, and the term

$$\mathbf{N}^{01}(\mathbf{y})u^{(0)}_{1,1x}(x_1), \tag{11.5}$$

where $\mathbf{N}^{01}(\mathbf{y})$ is determined from solution to the periodicity cell problem (9.37), which is three-dimensional elasticity theory problem. Since function $\mathbf{N}^{01}(\mathbf{y})$ is periodic in $y_1 = x_1/\varepsilon$, it plays a role of local perturbation, and does not change the macroscopic displacements of the beam. The term (11.5) significantly change the local stress-strain state and the local energy of the beam. As a result, it significantly changes the axial stiffness of the beam.

Comparing the hypotheses of the Euler-Bernoulli beam theory and the Kirchhoff-Love plate theory with the asymptotic calculations, we find that the representations of displacements in the classical theories and the asymptotic representations for the displacements coincide. A subtle point is that the coincidence depends on the order of the terms of the asymptotes. This point can be illustrated by an example of torsion of a rod. In the asymptotic expansion for the torsion problem, the first-order asymptotic corrector (8.45) of the beam displacements corresponds to the Coulomb hypothesis (Coulomb 1784) (this hypothesis assumes the pure rotation in the beam) and the Bernoulli hypothesis (this hypothesis assumes that the cross-sections of the beam remains flat). We see de Saint-Venant combination of rotation and axial displacements only in the second-order asymptotic corrector (8.51).

From the author's point of view, the comparison of classical hypotheses with asymptotic calculations does not give any arguments for considering classical (engineering) theories of plates and beams as "simplifications". From the point of view of asymptotic analysis, the classical (engineering) theories of plates and beams look like a step-by-step construction in terms of the asymptotic solution to the elasticity theory problem in thin and small-diameter regions.

We do not discuss the arguments in favor of one or another method for constructing the models of plates and beams. The corresponding discussions can be found in classical scientific works and numerous textbooks. The author would like only to note that small thickness or small diameter have always been among the main arguments and assumptions in the classical theories of plates and beams. However, the use of asymptotic techniques directly and substantially using a small parameter technique began in works devoted to the "asymptotic integration" of elasticity problems in thin domains in the 1960 and 1970s. As it turned out in the end, the classical Euler-Bernoulli and the Kirchhoff-Love hypotheses are closest to the asymptotic expansions obtained within the framework of the homogenization theory and the two-scale method.

Mention a comparison of classical and asymptotic solutions for the problem of dynamics of strings proposed by Kirchhoff (1876) analyzed in Andrianov and Koblik (2024). In the mentioned paper, it is shown that quasi-static Kirchhoff hypothesis is asymptotically justified.

It seems the theory of Euler-Bernoulli beams and the theory of Kirchhoff-Love plates are not simplifications of solutions of elasticity theory, but predictions and foresight of the terms of asymptotic solutions to elasticity theory problems in thin regions and regions of small diameter.

References

Andrianov IV, Koblik SG (2024) Asymptotic analysis of geometrically nonlinear beam vibrations: Kirchhoff and Bolotin equations. Z Angew Math Mech 104(10), e202400341

Coulomb CA (1784) Recherches theoriques et experimentales sur la force de torsion et sur l'elasticite des fils de metal. In: Histoire de l'Academie Royale des Sciences, pp 229–269

Kirchhoff G (1876) Vorlesungen iiber Mathematische Physik (Mechanik). B.G. Teubner, Leipzig

Index

A. Kolpakov, *Thin-Walled Structures with Initial Stresses*, Mathematical Engineering,
https://doi.org/10.1007/978-3-032-07247-4